ASTROPHYSICS OF PLANET FORMATION

The study of planet formation has been revolutionized by recent observational breakthroughs, which have allowed the detection and characterization of extrasolar planets, the imaging of protoplanetary disks, and the discovery of the Solar System's Kuiper Belt.

Written for beginning graduate students, this textbook provides a basic understanding of the astrophysical processes that shape the formation of planetary systems. It begins by describing the structure and evolution of protoplanetary disks, moves on to the formation of planetesimals, terrestrial and gas giant planets, and concludes by surveying new theoretical ideas for the early evolution of planetary systems.

Covering all phases of planet formation – from protoplanetary disks to the dynamical evolution of planetary systems – this introduction can be understood by readers with backgrounds in planetary science, and observational and theoretical astronomy. It highlights the physical principles underlying planet formation and the areas where more research and new observations are needed.

PHILIP J. ARMITAGE is a Professor in the Department of Astrophysical and Planetary Sciences at the University of Colorado, Boulder, and a Fellow of JILA. His research focuses on theoretical and computational studies of protoplanetary disks, planet formation, and black hole astrophysics. He has extensive teaching experience at the advanced undergraduate and graduate level.

ASTROPHYSICS OF PLANET FORMATION

PHILIP J. ARMITAGE

University of Colorado, Boulder

CAMBRIDGE
UNIVERSITY PRESS

CAMBRIDGE
UNIVERSITY PRESS

32 Avenue of the Americas, New York NY 10013-2473, USA

Cambridge University Press is part of the University of Cambridge.

It furthers the University's mission by disseminating knowledge in the pursuit of
education, learning and research at the highest international levels of excellence.

www.cambridge.org
Information on this title: www.cambridge.org/9781107653085

First published 2010
First paperback edition 2013

A catalogue record for this publication is available from the British Library

Library of Congress Cataloguing in Publication data
Armitage, Philip J., 1971–
Astrophysics of planet formation / Philip Armitage.
    p.   cm.
ISBN 978-0-521-88745-8 (hardback)
1. Planets – Origin.   2. Astrophysics.   I. Title.
QB603.074A76   2010
523.4 – dc22   2009036992

ISBN 978-0-521-88745-8 Hardback
ISBN 978-1-107-65308-5 Paperback

Contents

Preface

The study of planet formation has a long history. The idea that the Solar System formed from a rotating disk of gas and dust – the *Nebula Hypothesis* – dates back to the writings of Kant, Laplace, and others in the eighteenth century. A quantitative description of terrestrial planet formation was already in place by the late 1960s, when Viktor Safronov published his now classic monograph *Evolution of the Protoplanetary Cloud and Formation of the Earth and the Planets*, while the main elements of the core accretion theory for gas giant planet formation were developed in the early 1980s. More recently, a wealth of new observations has led to renewed interest in the problem. The most dramatic development has been the identification of extrasolar planets, first around a pulsar and subsequently in large numbers around main-sequence stars. These detections have furnished a glimpse of the Solar System's place amid an extraordinary diversity of extrasolar planetary systems. The advent of high resolution imaging of protoplanetary disks and the discovery of the Solar System's Kuiper Belt have been almost as influential in focusing theoretical attention on the initial conditions for planet formation and the role of dynamics in the early evolution of planetary systems.

My goals in writing this text are to provide a concise introduction to the classical theory of planet formation and to more recent developments spurred by new observations. Inevitably, the range of topics covered is far from comprehensive. The emphasis is firmly on the *astrophysical* aspects of planet formation, including the physics of the protoplanetary disk, the agglomeration of dust into planetesimals and planets, and the dynamical interactions between those bodies and the disk and between themselves. Planets are made of rock, ice, and gas, but the information that can be deduced from study of the chemical and geological make-up of those materials – the subject of *cosmochemistry* and much of traditional planetary science – is mostly ignored.

This book is an outgrowth of a graduate course that I teach at the University of Colorado in Boulder, for which the prerequisites are undergraduate classical physics and elementary mathematical methods. The primary readership is beginning graduate students, but most of the text ought to be accessible to undergraduates who have had some exposure to Newtonian mechanics and fluid dynamics. For the more sophisticated reader there is nothing here that is new, though the tone of the presentation – and in particular the emphasis on the coupling between turbulent processes in the disk and planet formation – focuses on what I consider to be important modern developments to a greater extent than older reviews. Despite recent progress one cannot disguise the fact that several critical problems in planet formation – foremost among them the nature of angular momentum transport within the protoplanetary disk and the formation mechanism of planetesimals – remain unsolved, and I have given extensive references to the technical literature to enable interested readers to explore these and other controversial topics further.

A number of colleagues have helped out in the preparation of this book. The discussion of the internal structure of the Solar System's gas giants draws heavily on the work of Tristan Guillot, who generously provided figures illustrating constraints on the core masses of Jupiter and Saturn. Keiji Ohtsuki was kind enough to provide figures showing the velocity evolution of planetesimals, while Sean Raymond prepared new figures depicting the late stages of terrestrial planet formation. My thanks also to Richard Alexander, Eric Feigelson, Dave Stevenson, Michele Trenti, Dimitri Veras, and Jared Workman, who shared their expertise on different topics and gave many of the chapters a critical reading.

Parts of the book were completed during a stay at UCLA, and I warmly thank Andrea Ghez and her colleagues in the Physics and Astronomy Department for their hospitality.

1

Observations of planetary systems

Planets can be defined informally as large bodies, in orbit around a star, that are not massive enough to have ever derived a substantial fraction of their luminosity from nuclear fusion. This definition fixes the maximum mass of a planet to be at the deuterium burning threshold, which is approximately 13 Jupiter masses for Solar composition objects ($1\ M_J = 1.899 \times 10^{30}$ g). More massive objects are called brown dwarfs. The lower mass cut-off for what we call a planet is not as well defined. Currently, the International Astronomical Union (IAU) requires a Solar System planet to be massive enough that it is able to clear the neighborhood around its orbit of other large bodies. Smaller objects that are massive enough to have a roughly spherical shape but which do not have a major dynamical influence on nearby bodies are called "dwarf planets." It is likely that some objects of planetary mass exist that are *not* bound to a central star, either having formed in isolation or following ejection from a planetary system. Such objects are normally called "planetary-mass objects" or "free-floating planets."

Complementary constraints on theories of planet formation come from observations of the Solar System and of extrasolar planetary systems. Space missions to all of the planets have yielded exquisitely detailed information on the surfaces (and in some cases interior structures) of the Solar System's planets, satellites, and minor bodies. A handful of the most fundamental facts about the Solar System are reviewed in this chapter, while other relevant observations are discussed subsequently in connection with related theoretical topics. By comparison with the Solar System our knowledge of individual extrasolar planetary systems is meager indeed – in many cases it can be reduced to a handful of imperfectly known numbers characterizing the orbital properties of the planets – but this is compensated in part by the large and rapidly growing number of known systems. It is only by studying extrasolar planetary systems that we can make statistical studies of the range of outcomes of the planet formation process, and avoid any bias introduced

Table 1.1. *The orbital elements (semi-major axis a, eccentricity e, and inclination i), masses, and equatorial radii of Solar System planets. The orbital elements are quoted for the J2000 epoch and are with respect to the mean ecliptic. Data from JPL.*

	a(AU)	e	i(deg)	M_p(g)	R_p(cm)
Mercury	0.3871	0.2056	7.00	3.302×10^{26}	2.440×10^{8}
Venus	0.7233	0.0068	3.39	4.869×10^{27}	6.052×10^{8}
Earth	1.000	0.0167	0.00	5.974×10^{27}	6.378×10^{8}
Mars	1.524	0.0934	1.85	6.419×10^{26}	3.396×10^{8}
Jupiter	5.203	0.0484	1.30	1.899×10^{30}	7.149×10^{9}
Saturn	9.537	0.0539	2.49	5.685×10^{29}	6.027×10^{9}
Uranus	19.19	0.0473	0.77	8.681×10^{28}	2.556×10^{9}
Neptune	30.07	0.0086	1.77	1.024×10^{29}	2.476×10^{9}

by the fact that the Solar System must necessarily be one of the subset of planetary systems that admit the existence of a habitable world.

1.1 Solar System planets

The Solar System has eight planets. Two are gas giants (Jupiter and Saturn) composed primarily of hydrogen and helium, although even their composition is substantially enhanced in heavier elements as compared to that of the Sun. Two are ice giants (Uranus and Neptune), composed of water, ammonia, methane, silicates, and metals, atop which sit relatively low mass hydrogen and helium atmospheres. Finally there are four terrestrial planets, two of which (Earth and Venus) are substantially more massive than the other two (Mars and Mercury). In addition there are a number of dwarf planets, including the trans-Neptunian objects Pluto, Eris, Haumea, and Makemake, and the asteroid Ceres. It is very likely that many more dwarf planets of comparable size remain to be discovered in the outer Solar System.

The orbital elements, masses and equatorial radii of the Solar System's planets are summarized in Table 1.1. With the exception of Mercury, the planets have almost circular, almost coplanar orbits. There is a small but significant misalignment of about 7° between the mean orbital plane of the planets and the Solar equator. Architecturally, the most intriguing feature of the Solar System is that the giant and terrestrial planets are clearly segregated in orbital radius, with the giants only being found at large radii where the Solar Nebula (the disk of gas and dust from which the planets formed) would have been cool and icy.

The planets make a negligible contribution ($\simeq 0.13\%$) to the mass of the Solar System, which overwhelmingly resides in the Sun. The mass of the Sun, $M_\odot = 1.989 \times 10^{33}$ g, is made up of hydrogen (fraction by mass in the envelope $X = 0.73$), helium ($Y = 0.25$), and heavier elements (described in astronomical parlance as "metals," with $Z = 0.02$). One notes that even most of the condensible elements in the Solar System are in the Sun. This means that if a significant fraction of the current mass of the Sun passed through a disk during the formation epoch the process of planet formation need not be 100% efficient in converting solid material in the disk into planets. In contrast to the mass, most of the angular momentum of the Solar System is locked up in the orbital angular momentum of the planets. Assuming rigid rotation at angular velocity Ω, the Solar angular momentum can be written as

$$J_\odot = k^2 M_\odot R_\odot^2 \Omega, \qquad (1.1)$$

where $R_\odot = 6.96 \times 10^{10}$ cm is the Solar radius. Taking $\Omega = 2.9 \times 10^{-6}$ s^{-1} (the Solar rotation period is 25 dy), and adopting $k^2 \approx 0.1$ (roughly appropriate for a star with a radiative core), we obtain as an estimate for the Solar angular momentum $J_\odot \sim 3 \times 10^{48}$ g cm^2 s^{-1}. For comparison, the orbital angular momentum associated with Jupiter's orbit at semi-major axis a is

$$J_J = M_J \sqrt{GM_\odot a} \simeq 2 \times 10^{50} \text{ g cm}^2 \text{ s}^{-1}. \qquad (1.2)$$

Even this value is small compared to the typical angular momentum contained in molecular cloud cores that collapse to form low-mass stars. We infer that substantial segregation of angular momentum and mass must have occurred during the star formation process.

The orbital radii of the planets do not exhibit any relationships that yield immediate clues as to their formation or early evolution.[1] Although the planets orbit close enough to perturb each other's orbits, the perturbations between the main planets are all nonresonant. Resonances occur when characteristic frequencies of two or more bodies display a near-exact commensurability. They adopt disproportionate importance in planetary dynamics because, in systems where the planets do not make close encounters, gravitational forces between the planets are generally much smaller (typically by a factor of 10^3 or more) than the dominant force from the star. These small perturbations are largely negligible unless special circumstances (i.e. a resonance) cause them to add up coherently over time. The simplest type of resonance, known as a *mean-motion resonance*, occurs when the periods P_1 and

---

[1] The Titius–Bode law, a well-known empirical relation between the orbital radii of the planets, is not thought to have any fundamental basis.

P_2 of two planets satisfy

$$\frac{P_1}{P_2} \simeq \frac{i}{j}, \tag{1.3}$$

where i and j are integers and use of the approximate equality sign denotes the fact that such resonances have a finite width. One can, of course, always find a pair of integers such that this equation is satisfied for arbitrary P_1 and P_2, so a more precise statement is that there are no dynamically important resonances among the major planets.[2] Nearest to resonance in the Solar System are Jupiter and Saturn, whose motion is affected by their proximity to a 5:2 mean-motion resonance known as the "great inequality" (the existence of this near resonance, though not its dynamical significance, was known even to Kepler). Among lower mass objects Pluto is one of a large class of Kuiper Belt Objects (KBOs) in 3:2 resonance with Neptune, and there are many examples of important resonances among satellites and in the asteroid belt.

1.1.1 The minimum mass Solar Nebula

The mass of the disk of gas and dust that formed the Solar System is unknown. However, it is possible to use the observed masses, orbital radii and compositions of the planets to derive a *lower limit* for the amount of material that must have been present, together with a crude idea as to how that material was distributed with distance from the Sun. This is called the "minimum mass Solar Nebula" (Weidenschilling, 1977a). The procedure is simple:

(1) Starting from the observed (or inferred) masses of heavy elements such as iron in the planets, augment the mass with enough hydrogen and helium to bring the augmented mixture to Solar composition.
(2) Divide the Solar System up into annuli, such that each annulus is centered on the current semi-major axis of a planet and extends halfway to the orbit of the neighboring planets.
(3) Imagine spreading the augmented mass for each planet across the area of its annulus. This yields a characteristic gas surface density Σ (units $g\,cm^{-2}$) at the location of each planet.

Following this scheme, one finds that out to the orbital radius of Neptune the derived surface density scales roughly as $\Sigma(r) \propto r^{-3/2}$. Since the procedure for constructing the distribution is somewhat arbitrary it is possible to obtain a number

---

[2] Roughly speaking, a resonance is typically dynamically important if the integers i and j (or their difference) are small. Care is needed, however, since although the 121:118 mean-motion resonance between Saturn's moons Prometheus and Pandora formally satisfies this condition (since the *difference* is small) one would not immediately suspect that such an obscure commensurability would be significant.

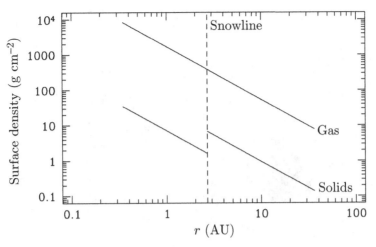

Fig. 1.1. The surface density in gas (upper line) and solids (lower broken line) as a function of radius in Hayashi's minimum mass Solar Nebula. The dashed vertical line denotes the location of the snowline.

of different normalizations, but the most common value used is that quoted by Hayashi (1981),

$$\Sigma(r) = 1.7 \times 10^3 \left(\frac{r}{1\ \text{AU}}\right)^{-3/2} \text{g cm}^{-2}. \tag{1.4}$$

Integrating this expression out to 30 AU the enclosed mass works out to be 0.01 $M_\odot$, which is comparable to the estimated masses of protoplanetary disks around other stars (though these have a wide spread). Hayashi (1981) also provided an estimate for the surface density of solid material as a function of radius in the disk,

$$\Sigma_s(\text{rock}) = 7.1 \left(\frac{r}{1\ \text{AU}}\right)^{-3/2} \text{g cm}^{-2} \text{ for } r < 2.7 \text{ AU}, \tag{1.5}$$

$$\Sigma_s(\text{rock/ice}) = 30 \left(\frac{r}{1\ \text{AU}}\right)^{-3/2} \text{g cm}^{-2} \text{ for } r > 2.7 \text{ AU}. \tag{1.6}$$

These distributions are shown in Figure 1.1. The discontinuity in the solid surface density at 2.7 AU is due to the presence of icy material in the outer disk that would be destroyed in the hotter inner regions.

Although useful as an order of magnitude guide, the minimum mass Solar Nebula (as its name suggests) provides only an approximate lower limit to the amount of mass that must have been present in the Solar Nebula. As we will discuss later, there are myriad reasons to suspect that both the gas and solid disks evolved substantially over time. There is no reason to believe that the minimum mass Solar Nebula reflects either the initial inventory of mass in the Solar Nebula, or the steady-state profile of the protoplanetary disk around the young Sun.

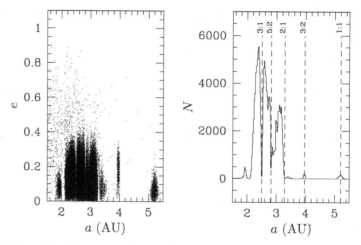

Fig. 1.2. The orbital elements of a sample of numbered asteroids in the inner Solar System. The left-hand panel shows the semi-major axes a and eccentricity e of asteroids in the region between the orbits of Mars and Jupiter. The right-hand panel shows a histogram of the distribution of asteroids in semi-major axis. The locations of a handful of mean-motion resonances with Jupiter are marked by the dashed vertical lines.

1.2 Minor bodies in the Solar System

In addition to the planets, the Solar System contains a wealth of minor bodies: asteroids, KBOs, comets and planetary satellites. Although the total mass in these reservoirs is now small[3] – estimates for the Kuiper Belt, for example, are of the order of 0.1 $M_\oplus$ (Chiang *et al.*, 2007) – the distribution of minor bodies is as important as study of the planets for the clues it provides to the early history of the Solar System. The first significant fact to note is that as a very rough generalization the Solar System is dynamically full, in the sense that most locations where small bodies could stably orbit for billions of years are, in fact, populated. In the inner Solar System, the main reservoir is the main asteroid belt between Mars and Jupiter, while in the outer Solar System the Kuiper Belt is found beyond the orbit of Neptune.

Figure 1.2 shows the distribution of a sample of numbered asteroids in the inner Solar System, taken from the *Jet Propulsion Laboratory*'s small-body database. Most of the bodies in the main asteroid belt have semi-major axes a in the range between 2.1 and 3.3 AU. However, the distribution of a is by no means smooth, and the crucial role of resonant dynamics in shaping the asteroid belt is obvious. There

---

[3] Indirect evidence suggests that the primordial asteroid and Kuiper belts were much more massive. A combination of dynamical ejection, and/or collisional grinding of bodies to dust that is then rapidly lost as a result of radiation pressure forces is likely to be responsible for their depletion.

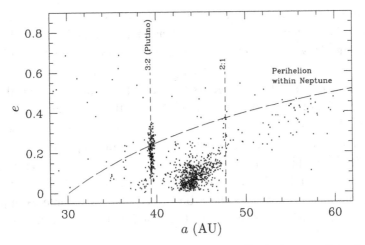

Fig. 1.3. The orbital elements of a sample of minor bodies in the outer Solar System beyond the orbit of Neptune. The dashed vertical lines indicate the locations of mean-motion resonances with Neptune. Objects with eccentricity above the long-dashed line have perihelia that lie within the orbit of Neptune.

are prominent regions, known as the Kirkwood (1867) gaps, where relatively few asteroids are found. These coincide with the locations of mean-motion resonances with Jupiter, most notably the 3:1 and 5:2 resonances. In addition to these locations – at which resonances with Jupiter are evidently depleting the population of minor bodies – there are *concentrations* of asteroids at both the co-orbital 1:1 resonance (the Trojan asteroids), and at the interior 3:2 resonance (the Hilda asteroids). Evidently different resonances can either destabilize or protect asteroid orbits (for a thorough analysis of the dynamics involved the reader should consult Murray & Dermott, 1999). Also notable is that the asteroids, unlike the major planets, have a distribution of eccentricity e that extends to moderately large values. Between 2.1 and 3.3 AU the mean eccentricity of the numbered asteroids is $\langle e \rangle \simeq 0.14$. As a result, collisions in the asteroid belt today typically involve relative velocities that are large enough to be disruptive. Indeed, a number of asteroid families (Hirayama, 1918) are known, whose members share similar orbital elements (a, e, i). These asteroids are interpreted as debris from disruptive collisions taking place within the asteroid belt, in some cases relatively recently (within the last few Myr, e.g. Nesvorný *et al.*, 2002).

Figure 1.3 shows the distribution of a sample of outer Solar System bodies, maintained by the IAU's *Minor Planet Center*. Outer Solar System bodies are divided into a number of dynamical classes. Resonant Kuiper Belt Objects orbit within one of Neptune's mean-motion resonances, most commonly the 3:2 resonance occupied by Pluto (such objects are called Plutinos). Some of these KBOs, like Pluto itself, cross Neptune's orbit and depend upon their resonant configuration to avert close encounters. The existence of this large resonant population of KBOs is believed to result from the outward migration of Neptune early in the Solar System's history. Classical KBOs comprise low-eccentricity bodies that are not in resonance with Neptune. Their orbits, when treated as test particles in the restricted three-body problem with Neptune as a perturber, are such that they will never cross the orbit of Neptune. The number of known classical KBOs drops rapidly for semi-major axes $a \gtrsim 47$ AU, reflecting either a physical edge to the population or, perhaps, a discontinuity in the physical properties of classical KBOs at this radius (Trujillo & Brown, 2001). Finally the scattered KBOs have typically highly eccentric and inclined orbits that do not cross the orbit of Neptune. A notable example is the large object Sedna, whose perihelion distance of 76 AU lies way beyond the orbit of Neptune.

Planetary satellites in the Solar System also fall into several classes. The regular satellites of Jupiter, Saturn, Uranus, and Neptune have relatively tight prograde orbits that lie close to the equatorial plane of their respective planets. This suggests that these satellites formed from disks, analogous to the Solar Nebula itself, that surrounded the planets shortly after their formation. The total masses of the regular satellite systems are a relatively constant fraction (about 10^{-4}) of the mass of the host planet, with the largest satellite, Jupiter's moon Ganymede, having a mass of 0.025 $M_\oplus$. The presence of resonances between different satellite orbits – most notably the *Laplace resonance* that involves Io, Europa, and Ganymede (Io lies in 2:1 resonance with Europa, which in turn is in 2:1 resonance with Ganymede) – is striking. As in the case of Pluto's resonance with Neptune, the existence of these nontrivial configurations among the satellites provides evidence for past orbital evolution that was followed by resonant capture. Orbital migration within a primordial disk, or tidal interaction with the planet, are candidates for explaining these resonances.

The giant planets also possess extensive systems of irregular satellites, which are typically more distant and which do not share the common disk plane of the regular satellites. These satellites were probably captured by the giant planets from heliocentric orbits. Finally the properties of the Moon seem most consistent with yet a third formation scenario – a giant impact early in the Earth's history which resulted in a heavy-element rich disk that condensed into the Moon (Hartmann & Davis, 1975; Cameron & Ward, 1976). It is possible that Pluto's large moon Charon formed in the aftermath of a similar impact.

1.3 Radioactive dating of the Solar System

Determining the ages of individual stars from astronomical observations is a difficult and usually imprecise exercise. For the Solar System, uniquely, the availability of apparently pristine meteorites allows accurate determination of its age and good constraints on the timing of some phases of the planet formation process.

The principle of radioactive dating of rock samples can be illustrated with a simple example. Consider a rock containing radioactive potassium (^{40}K) that solidifies from the vapor or liquid phases during the epoch of planet formation. One of the decay channels of ^{40}K is

$$^{40}\text{K} \rightarrow {}^{40}\text{Ar}. \tag{1.7}$$

This decay has a half-life of 1.25 Gyr and a branching ratio $\xi \approx 0.1$. (The branching ratio describes the probability that the radioactive isotope decays via a specific channel. In this case ξ is small because ^{40}K decays more often into ^{40}Ca.) If we assume that the rock, once it has solidified, traps the argon and that *there was no argon in the rock to start with*, then measuring the relative abundance of ^{40}Ar and ^{40}K suffices to determine the age. Quantitatively, if the parent isotope ^{40}K has an initial abundance $n_p(0)$ when the rock solidifies at time $t = 0$, then at later times the abundances of the parent isotope n_p and daughter isotope n_d are given by the usual exponential formulae that characterize radioactive decay,

$$n_p = n_p(0)e^{-t/\tau}$$
$$n_d = \xi n_p(0) \left[1 - e^{-t/\tau}\right], \tag{1.8}$$

where τ, the mean lifetime, is related to the half-life via $\tau = t_{1/2}/\ln 2$. The ratio of the daughter to parent abundance is

$$\frac{n_d}{n_p} = \xi \left(e^{t/\tau} - 1\right). \tag{1.9}$$

A laboratory measurement of the left-hand-side then fixes the age provided that the nuclear physics of the decay (the mean lifetime and the branching ratio) is accurately known. Notice that this method works to date the age of the rock (rather than the epoch when the radioactive potassium was formed) because minerals have distinct chemical compositions that differ – often dramatically so – from the average composition of the protoplanetary disk. In the example above, it is reasonable to assume that any ^{40}Ar atoms formed prior to the rock solidifying will not be incorporated into the rock, first because the argon will be diluted throughout the disk and second because it is an unreactive element that will not be part of the same minerals as potassium.

Radioactive dating is rarely as simple as the above illustration would suggest. A somewhat more representative example is the decay of rubidium 87 into strontium 87,

$$^{87}\text{Rb} \rightarrow {}^{87}\text{Sr}, \tag{1.10}$$

which occurs with a half-life of 48.8 Gyr. Unlike argon, strontium is not a noble gas, and we cannot assume that the rock is initially devoid of strontium. If we denote the initial abundance of the daughter isotope as $n_d(0)$, then measurement of the ratio (n_d/n_p) yields a single constraint on two unknowns (the initial daughter abundance and the age) and dating appears impossible. Again, the varied chemical properties of rocks allow progress. Suppose we measure samples from two different minerals within the same rock, and compare the abundances of ^{87}Rb and ^{87}Sr not to each other, but to the abundance of a separate stable isotope of strontium ^{86}Sr. Since ^{86}Sr is chemically identical to the daughter isotope ^{87}Sr that we are interested in, it is reasonable to assume that the ratio ^{87}Sr/^{86}Sr was initially constant across samples. The ratio ^{87}Rb/^{86}Sr, on the other hand, can differ between samples. As the rock ages, the abundance of the parent isotope drops and that of the daughter increases. Quantitatively,

$$n_p = n_p(0)e^{-t/\tau}$$

$$n_d = n_d(0) + \xi n_p(0)\left[1 - e^{-t/\tau}\right]. \tag{1.11}$$

Eliminating $n_p(0)$ between these equations and dividing by the abundance n_{ds} of the second stable isotope of the daughter species (^{86}Sr in our example) we obtain

$$\left(\frac{n_d}{n_{ds}}\right) = \left(\frac{n_d(0)}{n_{ds}}\right) + \xi \left(\frac{n_p}{n_{ds}}\right)\left[e^{t/\tau} - 1\right]. \tag{1.12}$$

The first term on the right-hand-side is a constant. We can then plot the relative abundances of the parent isotope (n_p/n_{ds}) and the daughter isotope (n_d/n_{ds}) from different samples on a ratio–ratio plot called an isochron diagram, such as the one shown schematically in Fig. 1.4. Inspection of Eq. (1.12) shows that we should expect the points from different samples to lie on a straight line whose slope (together with independent knowledge of the mean lifetime) fixes the age. Two samples are in principle sufficient to yield an age determination, but additional data provide a check against possible systematic errors – if the points fail to lie on a straight line something is wrong.

Radioactive dating of primitive meteorites known as chondrites using these techniques dates the formation of the Solar System to an epoch 4.57 Gyr ago. Knowing this age accurately is useful for calibrating Solar evolution models, but is otherwise of little interest for planet formation. More valuable are constraints on the time scale of critical phases of the planet formation process, where the questions we would like to answer are more subtle. For example, it would be valuable to be able to know whether the formation of km-sized bodies called planetesimals was sudden or spread out over many Myr. Addressing such questions is challenging using absolute chronometers based on long-lived isotopes (those with half-lives of the order of Gyr), so a complementary approach that derives *relative* ages from

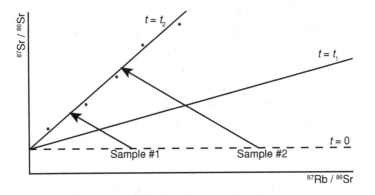

Fig. 1.4. Ratio–ratio plot for dating rocks using the radioactive decay ^{87}Rb $\rightarrow$ ^{87}Sr. The abundance of these isotopes is plotted relative to the abundance of a separate stable isotope of strontium ^{86}Sr. When the rock solidifies, different samples contain identical ratios of ^{87}Sr/^{86}Sr, but different ratios of ^{87}Rb/^{86}Sr. The ratios of different samples track a steepening straight line as the rock ages.

studying the decay products of short-lived isotopes is often employed. An example is the decay

$$^{129}\text{I} \rightarrow {}^{129}\text{Xe} \tag{1.13}$$

which has a half-life of only 17 Myr. If chondrites formed over a very extended period, the ratio of ^{129}Xe to a stable isotope of iodine ^{127}I could vary substantially from meteorite to meteorite, since the radioactive iodine present at early times when the first chondrites formed would largely have decayed away before the last chondrites formed. The observation that this ratio is relatively constant among chondrites constrains the interval of formation times to be no more than about 20 Myr. Similarly, the presence of daughter isotopes resulting from the decay of ^{26}Al – an *extremely* short-lived species with a half-life of 0.72 Myr – provides evidence that the Solar Nebula was polluted with recently synthesized radioactive isotopes. The most likely origin of these isotopes is nucleosynthesis within the cores of massive stars followed by ejection into the surrounding medium via either a supernova explosion or Wolf–Rayet stellar winds. In either case the implication is that the Sun formed in proximity to a rich star forming region (or within a cluster, now dissolved) that also contained massive stars.

The main caveat to be borne in mind with radioactive dating is that it depends upon there being no other processes besides radioactive decay that alter the abundance of either the parent or daughter isotopes. The radioactive age of a rock fixes the moment it solidified only if there has been no diffusion, or other alteration of the rock, during the intervening period. Even if the rock itself is pristine, high energy particles (cosmic rays) can induce nuclear reactions that may change the abundances of the critical isotopes. These problems are illustrated by the difficulty experienced in trying to date the formation of the Moon, whose age provides a critical constraint on theoretical models of terrestrial planet formation. Although lunar samples have been available since the Apollo landings in the 1960s, re-analysis in 2007 led to a substantial change in the age. The most recent results, based on an analysis of tungsten isotopes in lunar samples, suggest that the Moon formed 62^{+90}_{-10} Myr after the formation of the Solar System (Touboul *et al.*, 2007). It is clear that even the quoted errors on such determinations are larger than one would hope.

1.4 The snowline in the Solar Nebula

The presence of large quantities of liquid water on the Earth poses an interesting problem. For liquid water to exist now, the Earth's surface temperature must be 273 K $< T <$ 373 K, and this range of temperature is often taken to define the limits of the instantaneous habitable zone for planets around stars. Before the planets formed, however, the pressure in the protoplanetary disk was so low that water would have existed only in the vapor phase at temperatures above $T \simeq$ 150–170 K. The location in the disk beyond which the temperature falls below this value and condensation of water-rich minerals is possible is called the snowline. Somewhat paradoxically, then, it seems that for a planet to be habitable today it must have formed interior to the snowline in a region of the disk that would have been too hot for water-rich minerals to condense.

Determining the orbital radius of the snowline is obviously of critical importance for understanding the habitability of terrestrial planets. The snowline also plays a major role in the theory of planet formation, since the condensation of large amounts of icy solids adds substantially to the inventory of solid material available for planet building. Unfortunately, current astronomical observations do not allow for a precise determination of the radius of the snowline in disks detected around other stars. For the Solar System itself, compelling evidence that the snowline was located around 2.7 AU from the Sun at the epoch when planetesimals formed comes from combining meteoritic data with observations of asteroids. Meteorites of the class known as carbonaceous chondrites, which are water-rich, have properties (such as their reflectance spectra) that match those of asteroids found only in the outer asteroid belt beyond about 2.5 AU. Conversely, the ordinary and enstatite chondrites, which contain negligible amounts of water, appear to originate from the inner asteroid belt at a distance from the Sun of around 2 AU. This implies that – although the instantaneous location of the snowline surely changed as both the young Sun and the Solar Nebula evolved – it fell in the region of the modern-day asteroid belt at the critical epoch when large quantities of solid materials were condensing. As expected on the theoretical grounds discussed above, planetesimals at the orbital location where the Earth formed would have been dry.

Where then did the Earth's water come from? The two most obvious reservoirs are asteroids from the outer asteroid belt and comets, with asteroids being favored on account of the similarity between their deuterium to hydrogen (D/H) ratio and that of terrestrial water (Morbidelli *et al.*, 2000). The D/H ratio of water in the Earth's atmosphere and oceans is approximately 153 parts per million (ppm), comparable to the mean D/H ratio of carbonaceous chondrites of 159 ± 10 ppm but substantially less than the value of 309 ± 20 ppm inferred from the admittedly small sample of observed comets. This suggests that the bulk of the water on the Earth was probably delivered from the asteroid belt, though some cometary contribution is unavoidable. The mass of asteroids required is substantial. The mass of water in the crust of the Earth is $2.8 \times 10^{-4} M_{\oplus}$ (where the Earth mass, $M_{\oplus} = 5.974 \times 10^{27}$ g), to which must be added an uncertain but comparable mass of water locked in the mantle. If this water arrived via asteroids with compositions similar to the carbonaceous chondrites, which have mass fractions of water of $\approx 10\%$, the total mass required would have been a few $10^{-3} M_{\oplus}$. Although this mass is negligible on the scale of the impacts that assembled the Earth and formed the Moon, it still exceeds the total mass in the present-day asteroid belt by an order of magnitude.

1.5 Chondritic meteorites

In addition to their role in dating the Solar System and their importance for understanding water delivery, the chondritic meteorites also pose a major puzzle for theorists trying to understand the thermal history of the Solar Nebula. Both the ordinary and carbonaceous chondrites are undifferentiated, which in this context means that distinct phases that characteristically form at different temperatures are found mixed together in the same rock. Within these meteorites, one finds CAIs (inclusions enriched in refractory elements such as calcium and aluminum) and chondrules embedded within a surrounding matrix. Since the high temperatures and pressures found within large bodies would destroy such a mixture, the undifferentiated nature of these meteorites implies that they represent some of the earliest ("primitive") solids formed within the protoplanetary disk. This conclusion is supported by age dating, which shows that chondritic meteorites are not only old but also have a small dispersion in age.

Once extremely volatile elements are excluded the overall composition of chondritic meteorites is quite similar to that of the Sun. More remarkable are the properties of the three main individual phases:

- The calcium–aluminum inclusions (CAIs) are small macroscopic rocks with typical scales of a few mm. Their mineralogy is consistent with a high temperature origin ($T \simeq 2000$ K). The CAIs appear to have cooled relatively slowly (at a rate of the order of 10 K per hour) from the high temperature state.

- Chondrules are small (sub-mm to mm) solidified droplets of igneous rocks, whose detailed properties vary significantly from meteorite to meteorite even within a single class. Chondrules, like CAIs, represent a high temperature phase formed at $T \gtrsim 1300$ K, but unlike CAIs their properties are best explained as a consequence of very rapid cooling (at rates of 100–1000 K per hour).
- The matrix that binds together these inclusions is a low temperature phase made up of μm-sized particles of minerals that include more volatile elements.

The problem posed by the existence of the chondrules and CAIs in meteoritic material that originated from a few AU from the Sun is acute. In equilibrium, protoplanetary disk temperatures in excess of 1000 K would only have existed very close to the young Sun (at distances of the order of 0.1 AU and below). For the chondrules to have formed *in situ* requires that a substantial amount of matter in the disk further out must have been processed through intense and impulsive heating events – perhaps via shocks in the disk, lightning (!) or collisions between small bodies. Alternatively, the chondrules might have been heated very close to the inner edge of the disk and been transported outward prior to the formation of the meteorites or their parent bodies. Outward diffusion through the disk or ejection in an outflow originating close to the star are examples of mechanisms that have been proposed to accomplish this. These mechanisms (and many more details about the observed properties of primitive meteorites) are reviewed by Scott (2007). Despite the wealth of laboratory data there is no clear consensus as to how these meteorites formed and hence how to incorporate constraints derived from them into theoretical models of the disk.

Further evidence for mixing or redistribution of material within the Solar Nebula comes from the existence of crystalline silicates within cometary bodies that presumably formed in the neighborhood of the current Kuiper Belt. In particular, analysis of 1–10 μm particles collected by the *Stardust* spacecraft from the comet Wild 2 shows that many of them are made up of olivine or pyroxene – crystalline silicates that can be produced from amorphous precursors via annealing at temperatures of 800 K and above (Brownlee *et al.*, 2006). Evidently much of the disk – including very cold regions dominated by ices – was somehow polluted with at least a fraction of material processed through high temperatures. This observation too remains to be explained in detail.

1.6 Extrasolar planetary systems

The first extrasolar planetary system to be discovered was identified by Alex Wolszczan and Dale Frail via precision timing of pulses from the millisecond pulsar (a subclass of particularly rapidly rotating neutron stars formed during supernova explosions) PSR1257+12 (Wolszczan & Frail, 1992). The system contains at least three planets on nearly circular orbits within 0.5 AU of the neutron star, with the outer two planets having masses close to 4 $M_\oplus$ and the inner planet having a mass 0.02 $M_\oplus$ – comparable to the mass of the Moon! Although the precision of these measurements remains unsurpassed, the fact that surveys have failed to find large numbers of pulsar planets has largely stymied interpretation of the phenomenon. The substantial mass loss during a supernova explosion would unbind any pre-existing planets, so the planets in the PSR1257+12 system must have originated within a disk formed subsequent to the explosion. The observation that pulsar planets are not common implies that the conditions leading to their formation could involve moderately rare events.

Planet 51 Peg b, a Jupiter mass body orbiting a Solar-type star, was discovered by Michel Mayor and Didier Queloz in 1995 via a program of radial velocity monitoring of nearby stars (Mayor & Queloz, 1995).[4] Since then, several hundred more extrasolar planets have been found, the majority via radial velocity searches but with a growing contribution from transit searches and gravitational microlensing surveys. Two further techniques, astrometry and direct imaging, are likely to play a growing role in future planet discovery and characterization. These techniques are largely complementary – each method has its own biases and furnishes different information about the discovered planets.

1.6.1 Direct imaging

The most straightforward way to detect extrasolar planets is to image the planet as a source of light that is spatially separated from the stellar emission. It is also the most difficult, due to the extreme contrast between a planet and its host star. A planet of radius R_p, orbital radius a, and albedo A intercepts and reflects a fraction f of the incident starlight that is given by

$$f = \left(\frac{\pi R_p^2}{4\pi a^2}\right) A = 1.4 \times 10^{-10} \left(\frac{A}{0.3}\right) \left(\frac{R_p}{R_\oplus}\right)^2 \left(\frac{a}{1 \text{ AU}}\right)^{-2}. \qquad (1.14)$$

Recalling that magnitudes are defined in terms of the flux F via $m = -2.5 \log_{10} F + \text{const}$, one finds that Earth-like planets are expected to be 24–25 magnitudes fainter than their host stars. This faintness means that, quite apart

---

[4] In a striking illustration of the role of serendipity in astronomical discovery, the *precision* needed to detect 51 Peg b and similar close-in planets (which exhibit radial velocity signals in excess of 50 m s^{-1}) had in fact been available for a number of years. Campbell *et al.* (1988), for example, surveyed a small sample of stars for planetary or brown dwarf companions in the 1980s and attained mean errors of only 13 m s^{-1}. Had their sample of target stars included some with easily detectable planetary companions it is entirely possible that large numbers of detections would have followed soon after.

from the very unfavorable contrast ratio between planet and star, moderately deep exposures are needed to have even a chance of directly imaging planets in reflected light.[5] Alternatively, one may contemplate imaging extrasolar planets in their thermal emission. If we crudely approximate the emission at frequency ν from a planet with surface temperature T as a blackbody, then the spectrum is described by the Planck function

$$B_\nu(T) = \frac{2h\nu^3}{c^2} \frac{1}{\exp(h\nu/k_B T) - 1}, \qquad (1.15)$$

where h is Planck's constant, c the speed of light, and k_B Boltzmann's constant. The peak of the spectrum falls at $h\nu_{max} = 2.8 k_B T$, which lies in the mid-infrared for typical terrestrial planets (the wavelength corresponding to ν_{max} is $\lambda \approx 20$ µm for the Earth with $T = 290$ K). If the star, with radius R_*, also radiates as a blackbody at temperature T_*, the flux ratio at frequency ν is

$$f = \left(\frac{R_p}{R_*}\right)^2 \frac{\exp(h\nu/k_B T_*) - 1}{\exp(h\nu/k_B T) - 1}. \qquad (1.16)$$

For an Earth-analog around a Solar-type star the contrast ratio at the favorable 20 µm wavelength is $f \sim 10^{-6}$, which is some four orders of magnitude more favorable than the corresponding ratio in reflected light. This advantage of working in the infrared is, however, offset by the need for a larger telescope in order to spatially resolve the planet. The spatial resolution of a telescope of diameter D working at a wavelength λ is

$$\theta \sim 1.22 \frac{\lambda}{D}. \qquad (1.17)$$

A spatial resolution of 0.5 AU at a distance of 5 pc corresponds to an angular resolution of 0.1 arcsec, which is theoretically achievable in the visible ($\lambda = 550$ nm) with a telescope of diameter $D \approx 1.5$ m. At 20 µm, on the other hand, the required diameter balloons to $D \approx 50$ m, which is unfeasibly large to contemplate constructing as a monolithic structure in space.

These elementary considerations show that the difficulty of directly imaging planets depends strongly on their orbital radii. Giant planets orbiting at large or very large radii (tens of AU) are relatively easy to image, and convincing examples of such systems are now known (Kalas *et al.*, 2008; Marois *et al.*, 2008). The challenge of imaging terrestrial planets orbiting within the habitable zone is altogether more formidable, and one may even ask whether it is worth trying. If the goal is primarily to measure the abundance of terrestrial planets, together with their masses, radii and orbital properties, the answer is no – indirect methods can furnish those data

---

[5] To give a specific example, the *Advanced Camera for Surveys* on the *Hubble Space Telescope* could detect an **isolated** $m_V = 27$ object at a signal to noise ratio of 10 given an exposure time of a couple of hours.

much more easily. Imaging and spectroscopy are, however, indispensable tools for characterizing possibly habitable planets, determining the properties of their atmospheres, and looking for possible signatures of life such as the presence of oxygen, ozone, or methane. Given this astrobiological payoff, at least three classes of space-based technique are under study that promise to overcome the contrast problem by blocking or canceling the stellar contribution:

- A highly optimized coronagraph (a device that blocks starlight either directly with an occulting mask at the image plane or, more plausibly for this application, by modifying the shape or transmission of the pupil to achieve the same effect). Coronagraphs working in the optical can theoretically attain the necessary starlight suppression factor of the order of 10^{10}, though there are substantial practical difficulties related primarily to the need for very fine optical tolerances throughout the telescope system.
- A nulling interferometer working in the mid-infrared. The main challenge for this design is that, as noted previously, large baselines (measured in the tens of meters) are needed to achieve the required spatial resolution in the infrared. This requires that the interferometer elements be mounted on multiple free-flying spacecraft.
- An optical or ultraviolet telescope paired with a separate free-flying external occulter to block the starlight. This solution places the least stringent demands on the telescope optics, but requires that observations be coordinated with a second spacecraft – in practice a specially shaped occulting disk of size ~ 50 m flying 10^4–10^5 km from the telescope.

All of these architectures are optically possible but technically very challenging, and which approach is selected for implementation first will depend more on engineering considerations of cost and technical risk than on any simple physical principle. Over the longer term it is likely that observations in both the visible and mid-infrared wavebands will be needed, since detailed characterization of the possible existence of life on any nearby habitable planets will require spectroscopy over as wide a band as possible.

1.6.2 Radial velocity searches

A star hosting a planet describes a small orbit about the center of mass of the star–planet system. The radial velocity method for finding extrasolar planets works by measuring this stellar orbit via detection of periodic variations in the radial velocity of the star. The radial velocity is determined from the Doppler shift of the stellar spectrum, which can be measured to a precision of the order of a m s^{-1}.

1.6.2.1 Circular orbits

The principles of the radial velocity technique can be illustrated with the simple case of circular orbits. Consider a planet of mass $M_{\rm p}$ orbiting a star of mass M_* in a circular orbit of semi-major axis a. For $M_{\rm p} \ll M_*$ the Keplerian orbital velocity

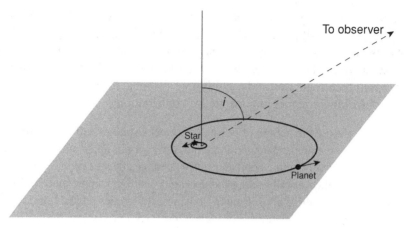

Fig. 1.5. The reflex motion (greatly exaggerated) of a star orbited by a planet on an eccentric orbit. The inclination of the system is the angle i between the normal to the orbital plane and the observer's line of sight.

of the planet is

$$v_{\mathrm{K}} = \sqrt{\frac{GM_*}{a}}. \tag{1.18}$$

Conservation of linear momentum implies that the orbital velocity v_* of the star around the center of mass is determined by $M_* v_* = M_{\mathrm{p}} v_{\mathrm{K}}$. If the planetary system is observed at an inclination angle i, as shown in Fig. 1.5, the radial velocity varies sinusoidally with a semi-amplitude

$$K = v_* \sin i = \left(\frac{M_{\mathrm{p}}}{M_*}\right)\sqrt{\frac{GM_*}{a}} \sin i. \tag{1.19}$$

Note that K is a directly observable quantity, as is the period of the orbit

$$P = 2\pi \sqrt{\frac{a^3}{GM_*}}. \tag{1.20}$$

If the stellar mass can be estimated independently but the inclination is unknown, as is typically the case, then we have two equations in three unknowns and the best that we can do is to determine a lower limit to the planet mass via the product $M_{\mathrm{p}} \sin i$. Since the average value of $\sin i$ for randomly inclined orbits is $\pi/4$ the statistical correction between the minimum and true masses is not large. The correction for individual systems is not normally known, however, and this can be an important source of uncertainty, for example when analyzing the dynamics and stability of multiple planet systems.

Consideration of the Solar radial velocity that is induced by the planets in the Solar System gives an idea of the typical magnitude of the signal. For Jupiter

$v_* = 12.5 \text{ m s}^{-1}$ while for the Earth $v_* = 0.09 \text{ m s}^{-1}$. For planets of a given mass there is a selection bias in favor of finding planets with small a. In an idealized survey in which the noise per observation is constant from star to star, Eq. (1.19) implies that the selection limit scales as

$$M_\text{p} \sin i|_\text{minimum} = Ca^{1/2}, \tag{1.21}$$

with C a constant. Planets with masses below this threshold are undetectable, as are planets with orbital periods that exceed the duration of the survey (since orbital solutions are generally poorly constrained when only a fraction of an orbit has been observed).

1.6.2.2 Eccentric orbits

For real applications it is necessary to consider the radial velocity signature produced by planets on eccentric orbits. The mathematics involved is straightforward, but requires the introduction of a thicket of celestial mechanics nomenclature. Here we state just the final results, the interested reader is referred to Murray & Dermott (1999) for further details.

Consider a planet on an orbit of eccentricity e, semi-major axis a, and period P. The orbital radius varies between $a(1 + e)$ (apocenter) and $a(1 - e)$ (pericenter). Suppose that passage of the planet through pericenter occurs at time t_peri. In terms of these quantities, the *eccentric anomaly*[6] E is defined implicitly via Kepler's equation,

$$\frac{2\pi}{P} \left(t - t_\text{peri} \right) = E - e \sin E. \tag{1.22}$$

Kepler's equation is transcendental and cannot be solved for E in terms of simple functions. However, it can readily be solved numerically. Once E is known, the *true anomaly* f is given by

$$\tan \frac{f}{2} = \sqrt{\frac{1 + e}{1 - e}} \tan \frac{E}{2}. \tag{1.23}$$

The true anomaly is the angle between the vector joining the bodies and the pericenter direction. Finally, in terms of these quantities, the radial velocity of the star is,

$$v_*(t) = K \left[\cos(f + \varpi) + e \cos \varpi \right] \tag{1.24}$$

where the longitude of pericenter ϖ is the angle in the orbital plane between pericenter and the line of sight to the system. The eccentric generalization of

---

[6] The eccentric anomaly has a rather complex geometrical interpretation, but for our purposes all that matters is that it is a monotonically increasing function of t that specifies the location of the body around the orbit.

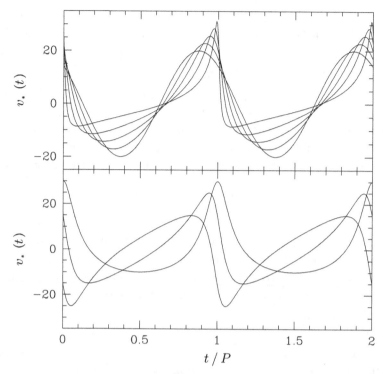

Fig. 1.6. Time dependence of the radial velocity of a star hosting a single planet. The upper panel shows the stellar radial velocity when the planet has a circular orbit (the symmetric sinusoidal curve), and when the planet has an eccentric orbit with $e = 0.2$, $e = 0.4$, $e = 0.6$; and $e = 0.8$. In all cases the longitude of pericenter $\varpi = \pi/4$. The lower panel shows, for a planet with $e = 0.5$, how the stellar radial velocity varies with the longitude of pericenter.

Eq. (1.19) in the same limit in which $M_p \ll M_*$ is

$$K = \frac{1}{\sqrt{1 - e^2}} \left(\frac{M_p}{M_*} \right) \sqrt{\frac{GM_*}{a}} \sin i. \qquad (1.25)$$

For a planet of given mass and semi-major axis, the amplitude of the radial velocity signature therefore increases with increasing e, due to the rapid motion of the planet (and star) close to pericenter passages.

Figure 1.6 illustrates the form of the radial velocity curves as a function of the eccentricity of the orbit and longitude of pericenter. Both e and ϖ can be measured given measurements of v_* as a function of time. Compared to the circular orbit of equal period, a planet on an eccentric orbit produces a stellar radial velocity signal of greater amplitude, but there are also long periods near apocenter where the gradient of v_* is rather small. These two properties of eccentric orbits mean that depending upon the observing strategy employed a radial velocity survey can be biased either in favor of or against finding eccentric planets. Such bias, however, is not a major concern for current samples, and the most important selection effects are those already discussed for circular orbits.

1.6.2.3 Noise sources

The amplitude of the radial velocity signal produced by Jovian analogs in extrasolar planetary systems is of the order of 10 m s^{-1}. High resolution astronomical spectrographs operating in the visible part of the spectrum have resolving powers $R \sim 10^5$, which correspond to a velocity resolution $\Delta v \approx c/R$ of a few km s^{-1}. The Doppler shift in the stellar spectrum due to orbiting planets therefore results in a periodic translation of the spectrum on the detector by a few *thousandths* of a pixel.[7] Detecting such small shifts reliably requires both exceptional instrumental stability (extending, for long period planets, over periods of many years) and careful consideration of the potential sources of noise in the measurement of the radial velocity signal.

Shot noise (i.e. uncertainty in the number of photons due purely to counting statistics) defines the ultimate radial velocity precision that can be attained from an observation of specified duration. An estimate of the shot noise limit can be derived by starting from a very simple problem: how accurately can the velocity shift of a spectrum be estimated given measurement of the flux in a single pixel on the detector? To do this, we follow the basic approach of Butler *et al.* (1996) and consider the spectrum in the vicinity of a spectral line, as shown in Fig. 1.7. Assume that in an observation of some given duration, N_{ph} photons are detected in the wavelength interval corresponding to the shaded vertical band. If we now imagine displacing the spectrum by an amount (expressed in velocity units) δv the change in the mean number of photons is

$$\delta N_{\text{ph}} = \frac{\text{d}N_{\text{ph}}}{\text{d}v} \delta v. \tag{1.26}$$

Since a 1σ detection of the shift requires that $\delta N_{\text{ph}} \approx N_{\text{ph}}^{1/2}$, the minimum velocity displacement that is detectable is

$$\delta v_{\text{min}} \approx \frac{N_{\text{ph}}^{1/2}}{\text{d}N_{\text{ph}}/\text{d}v}. \tag{1.27}$$

This formula makes intuitive sense – regions of the spectrum that are flat are useless for measuring δv while sharp spectral features are good. For Solar-type stars with photospheric temperatures $T_{\text{eff}} \approx 6000$ K the sound speed at the photosphere is

---

[7] For simplicity, we assume that one spectral resolution element corresponds to one pixel on the detector. A real instrument may over-sample the spectrum, but this practical point does not alter any of the basic results.

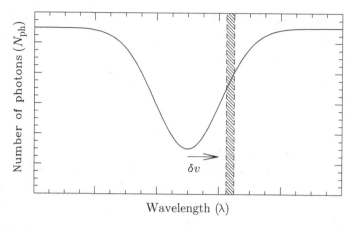

Fig. 1.7. Schematic spectrum in the vicinity of a single spectral line of the host star. The wavelength range that corresponds to a single pixel in the observed spectrum is shown as the vertical shaded band. If the spectrum shifts by a velocity δv the number of photons detected at that pixel will vary by an amount that depends upon the local slope of the spectrum.

around 10 km s^{-1}. Taking this as an estimate of the thermal broadening of spectral lines, the slope of the spectrum is at most

$$\frac{1}{N_{\text{ph}}} \frac{dN_{\text{ph}}}{dv} \sim \frac{1}{10 \text{ km s}^{-1}} \sim 10^{-4} \text{ m}^{-1}\text{s}. \qquad (1.28)$$

Combining Eq. (1.27) and (1.28) with knowledge of the number of photons detected per pixel yields an estimate of the photon-limited radial velocity precision. For example, if the spectrum has a signal to noise ratio of 100 (and there are no other noise sources) then each pixel receives $N_{\text{ph}} \sim 10^4$ photons and $\delta v_{\text{min}} \sim 100 \text{ m s}^{-1}$. If the spectrum contains N_{pix} such pixels the combined limit to the radial velocity precision is

$$\delta v_{\text{shot}} = \frac{\delta v_{\text{min}}}{N_{\text{pix}}^{1/2}} \sim \frac{100 \text{ m s}^{-1}}{N_{\text{pix}}^{1/2}}. \qquad (1.29)$$

Although this discussion ignores many aspects that are practically important in searching for planets from radial velocity data, it suffices to reveal several key features. Given a high signal to noise spectrum and stable wavelength calibration, photon noise is small enough that a radial velocity measurement with the m s^{-1} precision needed to detect extrasolar planets is feasible. The resolution of the spectrograph needs to be high enough to resolve the widths of spectral lines, but does not need to approach the magnitude of the planetary signal. In fact the intrinsic precision of the method depends first and foremost on the amount of structure that is present within the stellar spectrum, and the measurement precision will be degraded for stars whose lines are additionally broadened, for example by rotation.

Once precisions of the order of meters per second have been attained, intrinsic radial velocity jitter due to motions at the stellar photosphere presents a second fundamental limit. The vertical velocity at the photosphere of a star is not zero, due to the presence of both convection and p-mode oscillations (acoustic modes trapped in the star). Although p-mode oscillations are of great intrinsic interest – they are the signal that facilitates helioseismological investigation of the interior of the Sun and other stars – they represent noise for radial velocity searches for extrasolar planets. For the star HD 160691 the amplitude of numerous p-modes has been measured directly from radial velocity data (Bouchy *et al.*, 2005). Individual modes were found to have an amplitude of between 10 and 40 cm s^{-1}, summing up to a combined impact on the radial velocity signal of the order of a meter per second. Although, given enough data, the effect of resolved oscillations can be removed by careful filtering, the existence of intrinsic stellar radial velocity jitter means that attaining precision at the sub-meter per second level is challenging.

1.6.3 Astrometry

Astrometric measurement of the stellar reflex motion in the plane of the sky provides a complementary method for detecting planets. From the definition of the center of mass, the physical size of the stellar orbit is related to the planetary semi-major axis via $a_* = (M_p/M_*)a$. For a star at distance d from the Earth the angular displacement of the stellar photo-center during the course of an orbit has a characteristic scale,

$$\theta = \left(\frac{M_p}{M_*}\right)\frac{a}{d}. \tag{1.30}$$

Unlike radial velocity searches, which are biased toward detecting short period planets, astrometry favors large semi-major axes. A further difference is that astrometry measures two independent components of the stellar motion (versus a single component via radial velocity measurements), and this yields more constraints on the orbit. As a result there is no $\sin i$ ambiguity and all of the important planetary properties can be directly measured. Numerically the size of the signal is

$$\theta = 5 \times 10^{-4} \left(\frac{M_p}{M_J}\right)\left(\frac{M_*}{M_\odot}\right)^{-1}\left(\frac{a}{5\ \mathrm{AU}}\right)\left(\frac{d}{10\ \mathrm{pc}}\right)^{-1}\ \mathrm{arcsec}. \tag{1.31}$$

Even though the parameters adopted here are rather optimistic this is still a very small displacement, and none of the planets found in the first decade of discovery were identified this way. In principle, however, there are no fundamental obstacles

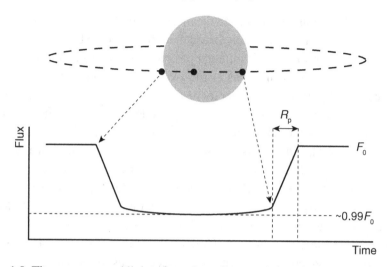

Fig. 1.8. The geometry and light curve of a stellar transit by a giant planet. During the transit the stellar flux, which is F_0 when the star is not eclipsed, is reduced by approximately 1%. The curvature of the light curve during transit is due to the phenomenon of limb darkening – the photosphere (defined as the optical depth $\tau = 2/3$ surface) is physically deeper looking toward the center of the disk, so we see hotter gas and greater intensity there than close to the limb.

to achieving astrometric accuracies of 1–10 μarcsec,[8] which is good enough to detect a wide range of hypothetical planets. A $10M_\oplus$ planet at 1 AU, for example, yields an astrometric signature of 3 μarcsec at $d = 10$ pc.

1.6.4 Transits

A planet whose orbit causes it to transit the stellar disk can be detected by monitoring the stellar flux for periodic dips. The amplitude of the transit signal is independent of the distance between the planet and the star, and provides a measure of the relative size of the two bodies. If a planet of radius R_p occults a star of radius R_*, with the geometry shown in Fig. 1.8, the fractional decrement in the stellar flux during the transit (assuming a uniform brightness stellar disk) is just

$$f = \left(\frac{R_p}{R_*}\right)^2. \tag{1.32}$$

Gas giant planets have a rather flat mass–radius relation, so it is reasonable to use Jupiter's radius $R_J = 7.142 \times 10^9$ cm as a proxy for all giant planet transits. The

---

[8] The European Space Agency's *GAIA* mission, for example, should attain an astrometric precision of 10 μarcsec for almost all stars with a visual magnitude $m_V \leq 15$.

amplitude of the signal for a gas giant transiting a Solar-radius star is $f = 0.01$. For rocky planets with the same mean density as the Earth,

$$f = 8.4 \times 10^{-5} \left(\frac{M_p}{M_\oplus}\right)^{2/3}. \tag{1.33}$$

The photometric precision of ground-based observations is limited by atmospheric fluctuations. Ground-based photometry accurate to one part in 10^3 is possible, and this is sufficient to detect massive extrasolar planets via their transit signatures. The regime relevant for terrestrial planet transits, on the other hand, is inaccessible from the ground and requires the higher precision photometry attainable above the atmosphere.

The probability that a planet will be observed to transit follows from elementary geometric arguments. For a planet on a circular orbit with semi-major axis a, the condition that some part of the planet will be seen to graze the stellar disk is that the inclination angle i satisfies $\cos i \leq (R_* + R_p)/a$. If an ensemble of such systems has random inclinations then the probability of observing transits is

$$P_{\text{transit}} = \frac{R_* + R_p}{a}. \tag{1.34}$$

For a planet similar to the Earth, the probability is $P_{\text{transit}} \approx 5 \times 10^{-3}$. If the geometry is favorable for observing transits, their maximum duration is roughly $2R_*/v_{\text{K}}$. More accurately (Quirrenbach, 2006),

$$t_{\text{transit}} = \frac{P}{\pi} \sin^{-1} \left(\frac{\sqrt{(R_* + R_p)^2 - a^2 \cos^2 i}}{a}\right), \tag{1.35}$$

where P is the orbital period. A planet similar to the Earth whose orbit is seen edge-on ($i = 90°$) transits its host star for 13 hours. Combining this result with the transit probability, one finds that if every star were to host an Earth-like planet, about 1 in 10^5 stars would be being transited at any given time. Clearly long duration surveys of large numbers of stars are needed to identify terrestrial planet transits. Locating gas giants in short period orbits is, of course, much easier. In either case, transit measurements yield the planetary radius and, once multiple transits have been observed, an accurate orbital period. The mass is not determined unless complementary radial velocity or astrometric information is available.

The most important sources of noise for transit searches are different for ground- and space-based experiments. From the ground, photometric error introduced by atmospheric fluctuations means that only giant planets with $R_p \sim R_J$ can be detected. The main practical difficulty is not noise per se, but rather confusion of planetary transit events with eclipses by more massive bodies. False alarms can be caused by grazing stellar eclipses, transits by very low mass stars (which have radii comparable to those of giant planets), and normal eclipsing binaries where the amplitude of the eclipse signal is diluted by light from unresolved background stars. Follow up radial velocity measurements can reject these contaminating events provided that the star is not so faint (or in such a crowded stellar field) as to make spectroscopy impossible.

For space-based transit searches, the ultimate limit is again set by photon statistics – enough photons must be received during $t_{transit}$ to detect the event at high significance despite the presence of shot noise. In practice this limit specifies a minimum size of telescope needed to detect planets of a particular class. Fairly modest apertures are sufficient, since even a 1 m diameter telescope receives more than enough photons to detect transits by Earth-radii planets around quite faint stars ($m_V = 12$). A second limitation is set by fluctuations in stellar brightness due to intrinsic stellar variability. For the Sun, sunspots and related phenomena introduce variability at the level of a tenth of a percent on the Solar rotation period. On the shorter time scales relevant to transit detection the variability is smaller. Analysis of data from the *Solar and Heliospheric Observatory* shows that the amplitude of the temporal power spectrum of Solar variability at time scales of a day is three orders of magnitude smaller than the amplitude at the Solar rotation period. This level is small enough that it should not prevent detection of Earth-like planets around stars whose activity level resembles that of the Sun.

Several variations of the basic transit method are also important:

- For giant planets in short period (typically a few days) orbits, the secondary eclipse when the planet moves behind the star may also be detectable. For example, a gas giant planet with a surface temperature of $T = 10^3$ K orbiting a Solar-type star yields a secondary eclipse with a fractional depth of $\sim 10^{-3}$ at $\lambda = 10$ μm. Measurement of this signal, which is reduced by at least an order of magnitude as compared to the transit signal, provides direct information on the atmospheric properties of the planet.
- The *reflected light* from giant planets in short period orbits that do not transit exhibits a phase modulation as the planet orbits the star. If the planet has an albedo A, the reflected light signal is a factor $\approx (4/A)(a/R_*)^2$ weaker than the corresponding transit signal. However, given sufficiently accurate photometry the geometric advantage of being able to detect planets across a wide range of orbital inclination can make this an efficient way to find close-in giant planets.

Once a planet has been detected in transit, accurate timing of the start of the transit can be analyzed to infer the presence of other perturbing planets in the same system. If the additional planet is in resonance with the transiting planet, this effect can be strong enough to permit detection of Earth-mass planets given transit times accurate at the 10^2 s level (Agol *et al.*, 2005; Holman & Murray, 2005). One can also usefully combine transit observations with high resolution radial velocity measurements. If the planet orbits in the equatorial plane of a rotating star, the transit first obscures a fraction of the stellar photosphere that is rotating toward the observer before subsequently obscuring a fraction of the receding photosphere just prior to egress. This time-dependent obscuration of different parts of the photosphere – known as the Rossiter–McLaughlin effect (Rossiter 1924) – results in a systematic shift in the apparent stellar radial velocity during transit. Detection of the effect allows for a measurement of any misalignment between the planet's orbital plane and the stellar equator. Such measurements are of interest because some theoretical explanations (such as the action of Kozai resonance, which we will discuss in Chapter 7) for the presence of massive planets in short-period orbits predict significant misalignments, while others such as gas disk migration predict near-perfect alignment of the orbital and equatorial planes.

1.6.5 Gravitational microlensing

Gravitational microlensing is an indirect but powerful method for detecting extra-solar planets that is based on entirely different physical principles from any of the aforementioned techniques. A photon that passes a star of mass M_* with impact parameter b is deflected by an angle

$$\alpha = \frac{4GM_*}{bc^2}.$$

(1.36)

This general relativistic result, which was famously confirmed by observations during Solar eclipses in the early twentieth century, is exactly twice the Newtonian prediction. If two stars lie at different distances along the same line of sight, consideration of the geometry illustrated in Fig. 1.9 implies that the image of the background star (the "source") is distorted by the deflection introduced by the foreground star (the "lens") into a ring. Writing the distance between the observer and the lens as d_L, the observer–source distance as d_S, and the lens–source distance as d_{LS}, the angular radius of the so-called *Einstein ring* is

$$\theta_E = \frac{2}{c}\sqrt{\frac{GM_* d_{LS}}{d_L d_S}}.$$

(1.37)

If the alignment between the source and the lens is nearly but not quite perfect, the axial symmetry is broken and the source is lensed into multiple images whose angular separation is also of the order of θ_E.

For planet detection, the configuration that is of greatest interest observationally is that where the source star lies in the Galactic bulge ($d_S = 8\,\text{kpc}$) and the lens star is in the disk ($d_L \approx 4\,\text{kpc}$). The Einstein ring radius for a Solar mass lens with these parameters is $\theta_E \sim 10^{-3}$ arcsec, so the multiple images cannot be spatially

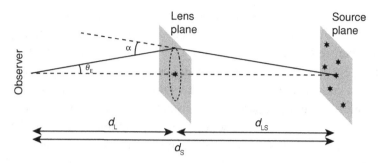

Fig. 1.9. Microlensing geometry. Light from a background star is deflected by a small angle α when it passes a foreground star of mass M_*. If the alignment between the foreground and background star is perfect, then by symmetry the observer sees the image of the background star distorted into a circular Einstein ring around the foreground star. The ring has an angular radius θ_E. When lensing is used as a tool to detect planets, the distances involved are such that the ring is unresolved. The observable is the change in brightness of the background star as the two stars first align, and then move apart.

resolved. Another less intuitive property of lensing, however, is that it conserves surface brightness. The area of the lensed image is greater than it would have been in the absence of lensing, and hence the flux from the lensed images exceeds that of the unlensed star. This means that if the alignment between the source and the lens varies with time, monitoring of the light curve of the source star can detect unresolved lensing via the time-variable magnification caused by lensing. In practice, simultaneous monitoring of $\sim 10^8$ stars in the Galactic bulge results in detection of ≈ 500 lensing events by disk stars every year. Light curves in which a single disk star lenses a background star are smooth, symmetric about the peak, and magnify the source achromatically. If the proper motion of the lens relative to that of the source is μ, the characteristic time scale of the lensing event is $t_E = \theta_E/\mu$. For events toward the Galactic bulge this time scale is about a month, scaling with the lens mass as $\sqrt{M_*}$.

This astrophysical digression is relevant to planet detection because microlensing light curves are altered when the lens star is part of a binary. For star–planet systems in which the mass ratio $q = M_p/M_* \ll 1$ the conceptually simplest case to consider is when the planet orbits close to the physical radius of the Einstein ring at the distance of the lens. If one of the multiple images of the source passes close to the planet during the lensing event the planet's gravity introduces an additional perturbation to the light curve. To an order of magnitude, the time scale of the perturbation is just $t_p \sim q^{1/2}t_E$, while the probability that the geometry will be such that a perturbation occurs is $P \sim Aq^{1/2}$, where A is the magnification of the source at the moment when the image passes near the planet. A second channel for planet detection occurs in rare high magnification events, during which planets close to the Einstein ring modify the light curve near peak regardless of their orbital position. This channel is valuable observationally since high magnification events can be anticipated based on photometric observations made well before the peak, allowing for detailed monitoring of those events that are most favorable for detecting planets.

The detailed analysis of microlensing light curves is highly technical, but the above discussion is enough to explain the attractive aspects of planet detection via lensing. First, the physical radius of the Einstein ring for disk stars lensing background stars in the Galactic bulge corresponds to a few AU. Lensing is therefore well suited to detect planets at relatively large orbital radii – comparable to the location of Jupiter in the Solar System – which are challenging to detect via radial velocity methods. Second, the weak $\sqrt{M_P}$ dependence of the time scale on planet mass allows for a wide range of planets to be detected. In particular, Jupiter mass planets yield a perturbation time scale of around a day, while Earth mass planets have a characteristic time scale of about an hour. Both time scales are quite accessible observationally. Moreover, *when* a planetary perturbation occurs its magnitude can be significant, even for Earth mass planets. Low mass planets which are challenging to find with other techniques can therefore be detected from the ground via gravitational microlensing.

1.7 Properties of extrasolar planets

Despite the explosive growth in the number of detections of extrasolar planets over the last decade we are still far from being able to conduct a fair census of extrasolar planetary systems. At the time of writing, robust statistical knowledge of the extrasolar planet population is confined to the corner of parameter space accessible to radial velocity surveys, whose selection function favors detection of relatively massive planets (typically gas giants) at orbital radii comparable to or smaller than that of Jupiter. Looking to the future, there are excellent prospects for surveying the region of parameter space occupied by potentially habitable planets via space-based transit surveys, and more limited possibilities for detecting extrasolar analogs to Saturn and the ice giants via astrometry and direct imaging surveys.

For the sample of planets detected via radial velocity surveys the available information is typically limited to those quantities derived directly from the radial velocity observables: a lower limit on the planet mass $M_p \sin(i)$, the orbital period P, the eccentricity of the orbit e, and the longitude of pericenter ϖ. In addition, the high resolution spectra acquired for radial velocity measurements permit an estimate of the host stars' masses and metal content.

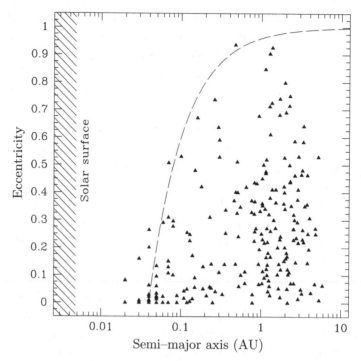

Fig. 1.10. The distribution of a sample of known extrasolar planets in semi-major axis and eccentricity, based on updated data from Butler *et al.* (2006). The dashed line shows the eccentricity of a planet whose pericenter distance is 0.04 AU.

The distribution of a sample of known extrasolar planets in $M_p \sin(i)$, a, and e (based on an updated version of the data published by Butler *et al.*, 2006) is shown in Figs. 1.10 and 1.11. Even bearing in mind the selection effects (which prevent detection of low mass planets and those planets with long orbital periods) the extrasolar planet distribution is strikingly different from what might be expected based on the architecture of the Solar System. To start with, a population of massive planets is detected at radii $a < 0.1$ AU that lie interior even to Mercury in the Solar System. These planets are often referred to as *hot Jupiters*. When they have been detected in transit the radii of hot Jupiters are consistent with a gas giant structure. The frequency of hot Jupiters around main sequence stars with spectral types similar to that of the Sun is approximately 1%. Most of these close-in planets have almost circular orbits, which may reflect either tidal circularization due to interaction between the planet and the star or an absence of processes that would excite orbital eccentricity for this class of planet. As the sensitivity of surveys improves, it has become clear that the mass distribution of close-in planets extends to masses below that of the ice giants in the Solar System. These planets are often dubbed "super-Earths."

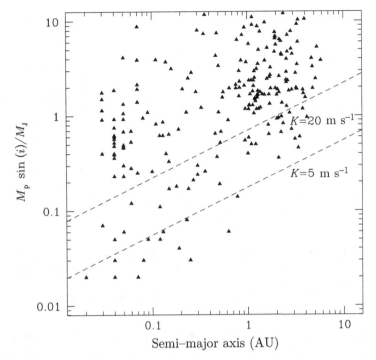

Fig. 1.11. The distribution of a sample of known extrasolar planets in semi-major axis and $M_p \sin(i)$, based on updated data from Butler *et al.* (2006). The dashed lines show the mass for which a planet around a Solar mass star induces a radial velocity signal of $K = 5 \text{ m s}^{-1}$ and $K = 20 \text{ m s}^{-1}$.

Although fascinating – particularly for studies of planetary and atmospheric structure – it is clear from Fig. 1.11 that the hot Jupiters are a relatively small subset of even detected giant extrasolar planets. The most notable aspect of planet detections at larger orbital radii is that typically these extrasolar giant planets do not have nearly circular orbits. For the planets in this sample with $M_p \sin(i) < 10 \ M_J$ and 1 AU $< a <$ 3 AU the median eccentricity is $\langle e \rangle = 0.27$. The eccentricity distribution is broad, and both Jovian analogs (planets at a few AU with close to circular orbits) and very eccentric planets with almost cometary orbits $e > 0.9$ are detected. There is no striking correlation between eccentricity and planet mass.

The distribution of planets in a plot of $M_p \sin(i)$ versus a primarily reflects the selection effects inherent to radial velocity surveys. The minimum detectable mass scales as $a^{1/2}$, Eq. (1.21), with the sample being reasonably complete for $K > 20 \text{ m s}^{-1}$. Within the currently detectable region of parameter space the planet mass function declines toward large masses, while the number of planets per logarithmic interval of orbital radius, $dN_p/d\log(a)$, increases toward large a up to the radius where selection effects cut in. Based on a sample of 1330 stars of

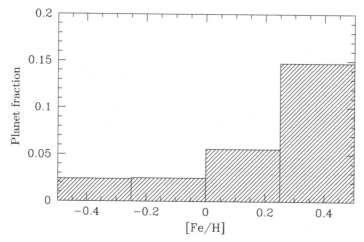

Fig. 1.12. The fraction of stars that host known extrasolar planets is plotted as a function of the metallicity of the host. The observed planet fraction increases strongly with metallicity. The currently observed fraction, plotted based on data from Fischer & Valenti (2005), is a lower limit to the true planet abundance.

spectral type FGKM, Marcy *et al.* (2005) derive a planet abundance within 5 AU of approximately 7%. This is certainly a lower limit as even many giant planets that are abundant at small orbital radii would be undetectable orbiting further out. Within this sample multiple planet systems are found to be common, with the planets in perhaps a third of the known multiple systems exhibiting evidence for the existence of mean-motion resonances.

Figure 1.12 shows the fraction of stars with known planetary companions as a function of the stellar metallicity,[9] using data from Fischer & Valenti (2005). Observed planet frequency rises rapidly with host metallicity, with even relatively modest increases in metallicity substantially enhancing the probability that currently detectable planets will be found around the star. Similar trends are seen in sub-samples of stars with either shallow or deep surface convection zones. This suggests that the trend of increasing planet frequency with metallicity is telling us something about planet formation – planets form more readily given a larger reservoir of solid material in the protoplanetary disk – rather than being a signature of stellar pollution by rocky debris scattered into the star subsequent to giant planet formation.

---

[9] Astronomical convention is to measure the abundance of heavy elements relative to hydrogen, with the resulting ratio being normalized to the ratio found in the Sun. A logarithmic scale is used, denoted by square brackets. Thus, a star with "[Fe/H] = 0" has the same fractional abundance of iron as the Sun, whereas one with [Fe/H] = 0.5 is enriched in iron by a factor ~ 3 as compared to the Sun. Iron is often taken to be a "typical" element when describing stellar metallicities though for very low metallicity stars, in particular, there can be large differences between e.g. [C/H] and [Fe/H].

1.8 Further reading

The properties of the planets, moons, and minor bodies of the Solar System are discussed in depth in any planetary science text. A good example is *Planetary Sciences* by Imke de Pater and Jack J. Lissauer (2001), Cambridge, UK: Cambridge University Press.

An extensive discussion of extrasolar planet detection methods can be found in the article by Andreas Quirrenbach in *Extrasolar Planets* (2006), Saas-Fee Advanced Course 31, D. Queloz, S. Udry, M. Mayor, and W. Benz (eds.), Berlin: Springer.

Data on the observed population of extrasolar planets continue to improve rapidly. For the latest results, it is best to consult contemporary reviews or download data compilations maintained at exoplanets.org and other websites.

2

Protoplanetary disk structure

Planets form from protoplanetary disks of gas and dust that are observed to surround young stars for the first few million years of their evolution. Disks form because stars are born from relatively diffuse gas (with particle number density $n \sim 10^5 \, \mathrm{cm}^{-3}$) that has too much angular momentum to collapse directly to stellar densities ($n \sim 10^{24} \, \mathrm{cm}^{-3}$). Disks survive as well-defined quasi-equilibrium structures because once gas settles into a disk around a young star its specific angular momentum *increases* with radius. To accrete, angular momentum must be lost from, or redistributed within, the disk gas, and this process turns out to require time scales that are much longer than the orbital or dynamical time scale.

In this chapter we discuss the structure of protoplanetary disks. Anticipating the fact that angular momentum transport is slow, we assume here that the disk is a static structure. This approximation suffices for a first study of the temperature, density, and composition profiles of protoplanetary disks, which are critical inputs for models of planet formation. It also permits investigation of the predicted emission from disks that can be compared to a large body of astronomical observations. We defer for Chapter 3 the tougher question of how the gas and solids within the disk evolve with time.

2.1 Disks in the context of star formation

Stars form in the Galaxy today from the small fraction of gas that exists in molecular form within relatively dense, cool, molecular clouds. Observationally, most star formation results in the formation of stellar clusters (which may subsequently disperse), within which most stars are part of binary or small multiple systems (Duquennoy & Mayor, 1991). Protoplanetary disks inherit their initial mass, size and chemical composition from this broader star formation environment, while their subsequent evolution may be influenced by environmental effects such as stellar radiation, ongoing gas accretion, or stellar flybys. The importance of environmental effects varies markedly depending upon the star formation environment. The Trapezium cluster at the core of the Orion Nebula, for example, has a stellar density in excess of 10^4 stars per pc^3, while the small cluster around the young massive star MWC 297 has a much lower density of $\rho_* \sim 10^2 \, pc^{-3}$ (Lada & Lada 2003). Evidently the common approximation that disks evolve in isolation will be better in the low density case. Surveys suggest that most stars form within rich clusters containing 100 or more stars, though even within such clusters the local environment can vary substantially (much of the Orion star forming region, for example, is much less dense than the Trapezium). Once disks have formed, the dominant environmental effects are normally due to close binary companions (if present) and external radiation produced by other stars within the cluster – especially massive ones that produce strong ultraviolet fluxes.

Molecular clouds are not homogeneous structures. Rather, their density and velocity fields exhibit structure across a wide range of spatial scales that is characteristic of turbulence (Larson, 1981; McKee & Ostriker, 2007). Any collapsing region will therefore possess nonzero angular momentum, leading to the formation of disks (and perhaps even binary companions) around newly formed stars. Within molecular clouds, stars form from dense knots of gas called molecular cloud cores. These cores have typical scales of ~ 0.1 pc and densities of $n \sim 10^5 \, cm^{-3}$. On these scales the *dynamical* importance of rotation is rather modest. An observational analysis by Goodman *et al.* (1993), for example, estimated the typical ratio of rotational to gravitational energy in dense cores to be

$$\beta \equiv \frac{E_{rot}}{|E_{grav}|} \sim 0.02. \tag{2.1}$$

Even these small values, however, correspond to very substantial reservoirs of angular momentum. If we crudely model a core as a uniform density sphere in solid body rotation, then we find that a $\beta = 0.02$ core with a mass of $M_\odot$ and a radius of 0.05 pc has an angular momentum

$$J_{core} \simeq 10^{54} \, g \, cm^2 \, s^{-1}, \tag{2.2}$$

which is three to four orders of magnitude in excess of the total angular momentum in the Solar System. Understanding how this angular momentum is lost or redistributed either during collapse, or subsequently, is at the heart of the *angular momentum problem* of star formation. Finessing this problem for now, we simply note that if a Solar mass star forms from such a cloud the mean specific angular momentum $l_{core} = J_{core}/M_\odot \approx 4 \times 10^{20} \, cm^2 \, s^{-1}$. Gas with this much angular momentum will circularize around the newly formed star at a radius r_{circ} where the specific angular momentum of a Keplerian orbit equals that of the core,

$$l_{\text{core}} = \sqrt{GM_\odot r_{\text{circ}}}. \tag{2.3}$$

For the parameters discussed above $r_{\text{circ}} \sim 100$ AU. The formation of disks with sizes comparable to or larger than the Solar System is thus an inevitable consequence of the collapse of molecular cloud cores.

2.1.1 Classification of Young Stellar Objects

A handful of young stars are to be found in the TW Hydrae Association at a distance of 50 pc from the Sun, but to obtain large samples of disk-bearing stars one must go out to richer star forming regions such as Ophiuchus ($d \approx 120$ pc), Taurus ($d \approx 150$ pc), or Orion ($d \approx 410$ pc). Spatially resolving even large disks at these distances is difficult, and as a consequence statistical measures of disk frequency and lifetime are more often derived using unresolved information from the spectral energy distribution (SED), which measures the distribution of flux as a function of frequency or wavelength. Conventionally, Young Stellar Objects (YSOs) are classified based on the slope of the SED in the infrared region of the spectrum, defined via a parameter

$$\alpha_{\text{IR}} \equiv \frac{\Delta \log(\lambda F_\lambda)}{\Delta \log \lambda}. \tag{2.4}$$

Conventions vary, but typically α_{IR} is measured by fitting observations that span a wavelength range between the near-IR (often the K band centered at 2.2 μm) and the mid-IR (often 10 μm or 24 μm).

Once the SED has been measured, YSOs are divided into four or five classes based on the magnitude of α_{IR} (Lada & Wilking, 1984; André *et al.*, 1993; André & Montmerle, 1994). Different observers use slightly different boundaries (depending, for example, on the particular wavelengths observed) but a typical classification scheme is:

- **Class 0**: the SED peaks in the far-IR or mm part of the spectrum, with no flux being detectable in the near-IR.
- **Class I**: the SED between near- and mid-IR wavelengths is approximately flat or rising, with $\alpha_{\text{IR}} \geq -0.3$. In modern literature this class is sometimes divided further into "flat-spectrum" sources with $\alpha_{\text{IR}} \simeq 0$ and true "Class I" sources with $\alpha_{\text{IR}} > 0.3$.
- **Class II**: the SED falls noticeably between near- and mid-IR wavelengths. The slope of the SED is $-1.6 \leq \alpha_{\text{IR}} < -0.3$.
- **Class III**: the IR SED is essentially that of a stellar photosphere, with $\alpha_{\text{IR}} < -1.6$.

The last two classes map closely on to an older classification scheme for optically visible pre-main-sequence stars that is based on the equivalent width of the Hα

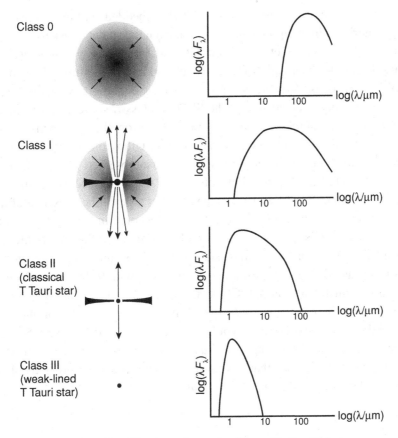

Fig. 2.1. Classification scheme for Young Stellar Objects.

line. Classical T Tauri stars[1] have an equivalent width of Hα that is in excess of 10Å. These are normally Class II sources. Weak-lined T Tauri stars have an Hα equivalent width that is less than 10Å. These are Class III objects.

Although the classification scheme for YSOs is fundamentally an empirical one, it is by now so entwined with a theoretical interpretation as an evolutionary sequence (Adams *et al.*, 1987) that it makes little sense to discuss the one without the other. Within the sequence, illustrated in Fig. 2.1, the Class 0 YSOs are identified with the least evolved objects observed during the earliest stages of cloud collapse. Any protostar present at the center of the cloud is so deeply embedded within optically thick gas and dust that it is not visible even at near-IR wavelengths. At this stage, rotationally supported material may be present, but its properties are largely unconstrained by observations. The first objects in which disks are detected

---

[1] Named after the prototype T Tauri, which was recognized as being part of an interesting class of variable stars associated with Galactic nebulae by Joy (1945).

are the Class I sources. These YSOs are still embedded within an envelope of infalling material, which reprocesses radiation from the star and disk toward longer wavelengths. Outflows or jets are often detected from these young sources with velocities that are high enough (of the order of 100 km s^{-1}) to imply launch points close to the star. Class II sources represent a later evolutionary phase when the envelope has largely been accreted. The SED of Class II sources can be modeled as the sum of that from the now optically visible star together with emission in the IR and mm from a surrounding circumstellar disk. Ultraviolet emission above the value predicted for a bare stellar photosphere is also observed in Class II sources, and this is attributed to accretion hotspots on the stellar surface as gas in the disk flows on to the star. The IR and UV excesses that signify the presence of a disk are common in the youngest stellar clusters, but become increasingly rare after a few million years (Haisch *et al.*, 2001), which appears to be the typical lifetime of protoplanetary disks around low mass stars. Once disks have been dissipated what remains is a Class III YSO, a pre-main-sequence star with little or no evidence for primordial circumstellar material. Class III YSOs can still be distinguished from ordinary stars by their location above the main sequence in a Hertzsprung–Russell diagram and by other signatures of youth such as strong X-ray activity. Some also display evidence for surrounding *debris disks* – gas poor disks whose emission is attributed to short-lived dust that is continually regenerated by erosive collisions of larger solid bodies in orbit around the star.

2.2 Vertical structure

The equilibrium structure of gas orbiting a star is determined in general by solving for a steady-state solution to the hydrodynamic equations and the Poisson equation for the gravitational potential. Such an exercise is nontrivial, and even when a solution can be found its dynamical stability is not guaranteed (Papaloizou & Pringle, 1984).

For protoplanetary disks two simplifications make the problem a great deal more straightforward. First, it is usually justified to assume that the total disk mass $M_{disk} \ll M_*$, and this allows us to neglect the gravitational potential of the disk and consider only stellar gravity. This approximation is accurate for disks with masses comparable to the minimum mass Solar Nebula ($M_{disk} \simeq 0.01 M_\odot$), becomes marginal for some of the most massive observed disks ($M_{disk} \simeq 0.1 M_\odot$), and fails at sufficiently early epochs when the disk mass may be comparable to that of the protostar. We will discuss some of the dynamical effects that are important for massive disks later on, but for now we arbitrarily define a protoplanetary disk as one whose mass is small enough ($M_{disk} \leq 0.1 M_\odot$) that the star dominates the

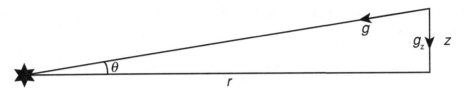

Fig. 2.2. The vertical structure of geometrically thin disks is set by a balance between the vertical component of the star's gravity g_z and a pressure gradient.

potential.[2] Second, the vertical thickness of the disk h is invariably a small fraction of the orbital radius. This follows from the fact that a disk has a large surface area and can cool via radiative losses rather efficiently. Efficient cooling implies relatively low disk temperatures and pressures, which are unable to support the gas against gravity except in a geometrically thin disk configuration with $h/r \ll 1$.

The structure of a geometrically thin protoplanetary disk follows from considering the vertical force balance at height z above the mid-plane in a disk orbiting at cylindrical radius r around a star of mass M_*. The vertical component of the gravitational acceleration (Fig. 2.2),

$$g_z = g \sin \theta = \frac{GM_*}{(r^2 + z^2)} \frac{z}{(r^2 + z^2)^{1/2}}, \tag{2.5}$$

must balance the acceleration due to the vertical pressure gradient in the gas $(1/\rho)(\mathrm{d}P/\mathrm{d}z)$. If we assume that the disk is vertically isothermal (a reasonable first guess if the temperature of the gas is set by stellar irradiation) and write the pressure as $P = \rho c_\mathrm{s}^2$, with c_s being the sound speed, we have

$$c_\mathrm{s}^2 \frac{\mathrm{d}\rho}{\mathrm{d}z} = -\frac{GM_* z}{(r^2 + z^2)^{3/2}} \rho. \tag{2.6}$$

The solution is

$$\rho = C \exp \left[\frac{GM_*}{c_\mathrm{s}^2 (r^2 + z^2)^{1/2}} \right], \tag{2.7}$$

where the constant of integration C is set by the mid-plane density. This expression is rarely used. Rather, we note that for a thin disk $z \ll r$ and $g_z \simeq \Omega^2 z$, where $\Omega = \sqrt{GM_*/r^3}$ is the Keplerian angular velocity. In this limit the vertical density profile has the simple form

$$\rho = \rho_0 e^{-z^2/2h^2}, \tag{2.8}$$

[2] More massive disks are sometimes described as protostellar rather than protoplanetary disks, but there is no consistent usage of these terms in the literature.

where the mid-plane density ρ_0 can be written in terms of the full-plane surface density Σ as

$$\rho_0 = \frac{1}{\sqrt{2\pi}} \frac{\Sigma}{h}, \tag{2.9}$$

and h, the vertical disk scale-height, is given by,

$$h \equiv \frac{c_s}{\Omega}. \tag{2.10}$$

Note that this means that the geometric thickness of the disk $h/r = \mathcal{M}^{-1}$, where the Mach number of the flow $\mathcal{M} = v_K/c_s$.

The assumptions that went into this calculation are generally self-consistent. Detailed disk models computed by Bell *et al.* (1997) show that for a disk accreting at a low rate on to a Solar mass star the mid-plane temperature at 1 AU is around $T \simeq 100$ K. The corresponding isothermal sound speed is

$$c_s^2 = \frac{k_B T}{\mu m_p}, \tag{2.11}$$

where k_B is the Boltzmann constant and μ is the mean molecular weight in units of the proton mass m_p. Taking $\mu = 2.3$ for a fully molecular gas of cosmic composition we have that $c_s \approx 0.6$ km s^{-1} and (at 1 AU around a Solar mass star) $h/r \approx 0.02$. The condition that the disk is geometrically thin is adequately satisfied. We should also verify that the vertical structure is not modified by the gravitational force from the disk itself. Representing the disk as a thin infinite sheet of mass with surface density Σ, a straightforward application of Gauss' theorem shows that the acceleration outside the sheet is constant with distance,

$$g_{z,\text{disk}} = 2\pi G\Sigma. \tag{2.12}$$

Equating this disk contribution to the acceleration due to the vertical component of stellar gravity at $z = h$, the condition that the self-gravity of the disk can be neglected when computing the vertical structure becomes

$$\Sigma < \frac{M_* h}{2\pi r^3}. \tag{2.13}$$

For the minimum mass Solar Nebula at 1 AU this condition is satisfied by more than an order of magnitude. More generally, if we write the enclosed disk mass as $M_{\text{disk}}(r) \sim \pi r^2 \Sigma$, self-gravity is ignorable provided that

$$\frac{M_{\text{disk}}}{M_*} < \frac{1}{2}\left(\frac{h}{r}\right). \tag{2.14}$$

At large radii $h/r \sim 0.1$, so for a massive disk with $M_{\text{disk}}/M_* = 0.1$ the neglect of disk self-gravity in the calculation of the vertical structure is not justified. Such disks require additional consideration, however, since the condition $M_{disk}/M_* \sim h/r$ also describes the mass above which disk self-gravity results in the disk becoming unstable to the development of nonaxisymmetric structure in the form of spiral waves. We discuss this separate (and more important) effect of self-gravity in Chapter 3.

The shape of the disk depends upon $h(r)/r$. If we parameterize the radial variation of the sound speed via

$$c_s \propto r^{-\beta}, \tag{2.15}$$

then the aspect ratio varies as

$$\frac{h}{r} \propto r^{-\beta+1/2}. \tag{2.16}$$

The disk will flare – i.e. h/r will increase with radius giving the disk a bowl-like shape – if $\beta < 1/2$. This requires a temperature profile $T(r) \propto r^{-1}$ or shallower. In an unwarped flaring disk the star is visible from all points on the surface of the disk.

2.3 Radial force balance

The density profile of the disk in the radial direction cannot be derived without either considering the nature of angular momentum transport, or by appealing to observational constraints (such as the minimum mass Solar Nebula). The orbital velocity of disk gas, however, can be determined given a surface density and temperature profile. We start from the momentum equation for an unmagnetized and inviscid fluid,

$$\frac{\partial \mathbf{v}}{\partial t} + (\mathbf{v} \cdot \nabla)\mathbf{v} = -\frac{1}{\rho}\nabla P - \nabla \Phi, \tag{2.17}$$

where $\mathbf{v}$ is the velocity, ρ the density, P the pressure, and Φ the gravitational potential. Specializing to a stationary axisymmetric flow in which the potential is dominated by that of the star, the radial component of the momentum equation implies that the orbital velocity of the gas $v_{\phi,gas}$ is given by,

$$\frac{v_{\phi,gas}^2}{r} = \frac{GM_*}{r^2} + \frac{1}{\rho}\frac{dP}{dr}. \tag{2.18}$$

Since the pressure near the disk mid-plane normally decreases outward, the second term on the right-hand-side is negative and the azimuthal velocity of the gas is slightly less than the Keplerian velocity (Eq. 1.18) of a point mass particle orbiting at the same radius. To quantify the difference we write the variation of the mid-plane pressure as a power-law near some fiducial radius r_0,

$$P = P_0 \left(\frac{r}{r_0}\right)^{-n},$$ (2.19)

where $P_0 = \rho_0 c_s^2$. Substituting we find that,

$$v_{\phi,\mathrm{gas}} = v_K \left(1 - n\frac{c_s^2}{v_K^2}\right)^{1/2}.$$ (2.20)

Recalling the definition of the vertical scale-height (Eq. 2.10) we note that the deviation from Keplerian velocity is $\mathcal{O}(h/r)^2$, and hence small for geometrically thin disks. For example, if the disk has a constant value of $h(r)/r = 0.05$ and a surface density profile $\Sigma \propto r^{-1}$ we obtain $n = 3$ and

$$v_{\phi,\mathrm{gas}} \simeq 0.996 v_K.$$ (2.21)

When considering the motion of the gas alone this difference is utterly negligible and we can safely assume the gas moves at the Keplerian velocity. The slightly lower gas velocity is however important for the evolution of solid bodies within the disk, since it results in aerodynamic drag and resultant orbital decay.

2.4 Radial temperature profile of passive disks

The time scale for the disk to attain thermal equilibrium is generally much less than the time scale for evolution of either the disk or the star. The temperature profile of the disk is then set by the balance between cooling and heating, for which there are two main sources:

- Intercepted stellar radiation, which is absorbed (usually by dust) and subsequently reradiated at longer wavelengths. Disks for which this is the main heating process are described as passive disks.
- Dissipation of gravitational potential energy, as matter in the disk spirals in towards the star.

The heating per unit area from both of these sources drops off with increasing distance from the star. If the disk extends out to large enough radii, heating due to the ambient radiation field provided by nearby stars can become significant and prevent the disk temperature from dropping below some floor level. For normal star forming environments this floor temperature might be 10–30 K.

The relative importance of accretional heating versus stellar irradiation depends upon the accretion rate and the amount of intercepted stellar radiation. Globally, inspiralling matter will radiate an amount of energy per unit mass that is approximately given by the gravitational potential at the stellar surface, GM_*/R_*, while the disk will intercept a fraction $f < 1$ of the stellar radiation. Taking $f = 1/4$ (the correct result for a razor-thin disk that extends all the way to the stellar surface) the accretion rate $\dot{M}$ above which accretional heating dominates can be crudely estimated as

$$\frac{GM_*\dot{M}}{R_*} = \frac{1}{4}L_*. \tag{2.22}$$

For a young Solar mass star with a luminosity $L_* = L_\odot = 3.9 \times 10^{33}$ erg s^{-1} and a radius $R_* = 2R_\odot$,

$$\dot{M} \approx 2 \times 10^{-8} \, M_\odot \text{yr}^{-1}. \tag{2.23}$$

Measured accretion rates[3] for classical T Tauri stars (Gullbring *et al.*, 1998) range from an order of magnitude above this critical rate to two orders of magnitude below, so it is oversimplifying to treat protoplanetary disks as being either always passive or always active. Rather, the thermal structure of disks at early epochs (when accretion is strongest) is likely dominated by internal heating due to the accretion, whereas at later times reprocessing of stellar radiation dominates.

2.4.1 Razor-thin disks

The temperature profile and spectral energy distribution of a passive protoplanetary disk are determined by the shape of the disk (whether it is flat, flared, or warped) and by the mechanism by which the absorbed stellar radiation is re-emitted. We begin by considering the simplest model: a flat razor-thin disk in the equatorial plane that absorbs all incident stellar radiation and re-emits it locally as a single temperature blackbody. The back-warming of the star by the disk is neglected.

We consider a surface in the plane of the disk at distance r from a star of radius R_*. The star is assumed to be a sphere of constant brightness I_*. Setting up spherical polar coordinates such that the axis of the coordinate system points to the center of the star, as shown in Fig. 2.3, the stellar flux passing through this surface is

$$F = \int I_* \sin\theta \cos\phi \, d\Omega, \tag{2.24}$$

where $d\Omega$ represents the element of solid angle. We count the flux coming from the top half of the star only (and to be consistent equate that to radiation from only

---

[3] The accretion rate can be determined observationally by measuring the excess (above the "normal" stellar photospheric emission) luminosity produced when inflowing gas impacts the stellar surface. This accretion signature is strongest toward the ultraviolet part of the spectrum (i.e. at wavelengths that are shorter than the blackbody peak for relatively cool pre-main-sequence stars), though it is still difficult to separate from the stellar flux.

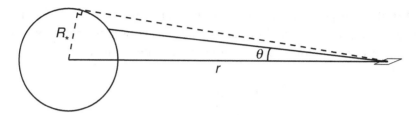

Fig. 2.3. Geometry for calculation of the radial temperature profile of a razor-thin protoplanetary disk. The flux impinging on the surface of the disk at distance r from the star is computed by integrating over the visible stellar surface.

the top surface of the disk), so the limits on the integral are

$$-\pi/2 < \phi \le \pi/2$$

$$0 < \theta < \sin^{-1}\left(\frac{R_*}{r}\right). \tag{2.25}$$

Substituting $d\Omega = \sin\theta d\theta d\phi$, the integral for the flux is

$$F = I_* \int_{-\pi/2}^{\pi/2} \cos\phi d\phi \int_0^{\sin^{-1}(R_*/r)} \sin^2\theta d\theta, \tag{2.26}$$

which evaluates to,

$$F = I_* \left[\sin^{-1}\left(\frac{R_*}{r}\right) - \left(\frac{R_*}{r}\right)\sqrt{1 - \left(\frac{R_*}{r}\right)^2}\right]. \tag{2.27}$$

For a star with effective temperature T_*, the brightness $I_* = (1/\pi)\sigma T_*^4$, with σ the Stefan–Boltzmann constant (e.g. Rybicki & Lightman, 1979). Equating F to the one-sided disk emission σT_{disk}^4 we obtain a radial temperature profile

$$\left(\frac{T_{\text{disk}}}{T_*}\right)^4 = \frac{1}{\pi}\left[\sin^{-1}\left(\frac{R_*}{r}\right) - \left(\frac{R_*}{r}\right)\sqrt{1 - \left(\frac{R_*}{r}\right)^2}\right]. \tag{2.28}$$

Integrating over radii, we obtain the total disk luminosity,

$$L_{\text{disk}} = 2 \times \int_{R_*}^{\infty} 2\pi r \sigma T_{\text{disk}}^4 dr$$

$$= \frac{1}{4}L_*. \tag{2.29}$$

We conclude that a flat passive disk extending all the way to the stellar equator intercepts a quarter of the stellar flux. The ratio of the observed bolometric luminosity of such a disk to the stellar luminosity will vary with viewing angle, but clearly a flat passive disk is predicted to be less luminous than the star.

The form of the temperature profile given by Eq. (2.28) is not very transparent. Expanding the right-hand-side in a Taylor series in the limit that $(R_*/r) \ll 1$ (i.e. far from the stellar surface), we obtain

$$T_{\text{disk}} \propto r^{-3/4}, \tag{2.30}$$

as the limiting temperature profile of a thin, flat, passive disk. For fixed molecular weight μ this in turn implies a sound speed profile

$$c_s \propto r^{-3/8}. \tag{2.31}$$

Assuming vertical isothermality, the aspect ratio given by Eq. (2.16) is

$$\frac{h}{r} \propto r^{1/8}, \tag{2.32}$$

and we predict that the disk ought to flare modestly to larger radii. If the disk does flare then the outer regions intercept a larger fraction of stellar photons, leading to a higher temperature. As a consequence, a temperature profile $T_{\text{disk}} \propto r^{-3/4}$ is probably the steepest profile we would expect to obtain for a passive disk.

2.4.2 Flared disks

The next step in sophistication is to consider a flared disk. If at cylindrical distance r from the star the disk absorbs stellar radiation at a height h_p above the mid-plane, the disk is described as flared if the ratio h_p/r is an increasing function of radius. Features of flared disks are, first, that all points on the surface of the disk have a clear line of sight to the star and, second, that as seen from the star the disk subtends a greater solid angle than a razor-thin disk. Flared disks absorb a greater fraction of the stellar radiation than flat disks, and thus produce stronger IR excesses. The temperature profile is also modified and this changes the shape of the resulting SED.

The temperature profile of a flared disk can be computed in the same way as for a razor-thin disk, namely by evaluating the flux (Eq. 2.24) by integrating over the part of the stellar surface visible from the disk surface at radius r. The exact calculation can be found in the appendix to Kenyon & Hartmann (1987), and while it is conceptually simple some messy geometry is required. Here we adopt an approximate treatment valid for $r \gg R_*$, and consider the star to be a point source of radiation. At cylindrical distance r, stellar radiation is absorbed by the disk at height h_p above the mid-plane. Note that h_p is *not* the same as the disk scale-height h, since the absorption of stellar radiation depends not just on the density but also on the opacity of the disk material to stellar photons (the height of the disk's photosphere – i.e. the surface at which the optical depth to the disk's own

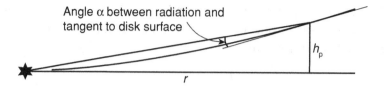

Fig. 2.4. Geometry for calculation of the radial temperature profile of a flared protoplanetary disk. At distance $r \gg R_*$, radiation from the star is absorbed by the disk at height h_p above the mid-plane. The angle between the tangent to the disk surface and the radiation is α.

thermal radiation is $\tau = 2/3$ – is yet a third different height). From consideration of the geometry in Fig. 2.4 the angle between the incident radiation and the local disk surface is given by[4]

$$\alpha = \frac{dh_p}{dr} - \frac{h_p}{r}.$$ (2.33)

The rate of heating per unit disk area at distance r is

$$Q_+ = 2\alpha \left(\frac{L_*}{4\pi r^2} \right),$$ (2.34)

where the factor of two comes from the fact that the disk has two sides and we have assumed that *all* of the stellar surface is visible from the surface of the disk (i.e. that the optically thick disk does not extend all the way to the stellar surface). Equating the heating rate to the rate of cooling by blackbody radiation,

$$Q_- = 2\sigma T_{\text{disk}}^4,$$ (2.35)

the temperature profile becomes

$$T_{\text{disk}} = \left(\frac{L_*}{4\pi\sigma} \right)^{1/4} \alpha^{1/4} r^{-1/2}.$$ (2.36)

Since $L_* = 4\pi R_*^2 \sigma T_*^4$ an equivalent expression is

$$\frac{T_{\text{disk}}}{T_*} = \left(\frac{R_*}{r} \right)^{1/2} \alpha^{1/4}.$$ (2.37)

The interior of an irradiated protoplanetary disk can reasonably be assumed to be isothermal, so this equation specifies the central sound speed and hence the vertical scale-height h via Eq. (2.10). If we additionally specify a relation between h and h_p – for example by assuming that $h_p \propto h$ which may be plausible if the disk is very optically thick – then Eq. (2.10), (2.11), (2.33), and (2.37) form a closed system of

[4] A more accurate approximation including a correction for the finite size of the star is $\alpha \simeq 0.4R_*/r + r d(h_p/r)/dr$.

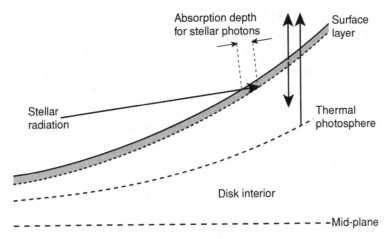

Fig. 2.5. The physics underlying radiative equilibrium models for protoplanetary disks. Stellar radiation is absorbed in a thin surface layer by dust, which reradiates in the infrared both upward into space and downward where it is absorbed once more and acts to heat the disk interior. The local emission from the disk is a superposition of radiation from the hot surface layer and the cooler interior.

equations whose solution determines α. At large radii, Kenyon & Hartmann (1987) find that the surface temperature approaches

$$T_{\rm disk}(r) \propto r^{-1/2}, \tag{2.38}$$

which is as expected substantially flatter than the $T_{\rm disk} \propto r^{-3/4}$ profile of a flat disk (Eq. 2.30).

2.4.3 Radiative equilibrium disks

Thus far we have assumed that the stellar radiation intercepted by the disk is reradiated as a single temperature blackbody. Additional complexity arises for real disks because the dominant opacity is provided by dust, which absorbs relatively short wavelength starlight (at around 1 μm) more efficiently than it emits longer wavelength thermal radiation. Figure 2.5 depicts the absorption and reradiation of stellar radiation for an optically thick disk in which dust is mixed evenly with the gas in the vertical direction. Short wavelength stellar radiation is absorbed by dust within a relatively thin surface layer for which the optical depth to grazing starlight $\tau \lesssim 1$. The layer is optically thin ($\tau \ll 1$) to the longer wavelength thermal radiation emitted by the hot dust. Approximately half of the incoming stellar flux is then reradiated directly to space, while the downward directed half is absorbed by and heats the interior of the disk. The local emission from the disk is then a superposition of a cool blackbody component from the disk interior with a warmer dilute blackbody generated by the surface layer.

The temperature of the dust in the surface layer can be calculated given knowledge of the radiative properties of the dust. We define the emissivity ϵ as the ratio of the efficiency with which the dust emits or absorbs radiation relative to a blackbody surface (by definition a perfect absorber and emitter with $\epsilon = 1$). Dust is generally a good absorber of radiation at wavelengths that are small compared to the particle size, with the emissivity dropping at longer wavelengths. For spherical dust particles of radius s we have, approximately,

$$\epsilon = 1, \qquad \lambda \leq 2\pi s$$

$$\epsilon = \left(\frac{\lambda}{2\pi s}\right)^{-1}, \quad \lambda > 2\pi s. \tag{2.39}$$

This is equivalent to assuming that the wavelength dependence of the monochromatic opacity $\kappa_\nu \propto \lambda^{-1}$. If a dust particle at radius r is exposed to a stellar flux $F_* = L_*/(4\pi r^2)$, its equilibrium temperature T_{dust} is determined by balancing heating at a rate $\pi s^2 \epsilon_* F_*$, where ϵ_* is the weighted emissivity for absorption of the stellar radiation spectrum, with cooling at a rate $4\pi s^2 \sigma T_{\text{dust}}^4 \epsilon_{\text{dust}}$, where ϵ_{dust} is the weighted emissivity for the dust particle's thermal emission. Equating,

$$\frac{T_{\text{dust}}}{T_*} = \left(\frac{\epsilon_*}{\epsilon_{\text{dust}}}\right)^{1/4} \left(\frac{R_*}{2r}\right)^{1/2}. \tag{2.40}$$

For particles that are small enough such that $\epsilon < 1$ for both absorption of stellar radiation and emission of thermal radiation, we can estimate the ratio $\epsilon_*/\epsilon_{\text{dust}}$ by evaluating the emissivity at the wavelength where thermal radiation peaks. We then have that $\epsilon \propto \lambda^{-1} \propto T$, so that $\epsilon_*/\epsilon_{\text{dust}} = T_*/T_{\text{dust}}$. Substituting in Eq. (2.40),

$$\frac{T_{\text{dust}}}{T_*} = \left(\frac{R_*}{2r}\right)^{2/5}. \tag{2.41}$$

Since the mismatch between the emissivity for absorption and emission becomes more severe at larger radii, the temperature of dust exposed to stellar radiation at the disk surface drops even more slowly with increasing distance than the temperature of a single temperature flared disk. Inserting fiducial values for the stellar radius $R_* = 2R_\odot$ and stellar effective temperature $T_* = 4000$ K we find that,

$$T_{\text{dust}} = 470 \left(\frac{r}{1 \text{ AU}}\right)^{-2/5} \text{ K}. \tag{2.42}$$

This scaling, and the analysis leading up to it, were derived by Chiang & Goldreich (1997).

This analysis assumes that the dust particles can be considered in isolation, whereas in reality they are embedded within gas whose temperature will not in general be equal to the dust temperature given by Eq. (2.40). If $T_{gas} < T_{dust}$, collisions between molecules and the dust particles will result in nonradiative cooling of the dust. If this cooling process is efficient enough it will invalidate the radiative equilibrium that was assumed in deriving Eq. (2.40).

To evaluate the conditions for which it is reasonable to assume that the dust and gas are thermally decoupled, we note that collisions between molecules and dust particles occur at a rate (per unit area of dust) $n_{gas} v_{th}$, where n_{gas} is the number density of molecules, and v_{th}, the thermal velocity of the molecules, is given by

$$v_{th} = \left(\frac{8k_B T_{gas}}{\pi \mu m_H}\right)^{1/2},$$ (2.43)

where μm_H is the mean particle mass. Each collision carries away of the order of kT_{dust} worth of energy. Collisions will be ignorable for thermal equilibrium of dust particles if

$$kT_{dust} n_{gas} v_{th} < \epsilon_{dust} \sigma T_{dust}^4.$$ (2.44)

Evaluating the thermal velocity of the gas assuming that its temperature is comparable to that of the dust, we find that collisions can be neglected provided that

$$n_{gas} < \left(\frac{\pi}{8}\right)^{1/2} \frac{(\mu m_H)^{1/2}}{k_B^{3/2}} \epsilon_{dust} \sigma T_{dust}^{5/2}.$$ (2.45)

The two-temperature disk model of Fig. 2.5 will therefore persist provided that the gas density in the region where dust is marginally optically thick to stellar radiation is low enough. To give a specific example, we can estimate whether a model in which the gas and dust temperatures are unequal is self-consistent at 1 AU, where we would predict that $T_{dust} = 470$ K. For dust particles whose size is such that $\epsilon_{dust} \simeq 0.1$ the above expression suggests that collisions with hydrogen molecules are ignorable for gas densities $n_{gas} \lesssim 2 \times 10^{13}$ cm^{-3}. Taking a minimum mass Solar Nebula surface density at 1 AU of 1.7×10^3 g cm^{-2} with $h/r = 0.05$, the *mid-plane* hydrogen number density is a few $\times 10^{14}$ cm^{-3}. The dust and the gas will therefore be thermally coupled if the dust has settled to the mid-plane and absorbs stellar radiation there. Conversely, n_{gas} will be below the threshold for thermal coupling if the dust is well-mixed with the gas and absorption occurs at a height $z \gtrsim 2.5h$. Evidently the extent of dust–gas energy exchange at 1 AU is rather subtle, and strictly it cannot be considered in isolation from questions of dust settling and agglomeration. At larger radii – where the surface density is smaller and the scale-height larger – the effect of collisions on the thermal state of irradiated dust can more safely be neglected.

2.4.4 The Chiang–Goldreich model

Although computation of disk models that include all of the aforementioned physics generally requires numerical techniques, useful analytic approximations are available. The most widely used is that developed by Chiang & Goldreich (1997). They considered a disk surrounding a 0.5 $M_\odot$ star with radius $R_* = 2.5\ R_\odot$ and effective temperature $T_* = 4000$ K. The disk has a surface density profile $\Sigma(r) \propto r^{-3/2}$, a surface density at 1 AU of 10^3 g cm^{-2}, and a dust to gas ratio of 10^{-2}. The resulting temperature of dust at the surface of the disk is

$$T_{\text{dust}} = 550 \left(\frac{r}{1\ \text{AU}}\right)^{-2/5} \text{K}. \tag{2.46}$$

The temperature of the disk interior T_i and the height of the visible photosphere h_p are approximated by piecewise polynomials. At 0.4 AU $< r <$ 84 AU,

$$T_i = 150 \left(\frac{r}{1\ \text{AU}}\right)^{-3/7} \text{K},$$

$$\frac{h_p}{r} = 0.17 \left(\frac{r}{1\ \text{AU}}\right)^{2/7}. \tag{2.47}$$

At 84 AU $< r <$ 209 AU,

$$T_i = 21 \text{ K},$$

$$\frac{h_p}{r} = 0.064 \left(\frac{r}{1\ \text{AU}}\right)^{1/2}. \tag{2.48}$$

At 209 AU $< r <$ 270 AU,

$$T_i = 200 \left(\frac{r}{1\ \text{AU}}\right)^{-19/45} \text{K},$$

$$\frac{h_p}{r} = 0.20 \left(\frac{r}{1\ \text{AU}}\right)^{13/45}. \tag{2.49}$$

These expressions, which can be generalized without difficulty to other stellar or disk properties, are quite useful in practice for giving an idea of the physical conditions within weakly accreting protoplanetary disks. Their main limitation arises from the fact that in ignoring heating due to accretion they tend to underestimate the temperature at the disk mid-plane. This deficiency can be remedied (e.g. Garaud & Lin, 2007), but there remains considerable uncertainty in *all* disk models due to the uncertain distribution (both radially and vertically) of dust and its evolution over time.

2.4.5 Spectral energy distributions

The spectral energy distribution of the disk can be computed by summing up the local disk emission at each radius, weighted by the area. The simplest case is where each annulus in the disk radiates as a blackbody at the local temperature $T_{\text{disk}}(r)$. If the disk extends from r_{in} to r_{out} the disk spectrum is

$$F_\lambda \propto \int_{r_{\text{in}}}^{r_{\text{out}}} 2\pi r \, B_\lambda[T_{\text{disk}}(r)] dr, \tag{2.50}$$

where B_λ is the Planck function,

$$B_\lambda(T) = \frac{2hc^2}{\lambda^5} \frac{1}{\exp[hc/\lambda k_B T] - 1}. \tag{2.51}$$

The behavior of the spectrum implied by Eq. (2.50) is easy to derive. At long wavelengths $\lambda \gg hc/k_B T_{\text{disk}}(r_{\text{out}})$ we recover the Rayleigh–Jeans form

$$\lambda F_\lambda \propto \lambda^{-3}, \tag{2.52}$$

while at short wavelengths $\lambda \ll hc/k_B T_{\text{disk}}(r_{\text{in}})$ there is an exponential cut-off that matches that of the hottest annulus in the disk,

$$\lambda F_\lambda \propto \lambda^{-4} \exp\left[\frac{-hc}{\lambda k_B T_{\text{disk}}(r_{\text{in}})}\right]. \tag{2.53}$$

For intermediate wavelengths,

$$\frac{hc}{k_B T_{\text{disk}}(r_{\text{in}})} \ll \lambda \ll \frac{hc}{k_B T_{\text{disk}}(r_{\text{out}})}, \tag{2.54}$$

the form of the spectrum depends upon the radial profile of the disk temperature distribution. For the razor-thin disk model with $T_{\text{disk}}(r) \propto r^{-3/4}$ we substitute

$$x \equiv \frac{hc}{\lambda k_B T_{\text{disk}}(r_{\text{in}})} \left(\frac{r}{r_{\text{in}}}\right)^{3/4}, \tag{2.55}$$

into Eq. (2.50). We then have, approximately,

$$F_\lambda \propto \lambda^{-7/3} \int_0^\infty \frac{x^{5/3} dx}{e^x - 1} \propto \lambda^{-7/3}, \tag{2.56}$$

and so

$$\lambda F_\lambda \propto \lambda^{-4/3}. \tag{2.57}$$

The overall spectrum, shown schematically in Fig. 2.6, is that of a "stretched" blackbody (Lynden-Bell & Pringle, 1974). This simple spectrum is too steep in the infrared to account for the observed properties of Class I and many Class II sources. The more realistic flared disk models, or those which incorporate a warm surface layer, yield flatter spectral energy distributions in the infrared which are able to reproduce the features of many YSO SEDs.

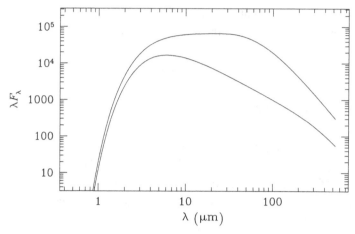

Fig. 2.6. The spectral energy distributions (λF_λ, plotted in arbitrary units on the vertical axis) of disks radiating as multicolor blackbodies with temperature profiles $T_{\text{disk}}(r) \propto r^{-1/2}$ (upper curve) and $T_{\text{disk}}(r) \propto r^{-3/4}$ (lower curve). For these toy models the disk was assumed to have a temperature of 1000 K at the inner edge at 0.1 AU, and to extend out to 30 AU.

2.5 Opacity

Dust is the dominant opacity source within protoplanetary disks everywhere except in the very innermost regions, where the temperature is high enough ($T \simeq 1500$ K) for dust particles to be destroyed leaving only molecular opacity. For an individual dust particle the cross-section to radiation of a given wavelength will depend upon the particle size, structure (for example the particle may have a spherical or more complex geometry), and composition. At a particular location within the disk, the total opacity due to dust will depend primarily upon the temperature and chemical composition of the disk (which determine which types of dust or ice are present) and the size distribution of the particles, which may change as coagulation proceeds. To a lesser extent, the gas density also influences the opacity by modifying the temperatures at which different species of dust particle are predicted to evaporate. Given a model for each of the above effects, it is possible to compute the frequency dependent opacity κ_ν of the gas and dust within the disk as a function of disk temperature and density.

Within the inner disk (where "inner" encompasses most of the radii of greatest interest for planet formation) it is typically true that the mean free path of thermal radiation from the disk interior is small compared to the disk scale-height. In this regime the radiation field is approximately isotropic and blackbody and the flux is proportional to the gradient of the energy density of the radiation $\mathbf{F}_{\text{rad}} = -c\nabla(aT^4)/(3\kappa_{\text{R}}\rho)$. The relevant opacity that enters into the equation for

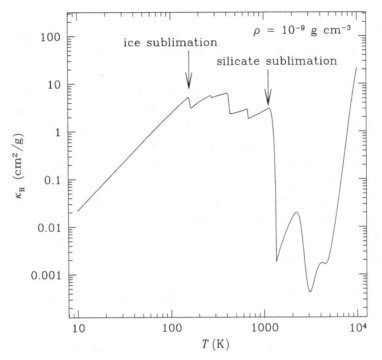

Fig. 2.7. The Rosseland mean opacity for dusty gas in the protoplanetary disk, as calculated by Semenov *et al.* (2003) for a gas density of 10^{-9} g cm^{-3}. The dust is assumed to be composed of homogeneous spherical particles whose size distribution follows a modified Mathis *et al.* (1977) law.

the flux is the Rosseland mean,

$$\frac{1}{\kappa_R} = \frac{\int_0^\infty (1/\kappa_\nu)(\partial B_\nu/\partial T)d\nu}{\int_0^\infty (\partial B_\nu/\partial T)d\nu}, \qquad (2.58)$$

where B_ν is the Planck function.

Figure 2.7 shows the temperature dependence of the Rosseland mean opacity for dusty gas of density 10^{-9} g cm^{-3}, as calculated by Semenov *et al.* (2003). The elemental abundances represent best estimates of the Solar values. The assumed size distribution of particles follows that proposed by Pollack *et al.* (1985),

$$n(s) = 1, \qquad\qquad\qquad s < 0.005 \ \mu m$$

$$n(s) = \left(\frac{s}{0.005 \ \mu m}\right)^{-3.5}, \qquad 0.005 \ \mu m \le s < 1 \ \mu m$$

$$n(s) = 4 \times 10^4 \left(\frac{s}{0.005 \ \mu m}\right)^{-5.5}, \quad 1 \ \mu m \le s < 5 \ \mu m$$

$$n(s) = 0, \qquad\qquad\qquad s \ge 5 \ \mu m. \qquad (2.59)$$

This distribution is itself a modified version of the Mathis *et al.* (1977) distribution that was derived based on observations of the wavelength dependence of stellar extinction. The specific model plotted assumes that the dust is composed of homogeneous spherical particles.

At low temperatures ($T \lesssim 150\,\mathrm{K}$), corresponding to regions of the disk beyond the snowline, water ice and volatile organic materials are the dominant particulates and the Rosseland mean opacity rises as T^2. At somewhat higher temperatures the opacity displays a sawtooth pattern that mirrors the temperatures at which successively hardier materials evaporate: first water ice, then in succession organics, troilite (FeS), iron, and silicates. Above the dust destruction temperature for silicates ($T \simeq 1500\,\mathrm{K}$) the opacity (which in this regime is the summed contribution of millions of molecular lines) plummets by at least two orders of magnitude. At still higher temperatures – which are occasionally of interest when modeling high accretion rate disks such as those encountered in FU Orionis objects – the opacity rises due to H^- scattering, enters a regime where bound–free and free–free absorption dominate, and finally plateaus at the electron scattering value.

2.5.1 Opacity in the optically thin outer disk

At large radii, lower temperatures and surface densities reduce the optical depth of the disk to its own thermal radiation. In the Chiang–Goldreich model, for example, for which $T_i \propto r^{-3/7}$ and $\Sigma \propto r^{-3/2}$ out to almost 100 AU, $\tau \sim \kappa_R(T_i)\Sigma/2 \propto r^{-33/14}$. Once the disk becomes optically thin the spectrum explicitly depends upon the wavelength dependence of the dust opacity. This occurs at mm wavelengths. If the mass of optically thin dust is M_d, Kirchoff's Law implies that the observed flux for a source at distance d is directly proportional to the opacity,

$$F_\nu = \frac{M_d}{d^2}\kappa_\nu B_\nu(T_i). \qquad (2.60)$$

At long wavelengths ($h\nu \ll k_B T_i$) the Planck function is well-approximated by the Rayleigh–Jeans form, $B_\nu(T) \approx 2k_B T \nu^2/c^2$, and we have

$$F_\nu \simeq \frac{2k_B}{c^2}\frac{M_d T_i}{d^2}\kappa_\nu \nu^2. \qquad (2.61)$$

This equation has two significant implications. First, provided that the disk temperature profile can be estimated observationally the observed flux in the mm regime provides a measure of the dust mass in the optically thin outer region of the disk. In practice the temperature profile can be constrained by comparing theoretical disk models to observations of the infrared SED which, since it is generated in the optically *thick* region of the disk, is sensitive to temperature but not to density.

Once the optically thin dust mass has been determined,[5] the total mass of the disk can be estimated given the expected gas to dust ratio for a disk of Solar (or some other) composition. Most quoted "disk masses" are derived in this manner.

The second consequence of Eq. (2.61) is that it allows for an observational determination of the frequency dependence of the opacity. If $\kappa_\nu \propto \nu^\beta$ then $F_\nu \propto \nu^{\beta+2}$. Observations of protoplanetary disks suggest that $\beta \approx 1$ (Beckwith & Sargent, 1991). This value is lower than that derived for dust in either diffuse interstellar clouds (Finkbeiner *et al.*, 1999) or molecular clouds (Goldsmith *et al.*, 1997), where $\beta_{ISM} \sim 2$ provides a better fit. (Note that β_{ISM} is the spectral index of the dust opacity in the Rayleigh–Jeans limit that would be observed in the interstellar medium.) It is plausible (Draine, 2006) that the lower values of β inferred for protoplanetary disks result from grain growth. Specifically, if the distribution of grain size follows a power-law $dn/ds \propto s^{-p}$ up to particles of size $s_{max} \gtrsim 3\lambda$ (i.e. an extended version of the small particle dust distribution described by Eq. 2.59), then Draine (2006) finds that the observed spectral index at wavelengths shorter than λ is

$$\beta \approx (p-3)\beta_{ISM}. \tag{2.62}$$

This is consistent with the observed $\beta \approx 1$ if $p \approx 3.5$ and $\beta_{ISM} \sim 2$.

2.5.2 Analytic opacities

Although numerical codes for computing low temperature opacities are widely available, analytic approximations can still be valuable, for example when computing analytic models of the vertical structure of actively accreting protoplanetary disks. Zhu *et al.* (2009) provide a set of analytic expressions for the opacity that improve upon the widely used Bell & Lin (1994) formulae for 10^2 K $< T < 10^5$ K. Writing the Rosseland mean opacity in the form

$$\log_{10} \kappa_R = A \log_{10} T + B \log_{10} P + C, \tag{2.63}$$

where P is the pressure and A, B, and C are constants, Zhu *et al.* (2009) give fits that describe the behavior of κ_R across the different regimes of grain, molecular, and atomic opacity. The fitting constants are listed in Table 2.1. Transitions between the different regimes occur when the values of the opacity given by neighboring pairs of expressions are equal, and are normally smoothed with an ad hoc function.

These and other analytic expressions are primarily fits to numerical results and as such do not capture the full subtlety of the actual opacity. Nonetheless there

---

[5] For a disk surface density profile that is shallower than $\Sigma(r) \propto r^{-2}$, most of the mass resides at large radius, so estimating the total dust mass from the optically thin emission is reasonable provided that the disk is large enough.

Table 2.1. *Fitting constants for analytic low temperature opacities.*

A	B	C	Opacity regime
0.738	0	−1.277	grains
−42.98	1.312	135.1	grain sublimation
4.063	0	−15.013	water
−18.48	0.676	58.93	empirical
2.905	0.498	−13.995	molecules
10.19	0.382	−40.936	H scattering
−3.36	0.928	12.026	bound–free & free–free
0	0	−0.48	electron scattering

are relatively few circumstances in which uncertainty in the opacity itself is the dominant source of error in theoretical calculations of planet formation – it is far more often the case that the amount and size distribution of dust are the biggest unknowns – and approximate formulae are thus useful in practice.

2.6 The condensation sequence

Since stars and their disks form from the collapse of the same material it is reasonable to assume that the elemental abundances in the disk (i.e. the ratio of the abundance of each element to the abundance of hydrogen) are very similar to those in the star. For the Solar System, the measured Solar photospheric abundances, supplemented by abundances derived from a small number of primitive chondritic meteorites, provide a proxy for the primordial Solar Nebula composition. Even once the raw abundances are known,[6] however, there remains the task of determining the chemical composition of the disk – i.e. how the elements are distributed among the possible chemical compounds and states – as a function of radius and time. This is not an easy task. Oxygen, for example, may be present in many forms, including carbon monoxide (CO), water (H_2O) in the form of vapor or ice, oxides such as Fe_3O_4, and silicates such as Mg_2SiO_4. The distribution of oxygen between these (and many other) forms will affect the opacity in the disk, the total surface density of solid materials, and the composition of larger bodies that form at a given radius.

The most general formulation of the problem of determining the composition of the disk is as an initial value problem. If the initial composition of the disk with

---

[6] In reality, determining the Solar abundances from spectroscopic observations of the photosphere is itself not an easy task, and there is ongoing debate about the precise abundances of important elements such as carbon and oxygen.

radius are known, a chemical network describing the rates of possible reactions can
be integrated forward in time to determine the composition at subsequent epochs.
This approach allows for the disk's chemical composition to be out of equilibrium
(for example in the outer disk, where the low temperatures mean that the rates of
energetically favorable reactions can be very low), and can be readily extended to
accommodate radial flow as the disk evolves.

In practice, solving for the evolution of a disk coupled to a fully consistent
treatment of the chemistry is challenging. It is therefore useful to pose a simpler question: for a given temperature and pressure what composition is the most
thermodynamically stable? Conceptually, we can imagine starting with a high temperature gas in which all of the elements are in the vapor phase and cooling it very
slowly so that the elements successively condense out into various solid phases.
Provided that the cooling is slow enough all reactions will have sufficient time to
reach equilibrium, and the resulting condensation sequence will depend only on
the abundances, the temperature, and the pressure. Under conditions of constant
temperature and pressure, the thermodynamically preferred state is the one that
minimizes the Gibbs free energy,

$$G \equiv H - TS, \tag{2.64}$$

where H is the enthalpy and S the entropy. If the chemical system comprises a
number of homogeneous (or ideal) phases, $i = 1, 2, \ldots$,

$$G = \Sigma \left(\mu_i f_i \right), \tag{2.65}$$

where f_i is the fraction in phase i with chemical potential μ_i. The composition
can then be determined by finding the f_i that minimize G, subject to the constraint
imposed by the specified overall abundance of the elements. In practice, matters
are complicated by the existence of *nonideal* solid solutions – solid phase mixtures
of two or more components in which there is both a change of entropy and enthalpy
on mixing. The Gibbs free energy is then written as,

$$G = \Sigma \left(\mu_i a_i \right) \tag{2.66}$$

where the a_i, known as the *thermodynamic activities*, are the quantities that need
to be determined. Although schematically straightforward, calculation of the condensation sequence is difficult in practice due both to the need for accurate thermodynamic data and because a very large number of different species need to be
considered simultaneously. One recent calculation of the condensation temperatures of the elements by Lodders (2003) includes no less than 2000 gaseous species
and 1600 different condensates.

Table 2.2 gives the values of a small number of condensation temperatures, as
calculated by Lodders (2003) at a fiducial pressure of $P = 10^{-4}$ bar. This pressure is

Table 2.2. *The condensation temperatures for a handful of important species, as computed by Lodders (2003) for a pressure of* 10^{-4} *bar.*

Species	Composition	Condensation temperature (K)
Methane	CH_4	41
Argon hydrate	$Ar.6H_2O$	48
Methane hydrate	$CH_4.7H_2O$	78
Ammonia hydrate	$NH_3.H_2O$	131
Water ice	H_2O	182
Magnetite	Fe_3O_4	371
Troilite	FeS	704
Forsterite	Mg_2SiO_4	1354
Perovskite	$CaTiO_3$	1441
Aluminum oxide	Al_2O_3	1677

somewhat higher than the typical mid-plane pressure in protoplanetary disks except in the innermost AU, and this slightly modifies the condensation temperatures (in particular, water ice is only stable below $T \simeq 150$–170 K under realistic disk conditions near the snowline). Nevertheless, the condensation sequence is primarily a function of temperature, and hence the sequence can be approximately mapped into a predicted radial distribution of different solid materials. There is a transition between purely rocky condensates in the hot inner disk and a combination of rocky and icy materials at greater radii. Depending upon the adopted abundances, Lodders (2003) finds a total condensate fraction between 1.3% and 1.9%, of which between 0.44% and 0.49% is in the form of rocky materials. These results imply a jump in the surface density of solids at the snowline that is substantial, but smaller than the factor of $\simeq 4$ implied by the minimum-mass Solar Nebula model of Hayashi (1981).

2.7 Ionization state of protoplanetary disks

The interior temperatures of typical protoplanetary disks range from a few thousand K in the immediate vicinity of the star to just tens of K in the outer regions. These temperatures are not high enough to ionize the gas fully. As a rule of thumb, an element with an ionization potential χ is ionized once the temperature exceeds

$$T \sim \frac{\chi}{10k_B}. \tag{2.67}$$

The ionization temperature for hydrogen is about 10^4 K and even potassium – one of the most easily ionized species with $\chi = 4.34$ eV – requires $T \gtrsim 5000$ K before it becomes ionized. As a first and excellent approximation we can therefore assume that protoplanetary disks are neutral. However, as we will see later when we discuss disk evolution, even apparently negligible ionization fractions suffice to couple magnetic fields dynamically to the gas, and this coupling in turn may be a critical element that permits angular momentum transport and accretion. An accurate understanding of the ionization fraction of protoplanetary disks as a function of radius and height above the mid-plane is thus important. This requires balancing ionization sources, which may be thermal or nonthermal (radioactive decay of short-lived nuclides, stellar X-rays, or cosmic rays), against recombinations in the gas phase or on the surfaces of dust grains. In more complex models it may be necessary to go beyond such equilibrium considerations and allow for the possibility that (relatively) well-ionized gas may be mixed by turbulent processes into neutral regions on a time scale shorter than the recombination time (Turner et al., 2007; Ilgner & Nelson, 2008).

2.7.1 Thermal ionization

Thermal ionization of the alkali metals dominates the ionization balance in the very innermost regions of the disk, usually well inside 1 AU. This regime can be treated straightforwardly. In thermal equilibrium the ionization state of a single species with ionization potential χ is described by the Saha equation (e.g. Rybicki & Lightman, 1979)

$$\frac{n^{\mathrm{ion}} n_{\mathrm{e}}}{n} = \frac{2U^{\mathrm{ion}}}{U} \left(\frac{2\pi m_{\mathrm{e}} k_{\mathrm{B}} T}{h^2} \right)^{3/2} \exp[-\chi / k_{\mathrm{B}} T]. \tag{2.68}$$

In this equation, n^{ion} and n are the number densities of the ionized and neutral species, and n_{e} ($= n^{\mathrm{ion}}$) is the electron number density. The partition functions for the ions and neutrals are U^{ion} and U respectively. The electron mass is m_{e}. The slightly modified temperature dependence (an extra factor of $T^{3/2}$) as compared to the Boltzmann factor, which governs the occupancy of atomic energy levels, arises because the ionized state is favored on entropy grounds over the neutral state.

In protoplanetary disks the first significant source of thermal ionization arises when the temperature becomes high enough to start ionizing the alkali metals. For potassium, the ionization potential $\chi = 4.34 \, \mathrm{eV}$. We write the fractional abundance of potassium relative to all other neutral species as $f = n_{\mathrm{K}}/n_{\mathrm{n}}$, and define the ionization fraction x as

$$x \equiv \frac{n_{\mathrm{e}}}{n_{\mathrm{n}}}. \tag{2.69}$$

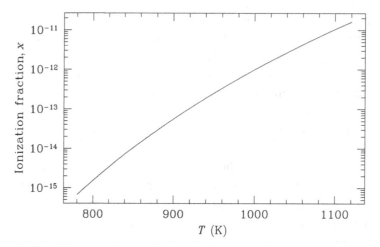

Fig. 2.8. The Saha equation prediction for the ionization state of the disk due to the thermal ionization of potassium (ionization potential $\chi = 4.34\,\mathrm{eV}$). The fractional abundance of potassium has been taken to be $f = 10^{-7}$, and the total number density of neutrals $n_\mathrm{n} = 10^{15}\,\mathrm{cm}^{-3}$.

While potassium remains weakly ionized, the Saha equation yields,

$$x \simeq 10^{-12} \left(\frac{f}{10^{-7}} \right)^{1/2} \left(\frac{n_\mathrm{n}}{10^{15}\,\mathrm{cm}^{-3}} \right)^{-1/2} \left(\frac{T}{10^3\,\mathrm{K}} \right)^{3/4}$$

$$\times \frac{\exp[-2.52 \times 10^4 / T]}{1.14 \times 10^{-11}}, \tag{2.70}$$

where the final numerical factor in the denominator is the value of the exponent at the fiducial temperature of 10^3 K. The predicted ionization fraction is shown as a function of temperature in Fig. 2.8. Ionization fractions that are large enough to be interesting for studies of magnetic field coupling are attained at temperatures of $T \sim 10^3$ K although the numbers are still *extremely* small – of the order of $x \sim 10^{-12}$ for our assumed parameters.

2.7.2 Nonthermal ionization

Outside the inner thermally ionized region, disk irradiation by stellar X-rays, cosmic rays, and radioactive decay of short-lived nuclides can all be significant sources of nonthermal ionization. Although the basic physics of each of these mechanisms is reasonably well-understood there is at least an order of magnitude uncertainty in the absolute value of the resulting ionization rates. To calculate the ionization *fraction*, which is the quantity of interest for most purposes, requires knowledge of both the ionization and recombination rates, and poor understanding of the latter introduces large additional uncertainties.

T Tauri stars are extremely luminous X-ray sources as compared to main-sequence stars of the same mass. Many disk-bearing classical T Tauri stars are detected in X-ray surveys of star forming regions, often with a soft X-ray luminosity in the range between $10^{28.5}$–$10^{30.5}$ erg s^{-1} (Feigelson & Montmerle, 1999; Feigelson *et al.*, 2007). Flares are observed with substantially higher luminosities. The spectra in the X-ray band are complex and include both line and continuum emission, which can be represented as a superposition of optically thin thermal bremsstrahlung components. The hardest spectral components have $T_X \sim$ few keV, which implies a tail of emission extending out to energies of 10 keV or more. Analysis of the X-ray properties supports an interpretation of the emission as originating in a magnetically powered corona that is a scaled-up version (both in power and in the size of the X-ray emitting magnetic field loops) of the Sun's. The existence of many other diagnostics of stellar magnetic activity, including Zeeman splitting of optical spectral lines and nonthermal radio emission, support this conclusion.

Some fraction of the stellar X-ray luminosity will be intercepted by the disk, either directly or following scattering in the disk atmosphere or in a stellar wind. If the gas is depleted of heavy elements (i.e. grains have condensed) the photo-ionization cross-section can be fit with a power-law (Igea & Glassgold, 1999),

$$\sigma(E) = 8.5 \times 10^{-23} \left(\frac{E}{\text{keV}} \right)^{-2.81} \text{cm}^2. \tag{2.71}$$

The steep power-law dependence of the cross-section with energy means that the penetrating power of X-rays is strongly energy dependent. X-rays with $E = 1$ keV are absorbed by a column $\Delta\Sigma \ll 1$ g cm^{-2}, while at 5 keV the stopping depth is a few g cm^{-2}. The outermost skin of the disk will therefore be ionized by soft X-rays, while the hard tail of the spectrum will penetrate further toward the interior. Expressions for computing the attenuation are given in Krolik & Kallman (1983) and in Fromang *et al.* (2002). For our purposes, it suffices to consider a rough fit to numerical results published by Igea & Glassgold (1999). For an X-ray luminosity of $L_X = 2 \times 10^{30}$ erg s^{-1} and a 5 keV thermal spectrum, Turner & Sano (2008) fit an ionization profile of the form

$$\zeta_X = 2.6 \times 10^{-15} \left(\frac{r}{1 \text{ AU}} \right)^{-2} \exp[-\Delta\Sigma/\Sigma_{\text{stop}}] \text{ s}^{-1}, \tag{2.72}$$

where $\Delta\Sigma$ is the column density measured downward from the disk surface and $\Sigma_{\text{stop}} = 8$ g cm^{-2} is an approximate stopping depth.

The high column density of gas in the inner regions of protoplanetary disks means that even the hardest stellar X-rays measured to date will be unable to penetrate to the disk mid-plane. Interstellar cosmic rays have a spectrum that extends up to much higher energies, with correspondingly greater penetrating power. Adopting the fiducial value for the interstellar cosmic ray flux (Spitzer & Tomasko, 1968) the resulting ionization rate is

$$\zeta_{CR} = 10^{-17} \exp[-\Delta\Sigma/\Sigma_{stop}] \, s^{-1}, \tag{2.73}$$

where in this case $\Sigma_{stop} = 10^2 \, g \, cm^{-2}$ (Umebayashi & Nakano, 1981). The main difficulty in applying this expression to protoplanetary disks comes from the uncertain effect of stellar winds on the cosmic ray flux. The Solar wind – which is probably much less powerful than typical T Tauri winds – is able to modulate the flux of Galactic cosmic rays measured at Earth, and it is quite conceivable that T Tauri winds (from either the star or the disk) may be able to exclude the bulk of the interstellar cosmic ray flux. Equation (2.73) is therefore realistically an upper limit to the ionization expected from cosmic rays, and the actual cosmic ray ionization rate could be negligibly small.

Radioactive decay provides a final source of ionization. It takes about $\Delta\epsilon = 36 \, eV$ to produce an ion pair in molecular hydrogen, so a decay that yields an amount of energy E will create about $E/\Delta\epsilon$ ions. If the nuclide under consideration has a fractional abundance (relative to hydrogen) f and a decay constant λ,[7] the ionization rate is (Stepinski, 1992)

$$\zeta_R = \frac{\lambda f E}{\Delta\epsilon}. \tag{2.74}$$

Abundant short-lived nuclides yield the highest ionization rates. Provided that it is originally present and has not yet had time to decay, ^{26}Al is predicted to dominate. Note that ^{26}Al has a very short half-life of 0.72 Myr ($\lambda \simeq 3 \times 10^{-14} \, s^{-1}$), an energy per decay of $E = 3.16$ MeV, and an estimated Solar System abundance of $f \sim 10^{-10}$. With these parameters $\zeta_R \simeq 2.6 \times 10^{-19} \, s^{-1}$, though this value will drop substantially over the disk lifetime due to the ongoing exponential decay. The rate may be further suppressed if the radioactive material is locked up within small pebble-sized solid bodies, and may be zero if the disk in question was not polluted with material from a relatively recent supernova.

Figure 2.9 shows the ionization rate as a function of column density for the simple models of X-ray, cosmic ray, and radioactive ionization discussed above. Despite the numerous uncertainties it is safe to conclude that X-ray ionization will dominate all other sources close to the disk surface, down to a column of 10–100 g cm^{-2}. Close to the mid-plane in the inner disk – where $\Sigma > 10^3 \, g \, cm^{-2}$ – either cosmic rays or ^{26}Al decays can potentially furnish a lower level of ionization.

---

[7] Related to the half-life via $t_{1/2} = \ln 2/\lambda$.

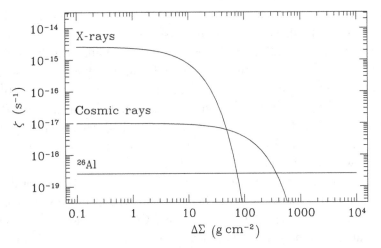

Fig. 2.9. The dependence of the ionization rate ζ on column density $\Delta\Sigma$, measured from the disk surface. Approximate contributions from the three main sources of nonthermal ionization are plotted, based on the simplified model described in the text: (a) cosmic rays, assuming that the interstellar flux is *not* shielded by a stellar wind; (b) stellar X-rays, evaluated at 1 AU using a single exponential fit (Turner & Sano, 2008) to numerical results by Igea & Glassgold (1999); and (c) radioactive decay of ^{26}Al present at a fractional abundance (relative to hydrogen) of $f = 10^{-10}$ (Stepinski, 1992).

The equilibrium ionization fraction x (Eq. 2.69) is obtained by balancing ionization against recombination. If we ignore for the moment both metal ions and dust particles, the dominant channel for absorbing free electrons would be dissociative recombination reactions with molecular ions. In the case of HCO^+, for example, the reaction is

$$e^- + HCO^+ \rightarrow H + CO. \tag{2.75}$$

Denoting the number density of molecular ions as n_{m^+}, ionization equilibrium requires that

$$\frac{dn_e}{dt} = \zeta n_n - \beta n_e n_{m^+} = 0. \tag{2.76}$$

A generic expression for the temperature dependence of the rate coefficient β is

$$\beta = 3 \times 10^{-6} T^{-1/2} \text{ cm}^3 \text{ s}^{-1}, \tag{2.77}$$

which yields an equilibrium electron abundance

$$x = \sqrt{\frac{\zeta}{\beta n_n}}. \tag{2.78}$$

We can evaluate the predicted electron abundance for conditions typical of protoplanetary disks. In the inner disk, the disk mid-plane is shielded from cosmic rays and a representative ionization rate due to radioactive decay might be $\zeta \sim 10^{-19}\,\mathrm{s}^{-1}$. For $n_n = 10^{15}\,\mathrm{cm}^{-3}$ and $T = 300\,\mathrm{K}$ we obtain $x \simeq 2.4 \times 10^{-14}$. For the same temperature close to the disk surface we may have $n_n = 10^{10}\,\mathrm{cm}^{-3}$ and a substantially enhanced ionization rate due to stellar X-rays, $\zeta_X \sim 10^{-15}\,\mathrm{s}^{-1}$. These parameters yield $x \simeq 7.6 \times 10^{-10}$. Although both values may appear negligibly small, in fact the difference is sufficient to imply substantially different coupling between the gas and magnetic fields.

The actual ionization equilibrium of disks is very probably a great deal more complex than this toy model suggests. In more realistic disk models, H_2^+ can charge exchange either on to other molecules such as HCO^+, or on to metal ions such as Mg^+. Radiative recombination with metal ions is much slower than dissociative recombination, so the presence of such charge exchange reactions can substantially alter the resulting electron fraction (Fromang *et al.*, 2002). A chemical network is required to assess the importance of these effects (e.g. Sano *et al.*, 2000; Ilgner & Nelson, 2006). Dust is also important: as a reservoir for metal atoms, as a charge carrier in its own right, and as a surface on which electrons and ions can recombine. Although all of these processes have been studied in some detail, it is unclear whether our knowledge of the composition of disks and of the distribution of solid particles is yet good enough to facilitate a comprehensive model for the ionization state.

2.8 Further reading

The structure of protoplanetary disks is reviewed in the article "Models of the structure and evolution of protoplanetary disks," C. P. Dullemond, D. Hollenbach, I. Kamp, & P. D'Alessio in *Protostars and Planets V* (2007), B. Reipurth, D. Jewitt, & K. Keil (eds.), Tucson: University of Arizona Press.

3

Protoplanetary disk evolution

Observationally it is clear that protoplanetary disks are not static structures, but rather evolve slowly over time. The observational manifestations of disks – phenomena such as an infrared excess or detectable mm flux – are almost always associated with stars in or near star forming regions, and are absent in (still youthful) clusters such as the Pleiades. Explaining theoretically why disks should evolve is not, however, an easy task. For a geometrically thin disk the angular velocity is essentially that of a Keplerian orbit (Eq. 2.20), and the specific angular momentum

$$l = r v_{\phi,\text{gas}} = \sqrt{GM_*r}, \tag{3.1}$$

is an *increasing* function of radius. For gas in the disk to flow inward and be accreted by the star, it therefore needs to lose angular momentum. Understanding the mechanisms that can result in angular momentum loss is the central problem in the theory of *accretion disks*, not just in the protoplanetary context but also for disks around black holes and other compact objects. It is a difficult problem because the effect of interest is subtle. Protoplanetary disks have an observed lifetime of several million years, which equates to $\sim 10^4$ dynamical times at the outer edge of the disk. In other words they are almost, but not quite, stable. In this chapter we discuss the evolution of disks, the origin of angular momentum transport, and the processes that disperse the gas at the end of the disk lifetime. The focus is on the evolution of the gas, which by mass is the dominant disk component. The evolution of the solid component, which is partially coupled to the gas but which also involves distinct physical processes, is discussed in Chapter 4.

3.1 Observations of disk evolution

Protoplanetary disks can be detected or observed via a number of complementary techniques, which include:

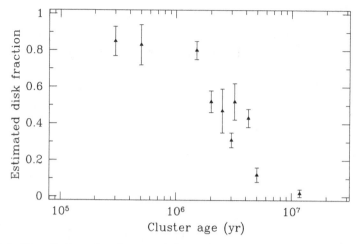

Fig. 3.1. Compilation of estimates of the optically thick disk fraction in young clusters as a function of the estimated cluster age. The disk fraction is estimated from either ground-based IR photometry (JHKL bands), or from *Spitzer* photometry. The plotted clusters are NGC 2024, the Trapezium, NGC 2264, NGC 2362 (all using data from Haisch *et al.*, 2001), NGC 1333 (disk fraction from Gutermuth *et al.*, 2008, plotted with an adopted age of 0.5 Myr), σ-Ori (Hernandez *et al.*, 2007), IC 348 (Lada *et al.*, 2006), Tr 37, NGC 7160 (Sicilia-Aguilar *et al.*, 2006), and Chamaeleon (Damjanov *et al.*, 2007). Note that there are substantial uncertainties in the absolute ages, especially for the younger clusters.

- Detection of infrared excesses (over the stellar photospheric flux) in the near- or mid-IR. The excess indicates the presence of warm dust close to the star, normally (depending upon the wavelength) on scales of the order of an AU or smaller.
- Detection of an accretion signature indicative of gas being accreted on to the star. Observational indicators include an ultraviolet excess, Hα emission with a large equivalent width, and a number of other emission lines such as the Ca II triplet in the infrared.
- Observation of mm or sub-mm flux arising from cool dust in the outer disk.
- Disk imaging, either in scattered visible light from the central star, or in silhouette against a bright background nebula.
- Detection of line emission from molecular species such as CO or NH_3, either as an unresolved source or as a spatially resolved image.

Currently the strongest constraints on disk evolution come from measurements of the fraction of young stars that exhibit IR excesses, but this reflects only the strengths of present observational technology – wide-field IR imaging is straightforward and exquisitely sensitive to small amounts of dust, while molecular line observations struggle to attain useful sensitivity on all but the brightest sources. Future instruments will likely change this balance.

Figure 3.1 shows the fraction of young stars in different clusters that show evidence for protoplanetary disks in the form of IR excesses. This "disk fraction" is plotted against the estimated age of the clusters, which are derived by comparing the locations of stars in the Hertzsprung–Russell diagram against theoretical stellar evolution tracks for pre-main-sequence stars.[1] The data sample includes the clusters compiled by Haisch *et al.* (2001), together with a selection of more recent results based on photometry obtained by the *Spitzer Space Telescope*. Where possible, the disk fraction refers to T Tauri stars (i.e. stars of around a Solar mass) that host optically thick disks. This last qualifier is intended to distinguish so-called *primordial disks* – in which substantial amounts of dust are mixed in with gas around the youngest stars – from the generally weaker *debris disks* that are observed around older stars (the disk around β Pictoris being one well-known example). Although there is no unambiguous observational test to distinguish primordial from debris disks most debris disks, when studied in detail, do not show evidence for any significant gaseous component. It is likely that the dust in these disks arises from erosive collisions between larger solid bodies.

The infrared excess measurements of disk frequency provide a robust measure of disk evolution. The disk frequency is close to 100% for clusters whose mean stellar age is less than about 1 Myr. For older clusters, the disk frequency drops steadily, reaching 50% at around 3 Myr and falling almost to zero by about 6 Myr. Evidence for primordial disks is essentially completely absent for stars of age 10 Myr or above. From these observations, one infers a disk lifetime of about 3–5 Myr.

Strictly speaking, the aforementioned constraint on the "disk lifetime" refers only to the survival time of small dust grains in the innermost ($\sim$1 AU) region of the disk. Does the disk as a whole evolve on the same time scale? Somewhat surprisingly, there is evidence that it may. First, the accretion rate of *gas* on to the star – measured entirely independently via spectroscopic observations of the hot continuum radiation produced when infalling gas impacts the stellar surface – decays with time on a similar time scale (Hartmann *et al.*, 1998). Second, the presence of one disk signature (such as an IR excess) usually, though not invariably, implies that the other signatures (such as a mm excess) are also present, *even when the different signatures arise at very different disk radii*. This observation implies that disks are dispersed across a range of radii on a relatively short time scale (Skrutskie *et al.*, 1990; Wolk & Walter, 1996; Andrews & Williams, 2005). We will discuss further the import of this observation for models of disk dispersal, but for now we note only that it makes the assignment of a single lifetime for the disk a meaningful concept.

---

[1] The derived ages need to be viewed with considerable caution for two reasons. First, for a very young cluster the spread in ages of the stars may be comparable to the mean age of the cluster. This makes the assignment of a single age to the system of dubious worth. Second, the evolutionary tracks used to assign ages return essentially arbitrary *absolute* ages for stars less than about 1 Myr old (Baraffe *et al.*, 2002), due to the neglect of accretion physics that matters at such early times. Statements about stars or clusters that rely on accurate determinations of ages less than a Myr should generally be regarded skeptically.

3.2 Surface density evolution of a thin disk

Consider an axisymmetric protoplanetary disk whose gas surface density profile is given by $\Sigma(r, t)$. We assume that the radial velocity $v_r(r, t)$ of gas in the disk is small,[2] and note that the fact that the disk is geometrically thin ($h/r \ll 1$) implies that the predominant forces at work are rotational support and gravity (cf. Section 2.3). If the potential is time-independent, local conservation of angular momentum implies that in the absence of angular momentum transport or loss $\Sigma(r, t)$ cannot change with time. Accretion and disk evolution will occur in the presence of angular momentum transport (often described as "viscosity," or, even more loosely, as "friction"), which allows local parcels of gas to reduce their angular momentum and spiral toward the star (global angular momentum conservation implies, of course, that gas elsewhere in the disk must *gain* angular momentum and move outward). This redistribution of angular momentum due to stresses within the disk is quite distinct from angular momentum loss – due for example to a magnetically driven outflow from the disk surface – and the evolution of disks under the action of winds is different from that due to internal redistribution.

The qualitative evolution of disks in the presence of dissipative processes was understood in the 1920s by, among others, the well-known geophysicist and astronomer Harold Jeffreys. The modern theory of thin disks was described in now-classic papers by Shakura & Sunyaev (1973) and Lynden-Bell & Pringle (1974). This theory is not fully predictive as it largely bypasses the central question of how efficiently angular momentum is transported within a disk flow, but it nonetheless forms the indispensable core to any discussion of disk evolution.

The evolution of $\Sigma(r, t)$ can be derived by considering the continuity equation (expressing the conservation of mass) and the azimuthal component of the momentum equation (expressing angular momentum conservation). The rate of change of the mass within an annulus in the disk extending between r and $r + \Delta r$ is given by

$$\frac{\partial}{\partial t}(2\pi r \Delta r \Sigma) = 2\pi r \Sigma(r) v_r(r) - 2\pi(r + \Delta r)\Sigma(r + \Delta r)v_r(r + \Delta r). \tag{3.2}$$

Writing for example $\Sigma(r + \Delta r) = \Sigma(r) + (\partial\Sigma/\partial r)\Delta r$, and taking the limit for small Δr, the continuity equation yields

$$r\frac{\partial\Sigma}{\partial t} + \frac{\partial}{\partial r}(r\Sigma v_r) = 0. \tag{3.3}$$

Following the same procedure (e.g. Pringle, 1981) conservation of angular momentum gives

$$r\frac{\partial}{\partial t}\left(r^2\Omega\Sigma\right) + \frac{\partial}{\partial r}\left(r^2\Omega \cdot r\Sigma v_r\right) = \frac{1}{2\pi}\frac{\partial G}{\partial r}, \tag{3.4}$$

---

[2] The radial velocity is defined such that $v_r < 0$ corresponds to inflow.

where Ω, the angular velocity of the gas in the disk, is at this point unspecified and need not be the Keplerian angular velocity due to a point mass. The rate of change of angular momentum in the disk is determined by the change in surface density due to radial flows (the second term on the left-hand-side) and by the *difference* in the torque exerted on an annulus by stresses at the inner and outer edges. For a viscous fluid, the torque G can be written in the form

$$G = 2\pi r \cdot v \Sigma r \frac{d\Omega}{dr} \cdot r, \qquad (3.5)$$

where v is the kinematic viscosity. The torque on an annulus is the product of the circumference, the viscous force per unit length, and the lever arm r, and is proportional to the gradient of the angular velocity. Note that this dependence, which is characteristic of a viscous fluid, is only an assumption if the "viscosity" is not a true microscopic process but rather an effective viscosity resulting from turbulence. Proceeding, we eliminate v_r between Eq. (3.3) and Eq. (3.4) and specialize to a Keplerian potential for which $\Omega \propto r^{-3/2}$. We then obtain the evolution equation for the surface density of a geometrically thin disk under the action of internal angular momentum transport

$$\frac{\partial \Sigma}{\partial t} = \frac{3}{r} \frac{\partial}{\partial r} \left[r^{1/2} \frac{\partial}{\partial r} \left(v \Sigma r^{1/2} \right) \right]. \qquad (3.6)$$

The evolution equation is a diffusive partial differential equation for the surface density $\Sigma(r, t)$. It is linear if the viscosity v is not itself a function of Σ. The equation can also be derived directly from the Navier–Stokes equations for a viscous fluid in cylindrical polar coordinates.

3.2.1 The viscous time scale

The diffusive form of Eq. (3.6) can be made more transparent with a change of variables. Defining

$$X \equiv 2r^{1/2}, \qquad (3.7)$$

$$f \equiv \frac{3}{2} \Sigma X, \qquad (3.8)$$

and assuming that the viscosity v is a constant, the evolution equation takes the prototypical form of a diffusion equation

$$\frac{\partial f}{\partial t} = D \frac{\partial^2 f}{\partial X^2}, \qquad (3.9)$$

with a diffusion coefficient D given by

$$D = \frac{12v}{X^2}. \qquad (3.10)$$

Apart from its pedagogical value, this version of the evolution equation can be useful numerically, since even naive finite difference schemes preserve conserved quantities accurately when the equation is cast in this form. The diffusion time scale across a scale ΔX implied by Eq. (3.9) is just $(\Delta X)^2/D$. Converting back to the physical variables, we then find that the time scale on which viscosity will smooth out surface density gradients on a radial scale Δr is

$$\tau_\nu \sim \frac{(\Delta r)^2}{\nu}. \tag{3.11}$$

If the disk has a characteristic size r, the surface density at all radii will evolve on a time scale

$$\tau_\nu \approx \frac{r^2}{\nu}. \tag{3.12}$$

This last time scale is described as the *viscous time scale* of the disk. It can be estimated observationally by measuring, for example, the rate at which accretion on to the star decays as a function of stellar age. For protoplanetary disks around Solar-type stars it appears to be of the order of a million years.

3.2.2 Solutions to the disk evolution equation

A steady-state solution to Eq. (3.6) can be derived by setting $\partial/\partial t = 0$ and integrating the resultant ordinary differential equation for the surface density. Applying the requisite boundary conditions is easiest if we start with the angular momentum conservation equation (Eq. 3.4) which does not assume Keplerian angular velocity. Setting the time derivative to zero and integrating, we have

$$2\pi r \Sigma v_r \cdot r^2 \Omega = 2\pi r^3 \nu \Sigma \frac{d\Omega}{dr} + \text{constant}. \tag{3.13}$$

Noting that the mass accretion rate $\dot{M} = -2\pi r \Sigma v_r$ we can write this equation in the form

$$-\dot{M} \cdot r^2 \Omega = 2\pi r^3 \nu \Sigma \frac{d\Omega}{dr} + \text{constant}, \tag{3.14}$$

where the constant of integration, which has the form of an angular momentum flux, remains to be determined. To specify the constant, we note that at a location in the disk where $d\Omega/dr = 0$ the viscous stress vanishes, and the constant is simply equal to the flux of angular momentum advected inward along with the mass,

$$\text{constant} = -\dot{M} \cdot r^2 \Omega. \tag{3.15}$$

A simple case to consider is that where the protoplanetary disk extends all the way down to the surface of a nonrotating (or slowly rotating) star. The disk and

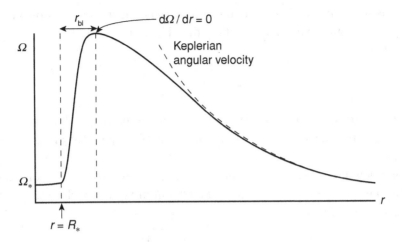

Fig. 3.2. Schematic illustration of the angular velocity profile in a disk that extends to the equator of a slowly rotating star. The viscous stress vanishes at a radial location $r = R_* + r_{bl}$ where Ω has a maximum.

the star form a single fluid system, and the angular velocity (shown schematically in Fig. 3.2) must make a continuous transition between $\Omega = 0$ in the star and $\Omega \propto r^{-3/2}$ within the disk. The viscous stress vanishes at a radius $R_* + r_{bl}$, where r_{bl} is the width of the *boundary layer* that separates the star from the Keplerian part of the disk. Within the boundary layer the angular velocity increases with radius, and the sub-Keplerian rotational support cannot balance the inward gravitational force.

The hydrodynamics (and magnetohydrodynamics) of protoplanetary disk boundary layers is quite complex, and the flow structure in this region remains rather uncertain. Elementary arguments, however, suggest that in most cases the radial extent of the boundary layer is a small fraction of the stellar radius. If magnetic fields and viscosity are negligible, the momentum equation (2.17) in axisymmetry can be written as

$$\frac{v_{\phi,\text{gas}}^2}{r} = \frac{GM_*}{r^2} + \frac{1}{\rho}\frac{dP}{dr} + v_r\frac{dv_r}{dr}, \qquad (3.16)$$

where P is the pressure. In the boundary layer – unlike in the disk itself – the rotational support (the term on the left-hand-side) cannot balance gravity. If, instead, radial pressure forces counteract the force of gravity[3] we require that

$$\frac{1}{\rho}\frac{dP}{dr} \sim \frac{c_s^2}{r_{bl}} \sim \Omega_K^2 r, \qquad (3.17)$$

---

[3] Similar arguments apply if we consider the influence of the $v_r dv_r/dr$ term rather than the pressure term, since causality requires that v_r be subsonic.

where c_s is some characteristic sound speed in the boundary layer region and Ω_K is the Keplerian angular velocity. Recalling that the vertical scale-height $h = c_s/\Omega_K$ we find that

$$\frac{r_{bl}}{r} \sim \left(\frac{h}{r}\right)^2. \tag{3.18}$$

Provided that the boundary layer (like the disk) is geometrically thin, we conclude that force balance mandates that the *radial* extent of the boundary layer must also be narrow.

We are now in a position to evaluate the constant in Eq. (3.14). For a narrow boundary layer, $R_* + r_{bl} \simeq R_*$ and the maximum in Ω occurs close to the stellar surface. We have that

$$\text{constant} \simeq -\dot{M}R_*^2\sqrt{\frac{GM_*}{R_*^3}}, \tag{3.19}$$

and the steady-state solution for the disk (within which the angular velocity is Keplerian) simplifies to

$$\nu\Sigma = \frac{\dot{M}}{3\pi}\left(1 - \sqrt{\frac{R_*}{r}}\right). \tag{3.20}$$

Once the viscosity is specified, this equation gives the steady-state surface density profile of a protoplanetary disk with a constant accretion rate $\dot{M}$. Away from the inner boundary one notes that $\Sigma(r) \propto \nu^{-1}$.

The solution given by Eq. (3.20) gives the surface density profile for a steady disk subject to a *zero-torque* boundary condition at the inner edge. Physically, this boundary condition is at least approximately realized for disks that extend to the equator of a slowly rotating star, and it is also the traditional choice in the more exotic circumstance of a disk of gas around a black hole. In classical T Tauri stars it is often the case that stellar magnetic fields truncate the disk before it reaches the stellar surface, and the resulting magnetic coupling between the star and the inner disk can violate the zero-torque assumption. More generally, one should note that the turnover in the surface density profile at small radii implied by Eq. (3.20) reflects the fact that for a Keplerian flow the only way in which the torque can vanish is if the surface density goes to zero. The turnover is therefore a purely formal result – in a real disk with a boundary layer the physical reason why the torque vanishes is because of the existence of a maximum in $\Omega(r)$ – and the inner boundary condition needs to be considered carefully if one needs the detailed form of the surface density very close to the inner edge of the disk.

Time-dependent analytic solutions to Eq. (3.6) can be derived for a number of simple forms for the viscosity and, although these forms are not particularly

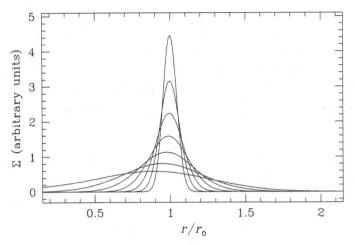

Fig. 3.3. The Green's function solution to the disk evolution equation with $\nu =$ constant, showing the spreading of a ring of mass initially orbiting at $r = r_0$. From top down the curves show the behavior as a function of the scaled time variable $\tau = 12\nu r_0^{-2}t$, for $\tau = 0.004$, $\tau = 0.008$, $\tau = 0.016$, $\tau = 0.032$, $\tau = 0.064$, $\tau = 0.128$, and $\tau = 0.256$.

realistic for protoplanetary disks, the resulting solutions suffice to illustrate the essential behavior implied by the disk evolution equation. If $\nu =$ constant, a Green's function solution to the evolution equation is possible.[4] Suppose that at $t = 0$, all of the gas lies in a thin ring of mass m at radius r_0

$$\Sigma(r, t = 0) = \frac{m}{2\pi r_0}\delta(r - r_0),$$ (3.21)

where $\delta(r - r_0)$ is a Dirac delta function. Given boundary conditions that impose zero-torque at $r = 0$ and allow for free expansion toward $r = \infty$ the solution is (Lynden-Bell & Pringle, 1974)

$$\Sigma(x, \tau) = \frac{m}{\pi r_0^2}\frac{1}{\tau}x^{-1/4}\exp\left[-\frac{(1 + x^2)}{\tau}\right]I_{1/4}\left(\frac{2x}{\tau}\right),$$ (3.22)

where we have written the solution in terms of dimensionless variables $x \equiv r/r_0$, $\tau \equiv 12\nu r_0^{-2}t$, and $I_{1/4}$ is a modified Bessel function of the first kind. Since the evolution equation is linear for $\nu = f(r)$, the time-dependent solution for arbitrary initial conditions can formally be written as a superposition of these solutions. Although this approach is rarely illuminating, the solution (Eq. 3.22), which is plotted in Fig. 3.3, illustrates several generic features of disk evolution. As t increases the initially narrow ring spreads diffusively, with the mass flowing toward $r = 0$ while simultaneously the angular momentum is carried by a negligible fraction of the

---

[4] Related solutions are known for the more general situation in which ν is a power-law in radius.

mass toward $r = \infty$. This segregation of mass and angular momentum is a general feature of the evolution of a viscous disk, and is necessary if accretion is to proceed without overall angular momentum loss from the system.

Often of greater practical utility than the Green's function solution is the self-similar solution also derived by Lynden-Bell & Pringle (1974). Consider a disk in which the viscosity can be approximated as a power-law in radius

$$\nu \propto r^{\gamma}. \tag{3.23}$$

Suppose that the disk at time $t = 0$ has the surface density profile corresponding to a steady-state solution (with this viscosity law) out to $r = r_1$, with an exponential cut-off at larger radii. Specifically, the initial surface density has the form

$$\Sigma(t = 0) = \frac{C}{3\pi \nu_1 \tilde{r}^{\gamma}} \exp\left[-\tilde{r}^{(2-\gamma)}\right], \tag{3.24}$$

where C is a normalization constant, $\tilde{r} \equiv r/r_1$, and $\nu_1 \equiv \nu(r_1)$. The self-similar solution is then

$$\Sigma(\tilde{r}, T) = \frac{C}{3\pi \nu_1 \tilde{r}^{\gamma}} T^{-(5/2-\gamma)/(2-\gamma)} \exp\left[-\frac{\tilde{r}^{(2-\gamma)}}{T}\right], \tag{3.25}$$

where

$$T \equiv \frac{t}{t_s} + 1, \tag{3.26}$$

$$t_s \equiv \frac{1}{3(2-\gamma)^2} \frac{r_1^2}{\nu_1}. \tag{3.27}$$

The evolution of related quantities such as the disk mass and accretion rate can readily be derived from the above expression for the surface density. The solution is plotted in Fig. 3.4. Over time, the disk mass decreases while the characteristic scale of the disk (initially r_1) expands to conserve angular momentum. This solution can be useful both for studying evolving disks analytically, and for comparing observations of disk masses, accretion rates, or radii with theory.

3.2.3 Temperature profile of accreting disks

The radial dependence of the effective temperature of an actively accreting disk can be derived by considering the net torque on a ring of width Δr. This torque – $(\partial G/\partial r)\Delta r$ – does work at a rate

$$\Omega \frac{\partial G}{\partial r} \Delta r \equiv \left[\frac{\partial}{\partial r}(G\Omega) - G\Omega'\right]\Delta r, \tag{3.28}$$

where $\Omega' = d\Omega/dr$. Written this way, we note that if we consider the whole disk (by integrating over r) the first term on the right-hand-side is determined solely by

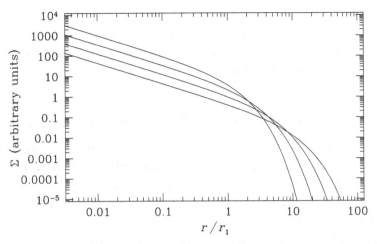

Fig. 3.4. The self-similar solution to the disk evolution equation is plotted for a viscosity $\nu \propto r$. The initial surface density tracks the profile for a steady-state disk ($\Sigma \propto r^{-1}$, Eq. 3.20) at small radius, before cutting off exponentially beyond $r = r_1$. The curves show the surface density at the initial value of the scaled time $T = 1$, and at subsequent times $T = 2$, $T = 4$, and $T = 8$.

the boundary values of $G\Omega$. We therefore identify this term with the *transport* of energy, associated with the viscous torque, through the annulus. The second term, on the other hand, represents the rate of loss of energy to the gas. We assume that this is ultimately converted into heat and radiated, so that the dissipation rate per unit surface area of the disk (allowing that the disk has two sides) is

$$D(r) = \frac{G\Omega'}{4\pi r} = \frac{9}{8}\nu\Sigma\Omega^2,\qquad(3.29)$$

where we have assumed a Keplerian angular velocity profile. For blackbody emission $D(r) = \sigma T_{\text{disk}}^4$. Substituting for Ω, and for $\nu\Sigma$ using the steady-state solution given by Eq. (3.20), we obtain

$$T_{\text{disk}}^4 = \frac{3GM_*\dot{M}}{8\pi\sigma r^3}\left(1 - \sqrt{\frac{R_*}{r}}\right).\qquad(3.30)$$

We note that apart from near the inner boundary ($r \gg R_*$) the temperature profile of an actively accreting disk is $T_{\text{disk}} \propto r^{-3/4}$. This has the same form as for a razor-thin disk that reprocesses stellar radiation (Eq. 2.30). Moreover, the temperature profile does *not* depend upon the viscosity. This is an attractive feature of the theory given that there are uncertainties regarding the origin and efficiency of disk angular momentum transport, though it also eliminates many possible routes to learning about the physics underlying ν via observations of steady disks.

Substituting a representative value for the accretion rate of $\dot{M} = 10^{-7}\,M_\odot\mathrm{yr}^{-1}$, we obtain for a Solar mass star at 1 AU an effective temperature $T_{\mathrm{disk}} = 150\,\mathrm{K}$. This is the *surface* temperature, as we will show shortly the central temperature is predicted to be substantially higher.

3.3 Vertical structure of protoplanetary disks

In an actively accreting disk it is intuitively plausible that the dissipation of gravitational potential energy into heat will be concentrated close to the disk mid-plane where the density is highest. At most disk radii there is a large optical depth between the surface and the mid-plane, so the heated gas cannot simply radiate directly to space. Rather, energy will need to be transported vertically toward the photosphere via radiative diffusion or turbulent transport, and this requires that there exist a temperature gradient between $z = 0$ and the disk photosphere. Our goal here is to calculate this gradient, and its effect on the vertical structure of the disk. This is dynamically important because – in almost all models – the magnitude of the viscosity that controls the evolution depends upon the *central* conditions in the disk rather than those at the surface. Given a model for the viscosity and knowledge of the vertical structure we can then derive an (in principle) self-consistent model for the disk structure and evolution.

The equations governing the vertical structure are essentially identical to those employed in stellar structure calculations except for a change in the geometry from spherically symmetric to plane parallel. If energy transport is due to radiative diffusion, the equations to be solved are: (a) hydrostatic equilibrium

$$\frac{\mathrm{d}P}{\mathrm{d}z} = -\rho g_z, \tag{3.31}$$

where g_z is the vertical gravity (Eq. 2.5); (b) an equation describing the vertical variation[5] of the flux F_z

$$\frac{\mathrm{d}F_z}{\mathrm{d}z} = \frac{9}{4}\rho\nu\Omega^2; \tag{3.32}$$

and (c) the equation of radiative diffusion describing the relation between the flux and the temperature gradient in an optically thick medium

$$\frac{\mathrm{d}T}{\mathrm{d}z} = -\frac{3\kappa_R\rho}{16\sigma T^3}F_z. \tag{3.33}$$

---

[5] For $h \ll r$ one may self-consistently assume that $F_z \gg F_r$, provided that the radial temperature profile remains smooth. The assumption of smoothness is violated if the disk is subject to thermal instabilities, but in this case one should be wary of using time-independent vertical structure models in any case.

These equations must be supplemented with an equation of state relating the pressure P to the fundamental variables ρ and T, with expressions for the Rosseland mean opacity κ_R, and with appropriate boundary conditions. The dependence of the viscosity on the local physical conditions is also required. If the resulting temperature gradient turns out to be steeper than the adiabatic gradient the disk is convectively unstable, and the radiative flux needs to be supplemented with the convective flux. This can be done in the usual manner using mixing length theory, or more crudely by simply setting the temperature gradient in convectively unstable regions equal to the adiabatic gradient. Given these ingredients, calculation of the disk vertical structure is a straightforward numerical exercise.

3.3.1 *The central temperature of accreting disks*

An approximation to the vertical temperature structure can be derived under the assumption that the energy dissipation due to viscosity is strongly concentrated toward $z = 0$. To proceed, we define the optical depth to the disk mid-plane

$$\tau = \frac{1}{2}\kappa_R \Sigma, \tag{3.34}$$

where κ_R is the Rosseland mean opacity and Σ is the disk surface density. The vertical density profile of the disk is $\rho(z)$. If the vertical energy transport occurs via radiative diffusion (in some regions convection may also be important), then for $\tau \gg 1$ the vertical energy flux $F(z)$ is given by the equation of radiative diffusion (Eq. 3.33 above)

$$F_z(z) = -\frac{16\sigma T^3}{3\kappa_R \rho}\frac{dT}{dz}. \tag{3.35}$$

Let us assume for simplicity that *all* of the energy dissipation occurs at $z = 0$. In that case $F_z(z) = \sigma T_{disk}^4$ is a constant with height. Integrating assuming that the opacity is constant,

$$-\frac{16\sigma}{3\kappa_R}\int_{T_c}^{T_{disk}} T^3 dT = \sigma T_{disk}^4 \int_0^z \rho(z')dz', \tag{3.36}$$

$$-\frac{16}{3\kappa_R}\left[\frac{T^4}{4}\right]_{T_c}^{T_{disk}} = T_{disk}^4 \frac{\Sigma}{2}, \tag{3.37}$$

where for the final equality we have used the fact that for $\tau \gg 1$ almost all of the disk gas lies below the photosphere. For large τ we expect that $T_c^4 \gg T_{disk}^4$, and the equation simplifies to

$$\frac{T_c^4}{T_{disk}^4} \simeq \frac{3}{4}\tau. \tag{3.38}$$

The implication of this result is that active disks with large optical depths are substantially hotter at the mid-plane than at the surface. For example, if at some radius the optical depth to the disk's thermal radiation is $\tau = 10^2$, then $T_c \approx 3T_{disk}$. This is important since it is the *central* temperature that largely determines, for example, which ices or minerals can be present. Relatively modest levels of accretion can thus affect the thermal and chemical structure of the disk substantially.

It will often be the case that *both* stellar irradiation and accretional heating contribute significantly to the thermal balance of the disk. If we define $T_{disk,visc}$ to be the effective temperature that would result from accretional heating in the absence of irradiation (i.e. the quantity called T_{disk}, with no subscript, above) and T_{irr} similarly to be the irradiation-only effective temperature, then application of Eq. (3.37) yields

$$T_c^4 \simeq \frac{3}{4}\tau T_{disk,visc}^4 + T_{irr}^4, \tag{3.39}$$

as an approximation for the central temperature, again valid for $\tau \gg 1$. Note that in disk models, such as those described in Section 2.4.4, in which the surface dust temperature exceeds the local blackbody temperature, the fraction of the irradiating flux that is thermalized within the disk is all that matters for the central temperature. The temperature T_{irr} that enters into the above formula is then the "interior" temperature T_i computed for passive radiative equilibrium disk models.

3.3.2 Shakura–Sunyaev α prescription

Although the effective temperature of an actively accreting disk (Eq. 3.30) is independent of the magnitude of the viscosity, the time scale on which evolution occurs (Eq. 3.11) and the profile of the surface density depend directly on the viscosity and its variation with radius. If we want to make progress in understanding the evolution of disks we cannot evade discussion of the physical origin of angular momentum transport forever! The first candidate to consider is molecular collisions, which generate a viscosity in a shear flow because of the finite mean-free path λ in the gas. Molecular viscosity, which is the normal mechanism considered in terrestrial fluids modeled with the Navier–Stokes equations, is given, approximately, by

$$\nu_m \sim \lambda c_s, \tag{3.40}$$

where the mean-free path in a gas with number density n is

$$\lambda = \frac{1}{n\sigma_{mol}}. \tag{3.41}$$

Here σ_{mol} is the cross-section for molecular collisions, which is very roughly equal to the physical size of molecules. Computing the molecular viscosity in detail is not at all a trifling task, but for our purposes high accuracy is not required as an order of magnitude estimate establishes that ν_m is far too small to matter in protoplanetary disks. Adopting

$$\sigma_{mol} \approx 2 \times 10^{-15} \, cm^2, \qquad (3.42)$$

as the collision cross-section of molecular hydrogen (Chapman & Cowling, 1970), and taking the sound speed at 10 AU to be $0.5 \, km \, s^{-1}$ and the number density to be $n = 10^{12} \, cm^{-3}$ we estimate that $\nu_m \sim 2.5 \times 10^7 \, cm^2 \, s^{-1}$. The implied viscous time scale (Eq. 3.11) is

$$t_\nu \simeq \frac{r^2}{\nu_m} = 3 \times 10^{13} \, yr. \qquad (3.43)$$

This is approximately ten million times longer than the observed time scale for disk evolution. Molecular viscosity is *not* the source of angular momentum transport within disks.

The consequence of a small molecular viscosity is, of course, a large Reynolds number, which can be defined as

$$Re \equiv \frac{UL}{\nu_m}, \qquad (3.44)$$

where U and L are characteristic velocity and length scales in the system. Taking $U = c_s$ and $L = h = 0.05r$, the fluid Reynolds number at 10 AU is

$$Re \sim 10^{10}, \qquad (3.45)$$

a staggeringly large number! One concludes that in the presence of a physical instability the protoplanetary disk will be highly turbulent, with dissipation of fluid motions occurring on a scale that is small compared to the disk scale-height.[6]

Let us assume for the time being that the protoplanetary disk *is* turbulent. The turbulent fluid motions will result in the macroscopic mixing of fluid elements at neighboring radii, which can act as an "effective" or "turbulent" viscosity. The magnitude of this turbulent viscosity can be estimated from dimensional arguments. If the turbulence is approximately isotropic, the outer scale of the turbulent flow is limited to be no larger than the smallest scale in the disk, which is generally the disk scale-height h. The velocity of the turbulent motions can be similarly limited to be no larger than the sound speed c_s, since supersonic motions result in shocks and rapid dissipation (this is true even if the fluid is magnetized). We therefore

---

[6] One will occasionally see the stronger statement that such a large Reynolds number implies that the disk is *necessarily* turbulent, even in cases where no explicit instability that would initiate a transition from laminar to turbulent flow can be identified. This type of reasoning, which derives from intuition gleaned from the study of terrestrial fluids, is suspect in astrophysical flows, and we do not advocate it here.

write the turbulent viscosity in the form

$$\nu = \alpha c_s h, \tag{3.46}$$

where α is a dimensionless quantity, known as the Shakura–Sunyaev α parameter, that measures the efficiency of angular momentum transport due to turbulence (Shakura & Sunyaev, 1973).

With the help of Eq. (3.46) it is possible to specify the viscosity in terms of local disk quantities, the sound speed c_s and the disk scale-height h, and thereby compute the local disk structure. We will follow this route shortly in Section 3.3.3. Before doing so, however, it is useful to note some of the limitations of any disk model constructed using the α prescription. First, the physical argument that we gave for Eq. (3.46) reasonably limits α to be less than unity, but does *not* give any basis for assuming that α is a constant. We will take α to be constant later only because it is impossible to proceed otherwise, not for any deeper reason. In fact, α may vary with the temperature, density, and composition of the disk gas, and there may even be regions which fail to satisfy the basic assumption by not developing turbulence at all. Second, although the turbulent viscosity has the same dimensions as a molecular viscosity, one should not forget that it arises from an entirely distinct physical process. In particular, studying the evolution of turbulent disks in two or three dimensions using the Navier–Stokes equations, valid for molecular viscosity, is to invite qualitatively wrong results. It is not to be recommended.

3.3.3 Vertically averaged solutions

Using the α prescription we can compute the viscosity ν as a function of r, Σ, and α. This in turn determines the steady-state surface density profile $\Sigma(r, \alpha, \dot{M})$ through Eq. (3.20), and can be employed together with Eq. (3.6) to calculate the time-dependent evolution of an arbitrary initial surface density profile. Here we will compute the viscosity in the vertically averaged or "one zone" approximation. This approximation amounts to replacing the equation of radiative diffusion (Eq. 3.33) with the approximate result given as Eq. (3.38), and replacing the vertical dependence of all other quantities by their central values.

Consider an annulus of the disk with surface density Σ at a radius where the Keplerian angular velocity is Ω (in thin disk solutions r and M_* enter only via the combination $\Omega = \sqrt{GM_*/r^3}$). The disk is characterized by eight variables: the mid-plane temperature T_c, effective temperature T_{disk}, sound speed c_s, density ρ, vertical scale-height h, opacity κ_R, viscosity ν, and optical depth τ. Apart from the effective temperature, which is defined at the photosphere, and the optical depth, which is evaluated *between* the surface and the mid-plane, all of these quantities are to be considered as the values at $z = 0$. Collecting together results that we have either already derived or which are trivial, we have the following set of equations:[7]

$$v = \alpha c_s h, \tag{3.47}$$

$$c_s^2 = \frac{k_B T_c}{\mu m_p}, \tag{3.48}$$

$$\rho = \frac{1}{\sqrt{2\pi}} \frac{\Sigma}{h}, \tag{3.49}$$

$$h = \frac{c_s}{\Omega}, \tag{3.50}$$

$$T_c^4 = \frac{3}{4} \tau T_{disk}^4, \tag{3.51}$$

$$\tau = \frac{1}{2} \Sigma \kappa_R, \tag{3.52}$$

$$v\Sigma = \frac{\dot{M}}{3\pi}, \tag{3.53}$$

$$\sigma T_{disk}^4 = \frac{9}{8} v \Sigma \Omega^2. \tag{3.54}$$

Once the opacity $\kappa_R(\rho, T_c)$ is specified, there are eight equations in eight unknowns to solve in order to determine the disk structure. If the opacity can be approximated as a power-law in density and temperature the solution is analytic, and one obtains an expression for v that is a power-law in r, α, and Σ.

As an explicit example, we may consider a disk in which the mid-plane opacity is due to icy particles. In this limit an approximate form for the opacity is

$$\kappa = \kappa_0 T_c^2, \tag{3.55}$$

with the constant having a numerical value in cgs units $\kappa_0 = 2.4 \times 10^{-4}$. For this opacity there is no dependence on density, and hence the equation for ρ is redundant. Eliminating variables in turn the remaining equations yield

$$\Sigma^3 = \frac{64}{81\pi} \frac{\sigma}{\kappa_0} \left(\frac{\mu m_p}{k_B} \right)^2 \alpha^{-2} \dot{M}. \tag{3.56}$$

For an accretion rate $\dot{M} = 10^{-7}\ M_\odot\ \mathrm{yr}^{-1}$ and an $\alpha = 0.01$, we find that $\Sigma \approx 140\ \mathrm{g\,cm}^{-2}$. The corresponding viscosity is $v \approx 5 \times 10^{15}\ \mathrm{cm}^2\,\mathrm{s}^{-1}$, and the viscous time at 30 AU is about 1.3×10^6 yr. This time scale is broadly consistent with protoplanetary disks evolving significantly over a few million years, and indeed

---

[7] For simplicity we assume that $r \gg R_*$ and neglect terms that depend upon the inner boundary conditions. Note also that different numerical factors can enter into, for example, the definition of the central density to be used in these equations. Given the overriding uncertainty that originates from our poor knowledge of angular momentum transport in disks, worrying about the "correct" value of such factors is rarely profitable.

most observational attempts to constrain α via measures of disk evolution return estimates $\alpha \sim 10^{-2}$ (e.g. Hartmann *et al.*, 1998).

3.4 Angular momentum transport mechanisms

The Shakura–Sunyaev α prescription furnishes a formula for the viscosity but leaves unanswered some of the most basic questions. What is the origin of the postulated turbulence? How large is α, and what is its dependence (if any) on the physical conditions within the disk? These issues remain at the forefront of research, both for protoplanetary disks and for other accreting systems. In this section and in Section 3.5.1 we sketch what currently appear to be the most important physical considerations, but the reader should bear in mind that some of the details are uncertain and subject to future revision.

3.4.1 The Rayleigh criterion

The stability of a rotating flow to infinitesimal hydrodynamic perturbations can be derived by linearizing the fluid equations, setting the time dependence of perturbations to be proportional to $e^{i\omega t}$, and searching for exponentially growing or decaying modes for which ω is imaginary. For a nonmagnetized, non-self-gravitating disk (roughly speaking, one for which $M_{disk}/M_* < h/r$) the appropriate stability criteria are due to Rayleigh (for a derivation see, e.g. Pringle & King, 2007). Such a disk flow is linearly stable to axisymmetric perturbations if and only if the specific angular momentum increases with radius. For instability we require

$$\frac{dl}{dr} = \frac{d}{dr}\left(r^2\Omega\right) < 0. \tag{3.57}$$

In a Keplerian disk the specific angular momentum is an increasing function of radius, $l \propto \sqrt{r}$ (Eq. 3.1), and the flow is predicted to be hydrodynamically stable. Despite the enormous value of the Reynolds number, there is then no ready justification for invoking hydrodynamic turbulence as the origin of the turbulent viscosity needed to drive accretion.

3.4.2 The magnetorotational instability

Coupling a magnetic field to the gas grants the fluid additional degrees of freedom that violently destabilize the disk. The condition for a weakly magnetized disk flow to be linearly unstable is that the angular velocity (rather than the specific angular momentum as in the fluid case) decrease with radius,

$$\frac{d}{dr}\left(\Omega^2\right) < 0, \tag{3.58}$$

and this condition *is* satisfied by Keplerian disks. The linear instability of a disk coupled to a magnetic field is known as the magnetorotational (MRI) or Balbus–Hawley instability. The mathematical analysis of the instability has a long history (Velikhov, 1959; Chandrasekhar, 1961) but its importance in the context of accretion disks was only recognized at a much later date (Balbus & Hawley, 1991).[8]

In the limit of ideal magnetohydrodynamics (MHD) the disk can be described by the equations of continuity,

$$\frac{\partial \rho}{\partial t} + \nabla \cdot (\rho \mathbf{v}) = 0, \tag{3.59}$$

momentum conservation,

$$\frac{\partial \mathbf{v}}{\partial t} + (\mathbf{v} \cdot \nabla) \mathbf{v} = -\frac{1}{\rho} \nabla \left(P + \frac{B^2}{8\pi} \right) - \nabla \Phi + \frac{1}{4\pi\rho} (\mathbf{B} \cdot \nabla) \mathbf{B}, \tag{3.60}$$

and magnetic induction,

$$\frac{\partial \mathbf{B}}{\partial t} = \nabla \times (\mathbf{v} \times \mathbf{B}). \tag{3.61}$$

The symbols have their conventional meanings, ρ is the density, $\mathbf{v}$ is the velocity, $\mathbf{B}$ is the magnetic field, P is the pressure, $B = |\mathbf{B}|$, and Φ is the gravitational potential. Magnetized disks have qualitatively different stability properties from fluid disks because of the presence of magnetic tension (described by the third term on the right-hand-side of Eq. 3.60), which exerts a force on the fluid that attempts to straighten curved field lines.

Demonstrating the existence of the MRI in the general case requires moderately lengthy algebra, which can be found in the review by Balbus & Hawley (1998). The essence of the instability, however, lies in the interplay of magnetic tension and flux freezing (Eq. 3.61) within a differentially rotating system, and this can be demonstrated in much simpler model systems. Consider an axisymmetric, incompressible disk that is threaded by a vertical magnetic field. In cylindrical polar coordinates (r, z, ϕ), the equations of motion of a small parcel of gas then read

$$\ddot{r} - r\dot{\phi}^2 = -\frac{d\Phi}{dr} + f_r, \tag{3.62}$$

$$r\ddot{\phi} + 2\dot{r}\dot{\phi} = f_\phi, \tag{3.63}$$

where the dots denote derivatives with respect to time, and f_r and f_ϕ are forces due to the coupling of the gas to the magnetic field which we will specify shortly.

---

[8] This long hiatus is something of a puzzle, as Chandrasekhar's work, at least, was certainly widely disseminated – the citation above is to his classic (if dense) textbook *Hydrodynamic and Hydromagnetic Stability*. Indeed, the significance of Chandrasekhar's result for the origin of turbulence within the protoplanetary disk *was* very nearly grasped by Safronov (1969), who noted that the MHD stability criterion does not reduce to the Rayleigh criterion as the magnetic field tends toward zero, and that "for a weak magnetic field the cloud should be less stable than we found earlier in the absence of the field." Safronov then, however, dismisses the MRI on the (incorrect) grounds that the instability requires that the viscosity and diffusivity are identically zero.

We now concentrate attention on a small patch of the disk at radius r_0, co-rotating with the overall orbital motion at angular velocity Ω. We define a local Cartesian coordinate system (x, y) via

$$r = r_0 + x, \tag{3.64}$$

$$\phi = \Omega t + \frac{y}{r_0}, \tag{3.65}$$

and substitute these expressions into Eq. (3.63) above. Discarding quadratic terms, the result is

$$\ddot{x} - 2\Omega\dot{y} = -x\frac{d\Omega^2}{d\ln r} + f_x,$$

$$\ddot{y} + 2\Omega\dot{x} = f_y. \tag{3.66}$$

The second term on the left-hand-side of these equations represents the Coriolis force. In the absence of magnetic forces these equations describe the epicyclic motion of pressure-less fluid perturbed from an initially circular orbit.

If the disk contains a vertical magnetic field B_z, perturbations to the fluid in the plane of the disk will be opposed by magnetic tension forces generated by the bending of the field lines. Consider in-plane perturbations varying with height z and time t as

$$\mathbf{s} \propto e^{i(\omega t - kz)}, \tag{3.67}$$

where $\mathbf{s}$ is the displacement vector. The corresponding perturbation to the magnetic field follows from the induction Eq. (3.61). It is

$$\delta\mathbf{B} = -ik B_z \mathbf{s}, \tag{3.68}$$

and this results in a magnetic tension force given by

$$\mathbf{f} = -(k v_A)^2 \mathbf{s}, \tag{3.69}$$

where $v_A = \sqrt{B_z^2/4\pi\rho}$ is the Alfvén speed. Using this expression for f_x and f_y, and recalling the $e^{i\omega t}$ time dependence, Eq. (3.66) becomes

$$-\omega^2 x - 2i\omega\Omega y = -x\frac{d\Omega^2}{d\ln r} - (k v_A)^2 x, \tag{3.70}$$

$$-\omega^2 y + 2i\omega\Omega x = -(k v_A)^2 y. \tag{3.71}$$

Combining these equations yields a dispersion relation (i.e. a relation between the wavenumber k and frequency ω of the perturbation) which is a quadratic in ω^2

$$\omega^4 - \omega^2 \left[\frac{d\Omega^2}{d\ln r} + 4\Omega^2 + 2(k v_A)^2 \right] + (k v_A)^2 \left[(k v_A)^2 + \frac{d\Omega^2}{d\ln r} \right] = 0. \tag{3.72}$$

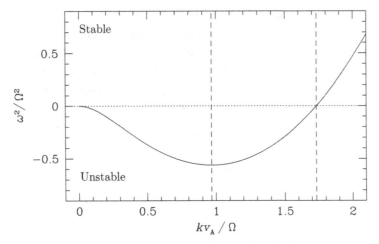

Fig. 3.5. The unstable branch of the dispersion relation for the magnetorotational instability (Eq. 3.72) is plotted for a disk with a Keplerian angular velocity profile. Both ω and kv_A have been plotted scaled to the orbital frequency Ω. The flow is unstable ($\omega^2 < 0$) for all spatial scales smaller than $kv_A < \sqrt{3}\Omega$ (rightmost dashed vertical line). The most unstable scale (shown as the dashed vertical line at the center of the plot) is $kv_A = (\sqrt{15}/4)\Omega$.

If $\omega^2 > 0$ then ω itself will be a real number, and the perturbation of the form $e^{i\omega t}$ will display oscillatory behavior. Instability occurs when $\omega^2 < 0$, since in this case ω is imaginary and the perturbation will have exponentially growing modes. The instability criterion is that

$$(kv_A)^2 + \frac{d\Omega^2}{d\ln r} < 0. \tag{3.73}$$

Taking the limit of a vanishingly weak field ($B_z \to 0$, $v_A \to 0$) we recover the aforementioned condition for the local linear instability of a differentially rotating disk in the presence of a weak magnetic field (Eq. 3.58). One may observe that the instability condition does not tend toward the Rayleigh criterion as the magnetic field goes to zero, but rather remains qualitatively distinct.

Two other important properties of the MRI – the growth rate of the instability and what it means for the magnetic field to be "weak" – can also be derived directly from Eq. (3.72). Specializing to a Keplerian rotation law $d\Omega^2/d\ln r = -3\Omega^2$ we find that the dispersion relation takes the form shown in Fig. 3.5. For a fixed value of the magnetic field strength (and hence a fixed Alfvén speed v_A) the flow is unstable for wavenumbers $k < k_{crit}$ (i.e. on sufficiently large spatial scales), where

$$k_{crit} v_A = \sqrt{3}\Omega. \tag{3.74}$$

As the magnetic field becomes stronger, the *smallest* scale $\lambda = 2\pi/k_{\text{crit}}$ which is unstable grows, until eventually it exceeds the disk's vertical extent which we may approximate as $2h$, where h is the scale-height. For stronger vertical fields no unstable MRI modes fit within the thickness of the disk, and the instability will be suppressed. Noting that $h = c_s/\Omega$, we can express the condition that the vertical magnetic field be weak enough to admit the MRI (i.e. that $\lambda < 2h$) as

$$B^2 < \frac{12}{\pi}\rho c_s^2,\tag{3.75}$$

where c_s is the disk sound speed. If we define the plasma β parameter as the ratio of gas to magnetic pressure

$$\beta \equiv \frac{8\pi P}{B^2},\tag{3.76}$$

Eq. (3.75) can be expressed alternatively as

$$\beta > \frac{2\pi^2}{3}.\tag{3.77}$$

A magnetic field whose *vertical* component approaches equipartition with the thermal pressure ($\beta \sim 1$) will therefore be too strong to admit the existence of linear MRI modes, but a wide range of weaker fields is acceptable.

The maximum growth rate can be determined by setting $d\omega^2/d(kv_A) = 0$ for the unstable branch of the dispersion relation shown in Fig. 3.5. The most unstable scale for a Keplerian disk is

$$(kv_A)_{\text{max}} = \frac{\sqrt{15}}{4}\Omega,\tag{3.78}$$

with a corresponding growth rate

$$|\omega_{\text{max}}| = \frac{3}{4}\Omega.\tag{3.79}$$

The main point to notice about this last result is that it implies an *extremely* vigorous growth of the instability, with an exponential growth time scale that is a fraction of an orbital period. Practically, this means that if a disk is unstable to the MRI it is hard to envisage other physical processes that will operate on a shorter time scale and prevent the MRI from dominating the evolution.

The physical origin of the MRI is fairly transparent for this simple configuration. Consider the effect of perturbing a weak vertical magnetic field threading an otherwise uniform disk, so that some fluid (tied to the magnetic field due to flux freezing) moves slightly inward and some slightly outward. Due to the differential rotation, the perturbed field line – which now connects adjacent annuli in the disk – will be sheared by the differential rotation of the disk into a trailing spiral pattern. Provided that the field is weak enough, magnetic tension is inadequate to snap the field lines back to the vertical. What tension there is, however, acts to reduce the angular momentum of the inner fluid element, and boost that of the outer fluid element. The transfer of angular momentum further increases the separation, leading to an instability.

The very rapid linear growth of the MRI implies, of course, that there is no physical way to set up a disk with initial conditions that resemble those used for analytic convenience above. Within a real disk it is the *nonlinear* properties of the instability that are of interest for angular momentum transport, and these can only be studied in detail via numerical MHD simulations. Disentangling physical from numerical effects in such calculations is a nontrivial exercise, but one consensus result is that the nonlinear evolution of the MRI leads to a state of sustained MHD turbulence which is able to maintain disk magnetic fields in the presence of dissipation. A flow with this property is described as a magnetic dynamo. Denoting the velocity fluctuations in the radial and azimuthal directions by δv_r and δv_ϕ respectively, we can define an equivalent Shakura–Sunyaev α parameter in terms of an average over the fluctuating velocity and magnetic fields

$$\alpha = \left\langle \frac{\delta v_r \delta v_\phi}{c_s^2} - \frac{B_r B_\phi}{4\pi \rho c_s^2} \right\rangle, \tag{3.80}$$

where the angle brackets denote a density-weighted average over time. This formula provides a way to connect the small-scale flow dynamics seen in numerical simulations with the gross angular momentum transport efficiency that is relevant for the evolution of the disk as a whole. Numerical work shows that the magnetic transport term (the second term on the right-hand-side, described as the Maxwell stress) is generically much larger than the fluid (or Reynolds) contribution in MHD turbulent flows initiated by the MRI, although both terms are positive. The predicted magnitude is less securely estimated – even in the ideal MHD limit – but could plausibly be consistent with the values $\alpha \sim 10^{-2}$ inferred from observations.

3.4.3 Disk winds and magnetic braking

Thus far we have assumed that the evolution of the disk is driven by redistribution of angular momentum within the protoplanetary disk. A qualitatively different idea holds that evolution is driven instead by angular momentum *loss* mediated by open magnetic field lines which thread the disk, as shown schematically in Fig. 3.6. If the magnetic field at the disk surface has vertical and azimuthal components B_z^s and B_ϕ^s respectively, then the torque per unit area exerted on the disk (counting both the upper and lower surfaces) is

$$T_m = \frac{B_z^s B_\phi^s}{2\pi} r. \tag{3.81}$$

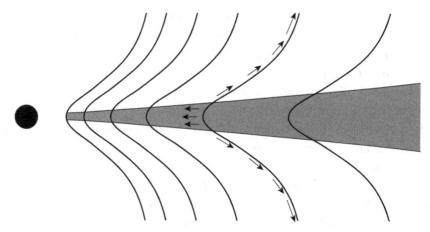

Fig. 3.6. Showing the geometry assumed in disk wind models for protoplanetary disk evolution. Open magnetic field lines connect the disk to an almost force-free disk corona. A small fraction of the accreting matter is accelerated outward along the field lines to form a magnetically launched disk wind, while the magnetic torque exerted by the field at the disk surface robs the disk of angular momentum and drives accretion.

One might worry about how to define the "surface" of the disk, but in practice the sharp drop in density with increasing height (Eq. 2.8) means that there is a reasonably well defined transition between the (normally) gas pressure dominated disk interior and the magnetically dominated outer layers. Often, it suffices to define the surface to lie at $z \pm h$. If the magnetic field is weak enough that the angular velocity remains Keplerian (i.e. magnetic pressure gradients are negligible) then the angular momentum per unit area is $\Sigma\sqrt{GM_*r}$, and the loss of that angular momentum via the magnetic torque results in a radial velocity

$$|v_{r,m}| = \frac{B_z^s B_\phi^s}{\pi \Sigma \Omega}. \tag{3.82}$$

This inflow velocity can be compared to that which results from internal redistribution of angular momentum due to viscosity,

$$v_{r,\text{visc}} = -\frac{3}{2}\frac{\nu}{r}. \tag{3.83}$$

Making use of Eq. (3.46) and (3.80), and noting that for MRI-generated transport the Maxwell stress is dominant over the Reynolds stress, one finds that,

$$\frac{|v_{r,m}|}{|v_{r,\text{visc}}|} \sim \frac{B_z^s B_\phi^s}{B_r B_\phi}\left(\frac{h}{r}\right)^{-1}, \tag{3.84}$$

where the magnetic fields in the denominator are the turbulent fields evaluated near the mid-plane and we have ignored numerical factors of the order of unity. One obtains the not terribly surprising result that if organized large scale fields thread the disk and have comparable strengths to the turbulent fields, a wind can carry away angular momentum more efficiently than internal processes can redistribute it, especially if the disk is thin.

It is straightforward to generalize the surface density evolution Eq. (3.6) to include the effect of angular momentum loss in a disk wind. If the angular velocity remains Keplerian,[9] the effect of the wind on the disk can be modeled as an additional advective term

$$\frac{\partial \Sigma}{\partial t} = \frac{3}{r}\frac{\partial}{\partial r}\left[r^{1/2}\frac{\partial}{\partial r}\left(\nu \Sigma r^{1/2}\right) \right] - \frac{1}{r}\frac{\partial}{\partial r}\left(r \Sigma v_{r,m}\right). \tag{3.85}$$

A full solution for the disk evolution requires coupling this equation to models for the mass loading and magnetic structure of the disk wind, which are complicated problems in their own right. The key physical question concerns the strength of the vertical magnetic field that threads the disk. For disk winds to play a major role in disk evolution we require a relatively strong vertical field. Since the flux threading the disk

$$\Phi = \oint B_z dS, \tag{3.86}$$

is a conserved quantity, development of net vertical field cannot occur as a result of local dynamo processes, but rather requires that magnetic flux of one sign either be dragged into or preferentially expelled from the disk. Although the molecular cloud cores that collapse to form disks unquestionably contain magnetic fields, it is not clear whether this field can be dragged inward efficiently if the disk is turbulent. If we consider an initially weak vertical field, inward dragging of field lines will occur on the viscous time scale $\tau_\nu = r^2/\nu$ (Eq. 3.12). This inward dragging will be opposed by outward diffusion, which for the geometry shown in Fig. 3.6 requires that the field reconnect across a vertical scale $\delta z \sim h \ll r$. For $B_z^s \sim B_r^s$, Lubow *et al.* (1994) find that the outward diffusion time scale is then given by

$$t_\eta \simeq \frac{r^2}{\eta}\left(\frac{h}{r}\right), \tag{3.87}$$

where η is the magnetic diffusivity within the disk. If the diffusivity and the viscosity are both generated from the same turbulent processes within the disk, we expect that to order of magnitude $\eta \sim \nu$, and hence $t_\eta \sim (h/r)t_\nu$. Outward diffusion of field lines then occurs on a shorter time scale than inward advection, and it is difficult to set up a self-consistent model for a magnetic wind driven disk.

---

[9] Discussion of the more general case can be found, for example, in Shu *et al.* (2007).

The argument above constitutes a theoretical impediment to the construction of models in which angular momentum loss in a wind dominates the internal transport of angular momentum within the disk. There are, however, ways around this problem. One possibility is that the magnetic field is always strong enough to suppress the linear growth of the MRI according to the condition given in Eq. (3.77). This circumvents the diffusion problem by removing the driver of the turbulent diffusivity itself. Another idea is that field dragging may be more efficient if the field close to the disk has a complex patchy structure rather than the simple geometry that is often assumed (Spruit & Uzdensky, 2005). Given the range of theoretical possibilities, one might hope to appeal to observations to settle the question of whether disk winds are important, but here too the available facts are meager. One indirect but arguably relevant observation is the measurement of the disk radius in dwarf novae – accreting white dwarf stars which display outbursts in which the accretion rate is much higher than in quiescence. In these systems, the disk is observed to expand during outbursts (Smak, 1984). Expansion is expected within viscous models for disk evolution (since the increased rate of angular momentum transport during outbursts injects angular momentum into the outer disk region), but is not a natural consequence of disk wind driven models. On the other hand, observations of accreting stellar mass black holes reveal a baffling range of variability, and it is frequently speculated that an additional degree of freedom in the form of a net vertical flux might help in explaining the observed phenomenology. Additional observations that are more directly relevant to protoplanetary disks are evidently needed.

3.4.4 Hydrodynamic turbulence

For disks in which the gas is well coupled to the magnetic field the MRI can operate and regardless of whether other angular momentum transport or angular momentum loss processes occur we expect the fluid to sustain MHD turbulence. Asking whether accretion disks would be turbulent in the *absence* of magnetic fields might therefore seem to be an academic exercise of the kind that brings academia into disrepute. As we will show in Section 3.5.1, however, there may well be regions within protoplanetary disks where the ionization fraction is so low that the magnetic field is effectively decoupled from the fluid dynamics of the gas. The MRI will not be able to function in such regions, and it is then a very valid question as to whether other – perhaps much weaker – angular momentum transport mechanisms exist that can operate in the effective absence of magnetic fields. Three possibilities have attracted the most attention: disk self-gravity, nonlinear or transient instabilities in purely hydrodynamic flows, and vortices.

Up to now we have considered disks whose own gravity is negligible compared to the force from the central star. Such disks are axisymmetric, which implies that no torque is exerted on gas at one radius due to the gravitational force from matter elsewhere in the disk. As the disk becomes more massive there is an increasing tendency for its own self-gravity to result in the formation of overdense "clumps." Gravitational forces between these clumps can result in angular momentum transport.

Self-gravity is an important potential mechanism for planetesimal formation, which is discussed in Chapter 4. We defer a formal mathematical analysis of the conditions for clump formation until then, in favor of a simple argument based on time scales. Within a disk, the self-gravity of the disk gas always has a tendency to form denser clumps, but this is opposed by both pressure forces and by shear, which tend to oppose clump formation. We can estimate the conditions under which self-gravity is strong enough to win out over these stabilizing effects by requiring that the time scale for collapse be shorter than the time scales on which sound waves can cross a clump, or shear destroy it.

Consider a forming clump of scale Δr and mass $m \sim \pi (\Delta r)^2 \Sigma$. In isolation, such a clump would collapse on the free-fall time scale

$$t_{ff} \sim \sqrt{\frac{\Delta r^3}{Gm}} \sim \sqrt{\frac{\Delta r}{\pi G \Sigma}}. \tag{3.88}$$

Stabilizing influences that may prevent collapse are pressure and shear. The time scale for a sound wave to cross the clump is

$$t_p \sim \frac{\Delta r}{c_s}, \tag{3.89}$$

where c_s is the disk sound speed, while the shear time scale (the time scale required for a clump to be sheared azimuthally by an amount Δr) is

$$t_{shear} = \frac{1}{r} \left(\frac{d\Omega}{dr} \right)^{-1} \sim \Omega^{-1}. \tag{3.90}$$

One observes that the shear time scale is independent of the clump size Δr, whereas the sound crossing time increases linearly with Δr. Pressure tends to stabilize small regions of the disk against gravitational collapse, while shear stabilizes the largest scales. The disk will be marginally unstable to clump formation if the free-fall time scale on the scale where shear and pressure support match is shorter than either t_p or t_{shear}. Setting all three time scales equal, we obtain a condition for instability in the form

$$\pi G \Sigma \gtrsim c_s \Omega. \tag{3.91}$$

At a given radius (i.e. at fixed Ω) the disk will be unstable if it is massive (large Σ) and/or cool and thin (small c_s). A more formal analysis (Toomre, 1964) gives the same result. The stability of a disk of either gas or stars is controlled by a parameter

$$Q \equiv \frac{c_s \Omega}{\pi G \Sigma}, \qquad (3.92)$$

known as the "Toomre Q" parameter.[10] Instability of the disk against self-gravity sets in once $Q \lesssim 1$. Using the fact that $h = c_s / \Omega$, and estimating the disk mass as $M_{disk} \sim \pi r^2 \Sigma$, we can write the instability threshold in the more intuitive form

$$\frac{M_{disk}}{M_*} \gtrsim \frac{h}{r}. \qquad (3.93)$$

This is (up to numerical factors) the same condition that we derived earlier (Eq. 2.14) for when disk gravity becomes important for the vertical structure. Since $h/r \approx 0.05$ is a representative number for protoplanetary disks, we require fairly massive disks before the effects of self-gravity can be expected to become important. Such disks are more likely to have existed at early epochs, possibly prior to the optically visible T Tauri phase of YSO evolution.

If self-gravity sets in within a disk, there are two possible outcomes:

- Collapse may continue unhindered, destroying the disk and forming one or more bound objects. Disk *fragmentation* via this process is a mechanism for planetesimal formation (in the case where the collapse occurs in the solid component of the disk material) and, perhaps, giant planet formation.
- Adiabatic heating as the clump contracts to higher density may yield enough pressure to prevent complete collapse. Numerical simulations show that the outcome is the development of spiral arms induced by the self-gravity within the disk. Gravitational forces set up by the spiral arms act much like magnetic tension in an MRI-unstable disk, and work to transport angular momentum outward and mass inward.

The boundary between these possibilities is set by the ability of the disk to radiate away its thermal energy. If the disk can cool on a short time scale, pressure cannot build up within contracting clumps, and fragmentation results. Slow cooling, which is physically more likely in most protoplanetary disks, results instead in stable angular momentum transport.

A priori it is far from obvious that angular momentum transport via self-gravity can be described using the local language of α disks that we have developed previously. If, however, one allows this sleight of hand, we can give a plausible argument for an important result derived more formally by Gammie (2001): namely

---

[10] If the disk is extremely massive, then the angular velocity profile itself may depart significantly from the Keplerian form. In that case, Ω in the stability criterion must be replaced by the *epicyclic frequency* κ, defined via $\kappa^2 = (1/r^3) d(r^4 \Omega^2)/dr$ (note that $\kappa = \Omega$ for a Keplerian disk).

that the equivalent α in a self-gravitating disk depends upon the cooling time of the gas. Let us assume that the disk does not fragment, in which case heating at a local rate

$$Q_+ = \frac{9}{4} \nu \Sigma \Omega^2, \tag{3.94}$$

must on average balance cooling. If the disk is optically thick, the local cooling rate is

$$Q_- = 2\sigma T_{\text{disk}}^4. \tag{3.95}$$

Writing the "gravitational viscosity" in the form,

$$\nu = \alpha_{\text{grav}} \frac{c_s^2}{\Omega} = \frac{\alpha_{\text{grav}}}{\Omega} \frac{k_B T_c}{\mu m_p}, \tag{3.96}$$

and setting $Q_+ = Q_-$, we find that the effective α depends upon the ratio $\sigma T_{\text{disk}}^4 / (\Sigma T_c)$. We recognize this ratio as being proportional to the cooling rate from the disk surface divided by the thermal energy. We therefore define

$$t_{\text{cool}} = \frac{U}{2\sigma T_{\text{disk}}^4}, \tag{3.97}$$

where U is the thermal energy content of the disk gas per unit area. Noting that, approximately, $U \simeq c_p \Sigma T_c$, where c_p is the heat capacity of the gas, the thermal balance condition yields

$$\alpha_{\text{grav}} \approx \frac{4}{9\gamma(\gamma - 1)\Omega t_{\text{cool}}}. \tag{3.98}$$

In this expression γ is the two dimensional adiabatic index (defined in terms of the two dimensional pressure P and internal energy U via $P = (\gamma - 1)U$) and we have reinserted the correct numerical factors from the analysis by Gammie (2001). One finds that the strength of angular momentum transport from self-gravity depends upon the cooling time of the disk, measured in units of the local dynamical time. This is intuitively reasonable. If the disk cools rapidly, strong gravitational turbulence, characterized by large density contrasts, is required in order to generate enough heat to balance the cooling. The strong density contrasts result in larger torques due to self-gravity. Eventually, if the cooling is too rapid, the disk cannot be stabilized against collapse, and fragmentation ensues. Simulations show that stable angular momentum transport is possible only for $\alpha_{\text{grav}} \lesssim 0.1$ (Gammie, 2001; Rice et al., 2005).

Returning now to non-self-gravitating protoplanetary disks, are there hydrodynamic instabilities that would initiate turbulence even in the absence of dynamically important magnetic fields? The question is notoriously controversial. The Rayleigh stability criterion – which implies that arbitrary perturbations to the linearized fluid equations asymptotically decay exponentially with time – leaves open two avenues by which fluid effects might be important within protoplanetary disks. One possibility is that disks might be unstable to nonlinear or finite amplitude perturbations, for which the $(\mathbf{v} \cdot \nabla)\mathbf{v}$ on the left-hand-side of the momentum Eq. (3.60) cannot be neglected. Neither numerical simulations (Hawley *et al.*, 1999) nor physical experiments (Ji *et al.*, 2006) have uncovered such instabilities, but there is also no analytic proof of the nonlinear stability of Keplerian disk flows. It is therefore conceivable that nonlinear instabilities might exist, provided that they manifest themselves only at Reynolds numbers larger than those accessible numerically or experimentally.

A related possibility is the existence of transiently growing disturbances in perturbed Keplerian disk flows. Although all linear modes eventually decay exponentially, it is possible to construct perturbations that evolve to large amplitudes *prior* to dying away. In the most optimistic scenario, this transient growth could trigger (unspecified) nonlinear instabilities and initiate turbulence. Even if this does not happen, it is certainly the case that there can be interesting coupling between turbulent and hydrodynamically stable regions of the disk, since turbulence can stir up disturbances that propagate into the formally stable areas. This may be relevant, for example, to disk models in which an MRI-active surface layer overlies a stable interior.

Vortices are a special class of hydrodynamic structure and are of particular interest for planet formation because of their ability to concentrate solid particles into dense regions at their core. This is quite independent of any possible role that they might play in angular momentum transport. Defining the fluid *vorticity* as

$$\omega = \nabla \times \mathbf{v}, \tag{3.99}$$

we obtain an equation for the evolution of the *vortensity* ω/ρ by taking the curl of the momentum equation. In the absence of magnetic fields and microscopic viscosity, the result is

$$\frac{D}{Dt}\left(\frac{\omega}{\rho}\right) = \left(\frac{\omega}{\rho}\right) \cdot \nabla\mathbf{v} - \frac{1}{\rho}\nabla\left(\frac{1}{\rho}\right) \times \nabla P. \tag{3.100}$$

This equation exposes two of the most important physical considerations concerning the role of vortices within protoplanetary disks. First, we may note that in a *barotropic* fluid (one in which $P = P(\rho)$ only) surfaces of equal pressure are always parallel to surfaces of equal density. In this limit the second term on the right-hand side vanishes, and any vortensity present within the disk is simply advected with the fluid motion. This illustrates why vorticity is interesting – once introduced it is an approximately conserved quantity that cannot be easily destroyed. Indeed, in two-dimensional models of protoplanetary disks, vortices are found to be stable long-lived structures whose interactions can act to transport angular momentum[11] (Godon & Livio, 1999; Johnson & Gammie, 2005). In three dimensions, the survival time of vortices within protoplanetary disks is less clear. Vortices appear to be destroyed rather quickly close to the disk mid-plane (Shen *et al.*, 2006), but have a better prospect of survival away from the mid-plane where the density is dropping rapidly with z and the fluid is more nearly two-dimensional (Barranco & Marcus, 2005). Second, vorticity can be generated within disks if $\nabla \rho \times \nabla P \neq 0$. In a thin disk geometry we expect that both the gradient in density and the gradient in pressure will be almost vertical, but small misalignments might occur due, for example, to the influence of stellar irradiation. This will result in a nonbarotropic source of vorticity. Similar effects may occur in the vicinity of planets, once those have formed within the gas disk.

3.5 Effects of partial ionization on disk evolution

As we noted in Section 2.7, the protoplanetary disk at the radii of greatest interest for planet formation (0.1 AU–50 AU) is expected to be predominantly neutral. Magnetic fields, which couple to charged particles, affect the neutral component of the fluid indirectly, via collisions between charged and neutral species. Accounting for the influence of partial ionization on the MRI (and, hence, on the likely properties of disk turbulence and angular momentum transport) requires us to consider the role of three new physical effects that, together, make up *nonideal* MHD:

- **Ohmic dissipation**. Magnetic fields are generated by currents, which can be dissipated as a result of collisions between the charge carriers and other species within the fluid. In a well-ionized fluid (for example in the Sun) the scale on which this dissipation occurs is typically much smaller than the macroscopic scales of interest, but this is not true in protoplanetary disks, where the Ohmic dissipation scale can approach, for example, the scales on which the MRI operates.
- **The Hall effect**. Charge carriers moving at an angle to a magnetic field experience a $(\mathbf{v} \times \mathbf{B})$ force which deflects their motion. This results in a *Hall current*, which, in turn, modifies the magnetic field.
- **Ambipolar diffusion**. The magnetic field couples to the charge carriers, which then exchange momentum with the neutral species via collisions. If the collision frequency is relatively low, there can be slippage of the neutrals relative to the charged particles, a process referred to as ambipolar diffusion.

Unfortunately from the perspective of economy, all three of these nonideal effects can be dominant at different locations within the disk. When one considers that

---

[11] This is in accord with the observation that vortices can be long lived in strongly stratified – and thus physically quasi-two-dimensional – atmospheres. Jupiter's Great Red Spot is a particularly spectacular example.

the nature of the primary charge carrier (electrons or charged dust particles) is also uncertain, a bewildering array of possible regimes presents itself! The linear growth of the MRI under protoplanetary conditions has been studied extensively (Blaes & Balbus, 1994; Balbus & Terquem, 2001; Desch, 2004; Salmeron & Wardle, 2005), while the nonlinear evolution including the effect of the Hall and Ohmic terms has been simulated by Sano & Stone (2002a). The discussion here follows that given by Sano & Stone (2002b).

In the presence of nonideal terms, the induction equation (Eq. 3.61) takes the form

$$\frac{\partial \mathbf{B}}{\partial t} = \nabla \times \left[\mathbf{v} \times \mathbf{B} - \eta \nabla \times \mathbf{B} - \frac{\mathbf{J} \times \mathbf{B}}{e n_e} + \frac{(\mathbf{J} \times \mathbf{B}) \times \mathbf{B}}{c \gamma \rho_i \rho} \right], \qquad (3.101)$$

where the second, third, and fourth terms on the right-hand-side represent Ohmic dissipation, the Hall effect, and ambipolar diffusion respectively. In this equation $\mathbf{v}$ is the velocity of the neutral fluid, $\mathbf{J}$ is the current density,

$$\mathbf{J} = \frac{c}{4\pi} (\nabla \times \mathbf{B}), \qquad (3.102)$$

while η, the magnetic diffusivity, is inversely proportional to the electric conductivity σ_c,

$$\eta = \frac{c^2}{4\pi \sigma_c}. \qquad (3.103)$$

Under the assumption that the charge carriers in the disk are electrons (with number density n_e, mass m_e, and charge e) and ions (with density ρ_i and mass m_i), we can write down expressions for the remaining terms in Eq. (3.101). The electrical conductivity (and, hence, the magnetic diffusivity) is determined by collisions between electrons and neutral species. If the cross-section for such collisions is σ, the collision frequency is

$$\nu_e = n_n \langle \sigma v \rangle_e, \qquad (3.104)$$

where n_n is the number density of neutrals and the brackets denote a velocity weighted average. The conductivity is

$$\sigma_c = \frac{e^2 n_e}{m_e \nu_e}. \qquad (3.105)$$

Numerically, the collision rate is given by (Draine *et al.*, 1983)

$$\langle \sigma v \rangle_e = 8.28 \times 10^{-10} T^{1/2} \, \text{cm}^3 \, \text{s}^{-1}. \qquad (3.106)$$

The strength of ambipolar diffusion depends upon the collision rate $\langle \sigma v \rangle_i$ between *ions* and neutral particles. Ambipolar diffusion is important if this collision rate is low enough. The drag coefficient γ in Eq. (3.101) is written as

$$\gamma = \frac{\langle \sigma v \rangle_i}{m_i + m_n}. \tag{3.107}$$

Here, $m_n = \mu m_H$ is the mass of the neutral particles, and the collision rate is

$$\langle \sigma v \rangle_i = 1.9 \times 10^{-9} \, cm^3 \, s^{-1}. \tag{3.108}$$

Given a model for the ionization state of the disk – which yields the masses and number densities of the charged and neutral species – these equations fully specify the magnitudes of the three nonideal terms. In a dust-free disk, Sano & Stone (2002b) find that the Hall term is usually the dominant nonideal effect. Ohmic dissipation is most important at small radii, while ambipolar diffusion can be significant at radii of the order of 100 AU. If dust is present, Ohmic dissipation typically dominates (Wardle & Ng, 1999).

3.5.1 Layered disks

We can estimate the minimum ionization fraction required for a protoplanetary disk to sustain the MRI by comparing the effect of the Ohmic dissipation term with the inductive term (the first term on the right-hand side of Eq. 3.101).[12] In the absence of dissipation, the MRI in the ideal limit results in a turbulent disk in which the magnetic field has a complex tangled structure. The development of a tangled field is opposed by the action of Ohmic dissipation, which acts diffusively to eliminate field gradients. We can estimate the critical ionization level needed for the MRI to operate by first calculating the magnitude of the magnetic diffusivity η for which the smoothing influence of Ohmic dissipation overcomes the competing tendency of the MRI to generate complex field structures.

To proceed, we first note that diffusion will erase small-scale structure in the magnetic field more readily than larger scale features. The largest scale MRI modes will therefore be the last to survive in the presence of Ohmic dissipation. Starting from the MRI dispersion relation (Eq. 3.72), we specialize to the case of a Keplerian disk and consider the weak-field/long wavelength limit ($k v_A / \Omega \ll 1$). The growth rate of the MRI in this limit is

$$|\omega| \simeq \sqrt{3}(k v_A). \tag{3.109}$$

---

[12] Since we have already noted that the Hall term is often the dominant nonideal effect in protoplanetary disks, one might worry that this is the wrong comparison – perhaps we should be comparing the Hall term to the inductive term? In fact, simulations suggest that the magnitude of the Hall effect does not substantially modify the conditions necessary for the MRI to operate, which is instead determined primarily by the strength of the purely dissipative Ohmic term (Sano & Stone 2002a). This is not obvious.

Writing this in terms of a spatial scale $\lambda = 2\pi/k$, we have that

$$|\omega| \simeq 2\pi\sqrt{3}\frac{v_A}{\lambda}. \tag{3.110}$$

Up to numerical factors, the MRI therefore grows on the Alfvén crossing time of the spatial scale under consideration. Equating this growth rate to the damping rate due to Ohmic diffusion,

$$|\omega_\eta| \sim \frac{\eta}{\lambda^2}, \tag{3.111}$$

and the condition for Ohmic dissipation to suppress the MRI on scale λ becomes

$$\eta \gtrsim 2\pi\sqrt{3}v_A\lambda. \tag{3.112}$$

To completely suppress the MRI, we demand that damping dominates growth on *all* scales, including the largest scale $\lambda \sim h$ available in a disk geometry. The limit on the diffusivity is then $\eta \gtrsim 2\pi\sqrt{3}v_A h$. It is instructive to express this result in a slightly different form. Recalling the definition of the fluid Reynolds number (Eq. 3.44), we define the *magnetic* Reynolds number Re_M via

$$Re_M \equiv \frac{UL}{\eta}. \tag{3.113}$$

where, as before, U is a characteristic velocity and L a characteristic scale of the system. Taking $U = v_A$ and $L = h$ for a disk, we can rewrite the condition for Ohmic dissipation to suppress the MRI in the form

$$Re_M \lesssim 1, \tag{3.114}$$

where numerical factors of the order of unity have been omitted.

The final step is to write the condition for the suppression of the MRI in terms of the ionization fraction $x \equiv n_e/n_n$. Making use of the expressions for the magnetic diffusivity and collision rate between electrons and neutrals (Eq. 3.103 through 3.106) the magnetic diffusivity is

$$\eta \simeq 2.3 \times 10^2 x^{-1} T^{1/2} \, \text{cm}^2 \, \text{s}^{-1}. \tag{3.115}$$

Under the assumption that Maxwell stresses dominate the transport of angular momentum, Eq. (3.80) implies that

$$\alpha \sim \frac{v_A^2}{c_s^2}, \tag{3.116}$$

which allows us to write the Alfvén speed in the definition of the magnetic Reynolds number in terms of α and the sound speed. Since $h = c_s/\Omega$, we obtain,

$$Re_M = \frac{v_A h}{\eta} = \frac{\alpha^{1/2} c_s^2}{\eta\Omega}. \tag{3.117}$$

Substituting for η and c_s^2, we obtain an expression for the predicted variation of the magnetic Reynolds number in a protoplanetary disk

$$\text{Re}_M \approx 1.4 \times 10^{12} x \left(\frac{\alpha}{10^{-2}}\right)^{1/2} \left(\frac{r}{1\,\text{AU}}\right)^{3/2} \left(\frac{T_c}{300\,\text{K}}\right)^{1/2} \left(\frac{M_*}{M_\odot}\right)^{-1/2}. \qquad (3.118)$$

This result is essentially identical[13] to that derived by Gammie (1996), whose analysis we have followed in this discussion. For the fiducial parameters, the critical ionization fraction below which Ohmic losses will suppress the MRI (i.e. for which $\text{Re}_M \lesssim 1$) is

$$x_{\text{crit}} \sim 10^{-12}. \qquad (3.119)$$

The first thing to observe is that this is a *very* small number. A tiny ionization fraction is enough to couple the magnetic field to the gas well enough to allow the MRI to operate within the disk.

Although x_{crit} is very small, it is far from obvious that even this level of ionization is attained throughout the disk. In Section 2.7 we found that thermal ionization of the alkali metals would yield $x > x_{\text{crit}}$ for $T_c \gtrsim 10^3$ K, but temperatures this high are only attained in the very inner disk. At larger distances from the star, the upper and lower surfaces of the disk will be well-enough ionized due to nonthermal processes (irradiation by the stellar X-ray flux, and/or cosmic rays) but the interior of the disk – shielded by the high column density from these external ionizing agents – may well have $x < x_{\text{crit}}$. We conclude that close to the mid-plane the MRI at the radii of greatest interest for planet formation may be suppressed by the low ionization fraction. Unless there are other nonmagnetic drivers of turbulence the disk may be quiescent with a low (or vanishing) efficiency of angular momentum transport.

Following this line of reasoning leads to the *layered* disk model proposed by Gammie (1996), which is shown schematically in Fig. 3.7. In the original version of the model, the protoplanetary disk at intermediate radii (from a few tenths of an AU out to of the order of 10 AU) has a sandwich structure. Accretion occurs only via an active surface layer which is well enough ionized by cosmic rays or stellar X-rays to sustain MHD turbulence driven by the MRI. The column density of this active layer is of the order of $10^2\,\text{g}\,\text{cm}^{-2}$. Closer to the mid-plane lies a quiescent "dead zone" within which the MRI is suppressed due to Ohmic dissipation. It is assumed that there are no other sources of turbulence or angular momentum transport in the dead zone, so this region is inactive and does not support accretion. The mass flux through the surface layer is an increasing function of radius, so as gas flows in

---

[13] The different temperature dependence arises from our use of a temperature dependent cross-section for electron–neutral collisions.

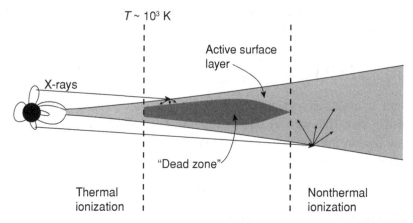

Fig. 3.7. Illustration of the layered disk model of the class proposed by Gammie (1996). In this model the innermost regions of the disk are hot enough such that thermal ionization suffices to couple the magnetic field to the gas well enough for the MRI to operate. At large radii, cosmic rays or stellar X-rays penetrate deep enough into the disk to provide the necessary level of ionization. At intermediate radii, it is hypothesized that accretion occurs primarily in an active surface layer ionized by these nonthermal processes, while the central "dead zone" is magnetically inactive.

from the outer disk (at radii where the column density is low enough that the entire disk is well enough ionized to sustain the MRI) it accumulates within the dead zone.

For the purposes of planet formation, the most important consequence of layered disk models is that they predict that the protoplanetary disk ought to be almost quiescent near $z = 0$ across most of the radii where terrestrial and gas giant planets form. This has immediate implications for several processes that we will discuss in depth in subsequent chapters, including dust settling, planetesimal formation, and low mass planet migration. Unfortunately, the basic question of whether layered disks actually exist remains open. As we have noted, there are significant uncertainties in both the ionization structure of protoplanetary disks and in the extent of residual hydrodynamic transport of angular momentum in the absence of magnetic fields. Even if the ionization level at the disk mid-plane is below x_{crit}, it is probable that a low level of turbulence and transport could persist there – driven for example by fluid stresses or magnetic fields that leak downward from the active surface layers (Fleming & Stone, 2003; Turner & Sano, 2008). A plausible hybrid model has the surface of the disk at AU scales being unstable to the MRI and fully turbulent, while the interior is less active but not entirely quiescent.

3.6 Disk dispersal

Depletion of the gaseous protoplanetary disk due to stellar accretion is predicted to be a gradual process. If we consider the self-similar solution (Eq. 3.25), for example, we find that for a time-independent viscosity scaling with radius as $\nu \propto r^{\gamma}$ the late-time behavior of the surface density is

$$\Sigma \propto t^{-(5/2-\gamma)/(2-\gamma)}, \tag{3.120}$$

which is a power-law in time ($\Sigma \propto t^{-3/2}$ for the case of $\nu \propto r$). A disk that evolved due only to accretion would steadily become optically thin over an increasing range of radii, in the process transitioning rather slowly from a Class II to a Class III source. Observationally this is *not* what is observed. Although a number of candidate transition disks are known – typically YSOs that lack a near-IR excess despite the presence of a robust mid-IR excess – the scarcity of such systems suggests that the dispersal phase of protoplanetary disks lasts of the order of 10^5 yr (Simon & Prato, 1995; Wolk & Walter, 1996). The relative brevity of the dispersal time scale suggests that additional physical processes beyond viscous evolution contribute to the loss of gas from the disk. Plausibly the additional evolutionary agent is *photoevaporation*, a process in which ultraviolet or X-ray radiation heats the disk surface to the point at which it becomes hot enough to escape the gravitational potential as a thermally driven wind. Photoevaporative flows from young stars with disks are observed in the Orion nebula cluster (O'Dell *et al.*, 1993; Johnstone *et al.*, 1998), where an intense ionizing radiation field is provided by massive stars. Weaker ultraviolet irradiation originating from low mass stars themselves would suffice to disperse disks more generally on time scales consistent with those inferred observationally.

3.6.1 Photoevaporation

The basic physics of photoevaporation is simple: ultraviolet radiation heats the disk surface to a temperature high enough that the thermal energy of the gas exceeds its gravitational binding energy. A pressure gradient drives this unbound gas away from the star, dispersing the disk. The details of photoevaporation depend upon the source of the irradiating flux (which can be the disk-bearing star itself or external stars in a cluster) and the energy of the photons. Extreme ultraviolet (EUV) radiation ($E > 13.6$ eV, $\lambda < 912$ Å) ionizes hydrogen atoms, producing a layer of hot gas whose temperature – around 10^4 K – is almost independent of the density of the disk at the radius under consideration. Far ultraviolet (FUV) radiation (6 eV $< E < 13.6$ eV) dissociates H_2 molecules, creating a neutral atomic

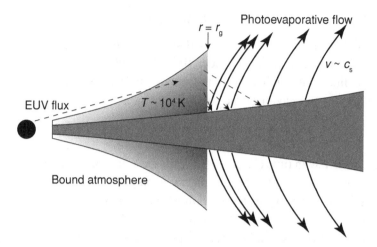

Fig. 3.8. The geometry assumed in simple models of disk photoevaporation by a central stellar EUV source (Hollenbach *et al.*, 1994). The EUV flux incident on the disk ionizes the surface layers, heating them to a temperature of around 10^4 K. Inside a critical radius r_g (set very approximately by the condition that the sound speed in the ionized gas be smaller than the escape velocity) the ionized layer forms a bound atmosphere that absorbs the stellar flux and reradiates recombination radiation. At larger radii the gas escapes as a thermally driven disk wind.

layer whose temperature depends upon the precise balance of heating and cooling processes. Typical values are 100–5000 K.

Since we have two different flavors of ionizing radiation (EUV and FUV) and two geometrically different sources (the central star or an ambient radiation field) there are four distinct regimes of photoevaporation. All four can have a significant impact upon disk evolution. Here we focus on disk photoevaporation due to EUV irradiation by the central star, since it is this regime that is probably most important for the loss of gas from the planet-forming region of the disk.

Let us consider a disk exposed to EUV flux from the central star, as shown in Fig. 3.8. The photoionized gas has a temperature $T \simeq 10^4$ K, which corresponds to a sound speed $c_s \simeq 10 \, \text{km s}^{-1}$. We define the gravitational radius r_g to be the location where the sound speed in the ionized gas equals the orbital velocity

$$r_g = \frac{GM_*}{c_s^2} = 8.9 \left(\frac{M_*}{M_\odot}\right) \left(\frac{c_s}{10 \, \text{km s}^{-1}}\right)^{-2} \text{AU}. \qquad (3.121)$$

In the simplest analysis, gas at $r < r_g$ is bound, and forms an extended atmosphere above the neutral disk surface with a scale-height $h \propto r^{3/2}$. Beyond r_g the ionized gas is unbound, and escapes the disk at a rate

$$\dot{\Sigma}_{\text{wind}} \sim 2\rho_0(r)c_s, \qquad (3.122)$$

where ρ_0, the density at the base of the ionized layer, is determined by the intensity of the irradiating EUV flux. Radiative transfer calculations by Hollenbach *et al.* (1994) suggest that while the inner disk is present, the flux at $r > r_g$ is dominated by photons generated from recombinations in the atmosphere at smaller radii, and that the base density scales with radius roughly as $\rho_0(r) \propto r^{-5/2}$. Most of the mass loss is then concentrated at radii near r_g.

More sophisticated analyses and numerical simulations result in a refinement of the simple picture of photoevaporation given above (Begelman *et al.*, 1983; Font *et al.*, 2004). These calculations suggest that mass loss starts at a radius of about $0.2\ r_g$ (about 2 AU for a Solar mass star), and that the gas escapes the disk at a modest fraction of the sound speed. The total mass loss rate from a disk exposed to an ionizing EUV flux Φ is estimated as

$$\dot{M}_{\text{wind}} \approx 1.6 \times 10^{-10} \left(\frac{\Phi}{10^{41}\ \text{s}^{-1}} \right)^{1/2} \left(\frac{M_*}{1\ M_\odot} \right)^{1/2} M_\odot\ \text{yr}^{-1}. \qquad (3.123)$$

An analytic fit to the radial dependence of the mass loss rate is given in Alexander & Armitage (2007).

For an ionizing photon flux $\Phi \sim 10^{41}\ \text{s}^{-1}$ the predicted mass loss rate from the disk is essentially negligible at early times (when the stellar accretion rate is three orders of magnitude greater), but comes to dominate after a few viscous times of disk evolution. This behavior is qualitatively consistent with that required to reproduce the observed evolution of T Tauri disks, provided that the low-mass stellar EUV flux – which must be two to three orders of magnitude in excess of the present-day output of the Sun for the model to work – is high enough. Although there is a dearth of direct measurements, such an enhanced EUV output for young stars seems likely given the observed strength of other activity indicators (for example the X-ray flux).

3.6.2 Viscous evolution with photoevaporation

Incorporating photoevaporation into time-dependent models of protoplanetary disk evolution is straightforward. The flow away from the disk carries the same specific angular momentum as the disk at the launch point, and hence the effect of photoevaporation can be captured as a simple mass sink in the disk evolution equation

$$\frac{\partial \Sigma}{\partial t} = \frac{3}{r} \frac{\partial}{\partial r} \left[r^{1/2} \frac{\partial}{\partial r} \left(\nu \Sigma r^{1/2} \right) \right] + \dot{\Sigma}_{\text{wind}}(r, t). \qquad (3.124)$$

The mass loss term must be specified via a model of the photoevaporation process, which in general will depend nonlocally on the disk properties (most pertinently,

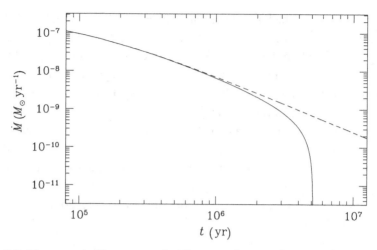

Fig. 3.9. The numerically computed stellar accretion rate for protoplanetary disk models with (solid line) and without (dashed line) photoevaporative mass loss. The model, which is based on that in Clarke *et al.* (2001), assumes a viscosity $\nu \propto r$, and photoevaporative mass loss that scales as $\dot{\Sigma}_{wind} \propto r^{-5/2}$ outside $r_g = 5$ AU. No mass loss is assumed to occur within r_g. The mass loss rate integrated from r_g to 25 AU is 10^{-9} $M_\odot$ yr^{-1}. The initial surface density profile matches the self-similar solution with an initial accretion rate of 3×10^{-7} $M_\odot$ yr^{-1} and a cut-off at 10 AU. The viscosity is normalized such that the viscous time scale at 10 AU is 3×10^5 yr. The presence of a wind from the outer disk results in a sharply defined epoch of disk dispersal.

if there is no disk gas at small radius there will be no bound atmosphere to absorb the stellar EUV flux, which instead will directly illuminate the inner disk edge).

The behavior of protoplanetary disks that evolve due to a combination of angular momentum redistribution and EUV photoevaporation has been studied by Clarke *et al.* (2001) and by Alexander *et al.* (2006). Three phases in the evolution can be distinguished. In the first phase, mass loss due to photoevaporation is negligible compared to the mass flux flowing through the disk as a result of viscous transport of angular momentum, and the disk evolves as if there were no mass loss. Eventually, the mass accretion rate drops to become comparable to the wind mass loss rate, and mass flowing in toward the star from large disk radii is diverted into the wind rather than reaching the inner disk. An annular gap develops in the disk near r_g, and the inner disk – now cut off from resupply – drains rapidly on to the star on its own viscous time scale of the order of 10^5 yr. Figures 3.9 and 3.10 show the evolution of the mass accretion rate and disk surface density up to this point. Finally, the disk interior to r_g becomes optically thin to the stellar EUV flux, which then directly illuminates the inner edge of the outer disk. This results in an increased rate of

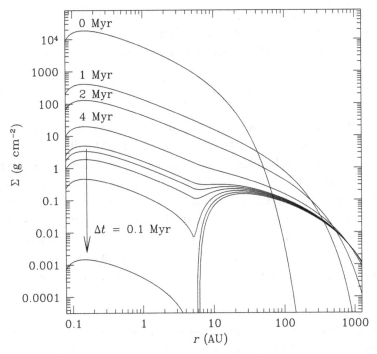

Fig. 3.10. The evolution of the surface density within a protoplanetary disk model that includes photoevaporative mass loss. As in Fig. 3.9, the model assumes a viscosity $\nu \propto r$, and photoevaporative mass loss that scales as $\dot{\Sigma}_{wind} \propto r^{-5/2}$ outside $r_g = 5$ AU. No mass loss is assumed to occur within r_g. The disk evolves slowly on Myr time scales until the stellar accretion rate drops to $\dot{M} \sim \dot{M}_{wind}$, at which point the inner disk becomes cut off from the outer disk and drains rapidly (on a 10^5 yr time scale) on to the star.

mass loss due to photoevaporation, which disperses the entire outer disk on a time scale of a few 10^5 yr.

3.7 Magnetospheric accretion

If the star is unmagnetized, the protoplanetary disk will extend inward until it meets the stellar surface. This is the boundary layer geometry of accretion discussed in Section 3.2.2 and used to justify the zero-torque boundary condition typically assumed for protoplanetary disks. Observations, however, show that T Tauri stars are not unmagnetized, with kG magnetic fields being quite common (Johns-Krull, 2007). These fields are strong enough to disrupt the disk close to the star, resulting in a *magnetospheric* accretion flow geometry in which gas at small radii flows in along field lines to strike the star in accretion hot spots away from the stellar equator.[14] The simplest geometry to consider is one in which the stellar magnetic field is dipolar, aligned with the stellar rotation axis, and perpendicular to the disk plane. The unperturbed field has a vertical component at the disk surface

$$B_z = B_* \left(\frac{r}{R_*} \right)^{-3},$$ (3.125)

which falls off rapidly with increasing distance (note that $B_* R_*^3 = \mu$, the stellar dipole moment). In the presence of a disk, the vertical field will thread the disk gas and – as long as the coupling between the magnetic field and the matter is good enough – the field will be distorted by differential rotation between the Keplerian disk and the star. The differential rotation will twist the field lines that couple the disk to the star, generating an azimuthal field component at the disk surface B_ϕ. Computing this field accurately is not an easy matter (it may depend, for example, on the nature of the turbulence within the disk), but to an order of magnitude it is reasonable to assume that the rapid differential rotation results in a field

$$B_\phi \sim B_z,$$ (3.126)

that is limited by the onset of instabilities that afflict very strongly twisted fields (Lynden-Bell & Boily, 1994). If we define the *co-rotation radius* r_{co} as the radius where the field lines – rotating with stellar rotation period P – have the same angular velocity as that of Keplerian gas in the disk

$$r_{co} = \left(\frac{GM_* P^2}{4\pi^2} \right)^{1/3},$$ (3.127)

we can identify two regions of star–disk magnetic interaction:

- Interior to co-rotation ($r < r_{co}$) the disk gas has a greater angular velocity than the field lines. Field lines that link the disk and the star in this region are dragged forward by the disk, and exert a braking torque that tends to remove angular momentum from the disk gas.
- Outside co-rotation ($r > r_{co}$) the disk gas has a smaller angular velocity than the field lines. The field lines are dragged backward by the disk, and the coupling results in a positive torque on the disk gas.

Young stars are typically rapid rotators, with periods of a week or less, so the co-rotation radius that separates these two regimes lies in the inner disk. For $P = 7$ days, for example, the co-rotation radius around a Solar mass star is at $r_{co} \simeq 15\,R_\odot$ or 0.07 AU.

---

[14] This accretion geometry was initially studied for accretion on to magnetized compact objects (Pringle & Rees, 1972; Ghosh & Lamb, 1979) and subsequently applied to accretion on to T Tauri stars (Königl, 1991).

The radius within which the external torque due to the star–disk magnetic interaction dominates over the internal viscous torque is estimated by comparing the time scales on which each mechanism removes the angular momentum of the disk gas. The stellar magnetosphere threading the disk will remove angular momentum (interior to co-rotation) on a time scale that is the angular momentum content per unit area divided by the magnetic torque (given by Eq. 3.81)

$$t_m \simeq 2\pi \frac{\Sigma \sqrt{GM_* r}}{B_z^s B_\phi^s r}, \tag{3.128}$$

with the magnetic fields being evaluated at the disk surface. Equating this to the viscous time scale ($t_\nu \simeq r^2/\nu$) defines the magnetospheric radius

$$r_m \simeq \left(\frac{3 B_*^2 R_*^6}{2 \dot{M} \sqrt{GM_*}} \right)^{2/7}. \tag{3.129}$$

In deriving this expression we have made use of the steady-state disk relation $\nu \Sigma = \dot{M}/(3\pi)$ and assumed the undistorted dipolar magnetic field geometry discussed above. The scaling of the magnetospheric radius with the magnetic field strength and mass accretion rate is robust – in fact the same scalings apply to the Alfvén radius derived for *spherical* accretion – but equally valid physical arguments result in different numerical factors. Adopting the expression above, the magnetospheric radius for fiducial T Tauri parameters ($B_* = 1\,\text{kG}$, $R_* = 2R_\odot$, $\dot{M} = 10^{-8}\,M_\odot\,\text{yr}^{-1}$, $M_* = M_\odot$) is

$$r_m \simeq 16 R_\odot. \tag{3.130}$$

One concludes that kG-strength stellar magnetic fields are strong enough to dominate the dynamics of protoplanetary disks in the innermost regions, and that a stellar magnetosphere that extends out to roughly the co-rotation radius is plausible.

The presence of magnetospheric rather than boundary layer accretion means that protoplanetary disks around magnetic stars are predicted to have inner edges that lie at $r \simeq r_m$ rather than at the stellar equator. At smaller radii there is a low density magnetospheric cavity within which gas is channeled on to the star along field lines at roughly the free-fall velocity. This geometry has a number of observational implications:

- The innermost region of the disk that would otherwise produce strong near-IR emission is missing. This alters the IR colors of classical T Tauri stars.
- The final accretion on to the stellar surface occurs at roughly the free-fall velocity. This explains the P Cygni profiles (indicative of infall) that are seen in some spectral lines (Hartmann *et al.*, 1994).

- The zero-torque inner boundary condition that is justified for boundary layer accretion need no longer apply. Instead, the star's spin angular momentum can be coupled to the disk angular momentum via the torque exerted by the magnetic field. This torque can modify the stellar rotation rate.

Although important in their own right most of these effects apply at radii that are small enough ($r \lesssim 0.1$ AU) to be relatively unimportant for planet formation. The change to the inner boundary condition formally alters the steady-state disk surface density profile everywhere, but the fractional change becomes small well away from the inner boundary at $r \simeq r_{\mathrm{m}}$. The presence of a low-density magnetospheric cavity is of greater interest for models of planet migration (discussed in Chapter 7), which commonly invoke angular momentum loss to the gaseous protoplanetary disk as the driver of orbital decay. Such migration is likely to slow and eventually halt once the planet enters the magnetospheric cavity.

3.8 Further reading

The most comprehensive reference to the classical theory of accretion disks is *Accretion Power in Astrophysics*, J. Frank, A. King, & D. Raine (2002), Cambridge, UK: Cambridge University Press. A concise and readable summary of the basics can be found in the review article "Accretion discs in astrophysics," J. E. Pringle (1981), *Annual Review of Astronomy & Astrophysics*, **19**, 137.

A rigorous exposition of the relationships between turbulence, dynamo theory, and accretion can be found in the review article "Instability, turbulence, and enhanced transport in accretion disks," S. A. Balbus & J. F. Hawley (1998), *Reviews of Modern Physics*, **70**, 1. Textbook level derivations of the relevant fluid and MHD instabilities are to be found in *Astrophysical Flows*, J. Pringle & A. King (2007), Cambridge, UK: Cambridge University Press.

4

Planetesimal formation

The formation of terrestrial planets from micron-sized dust particles requires growth through at least 12 orders of magnitude in size scale. It is conceptually useful to divide the process into three main stages that involve different dominant physical processes:

- **Planetesimal formation.** Planetesimals are defined as bodies that are large enough (typically of the order of 10 km in radius) that their orbital evolution is dominated by mutual gravitational interactions rather than aerodynamic coupling to the gas disk. With this definition it is self-evident that aerodynamic forces between solid particles and the gas disk are of paramount importance in the study of planetesimal formation, since these forces dominate the evolution of particles in the large size range that lies between dust and substantial rocks. The efficiency with which particles coagulate upon collision – loosely speaking how "sticky" they are – is also very important.
- **Terrestrial planet formation.** Once a population of planetesimals has formed within the disk their subsequent evolution is dominated by gravitational interactions. This phase of planet formation, which yields terrestrial planets and the cores of giant planets, is the most cleanly defined since the basic physics (Newtonian gravity) is simple and well-understood. It remains challenging due to the large number of bodies – it takes 500 million 10 km radius planetesimals to build up the Solar System's terrestrial planets – and long time scales involved.
- **Giant planet formation and core migration.** Once planets have grown to about an Earth mass, coupling to the gas disk becomes significant once again, though now it is *gravitational* rather than aerodynamic forces that matter. For $M_p \sim M_\oplus$ this coupling can result in eccentricity damping and exchange of orbital angular momentum with the gas (a process described as *migration*), while for $M_p \sim 10\,M_\oplus$ the interaction becomes strong enough that the planet can start to capture an envelope from the protoplanetary disk.

Although the boundaries between these regimes are somewhat arbitrary and inconsistently defined (the term "planetesimal," for example, can refer to anything from a 1–100 km scale body) it is useful to keep this ordering of the most important physics in mind as we discuss planet formation.

4.1 Aerodynamic drag on solid particles

Consider a spherical particle of solid material of radius s and material density ρ_m. The first step to understanding how such a particle evolves within the protoplanetary disk is to calculate the aerodynamic force experienced by the particle when it moves at a velocity v *relative to the local velocity* of the gas disk. In calculating the force there are two physical regimes to consider. If $s \lesssim \lambda$, the mean-free path of gas molecules within the disk, then the fluid on the scale of the particle is effectively a collisionless ensemble of molecules with a Maxwellian velocity distribution. The drag force in this regime – which is normally the most relevant for small particles within protoplanetary disks – is called Epstein drag. In the alternate Stokes drag regime, which applies for $s \gtrsim \lambda$, the disk gas flows as a fluid around the obstruction presented by the particle. In either regime the force scales with the frontal area πs^2 that the particle presents to the gas. This means that the acceleration caused by gas drag – which is proportional to the drag force divided by the particle mass – *decreases* with particle size (as s^{-1} for spherical particles) and eventually becomes negligible once bodies of planetesimal size have formed.

4.1.1 Epstein drag

Epstein drag is felt by solid particles that are smaller than the mean-free path of gas molecules within the disk. The form of the drag law in this regime can be derived by considering the frequency of collisions between the particle and gas molecules, given by elementary arguments as the product of the collision cross-section, relative velocity, and molecule number density. We model the solid particle as a sphere of radius s moving with velocity v relative to the disk gas. Within the gas the mean thermal speed of the molecules is

$$v_{th} = \sqrt{\frac{8k_B T}{\pi \mu m_H}}. \tag{4.1}$$

The gas temperature and density are T and ρ, and the mean molecular weight is μ. Up to factors of the order of unity, the frequency with which gas molecules collide with the "front" side of the particle is

$$f_+ \approx \pi s^2 (v_{th} + v) \frac{\rho}{\mu m_H}, \tag{4.2}$$

while the collision frequency on the back side is

$$f_- \approx \pi s^2 (v_{th} - v) \frac{\rho}{\mu m_H}. \tag{4.3}$$

Noting that the momentum transfer per collision is approximately given by $2\mu m_H v_{th}$, we find that the net drag force in the Epstein regime scales as

$$F_D \propto -s^2 \rho v_{th} v. \tag{4.4}$$

The force is linear in the relative velocity and proportional to the surface area of the particle and to the thermal speed of molecules in the disk gas. A more accurate derivation, valid for $s < \lambda$, $v \ll v_{th}$, and a Maxwellian distribution of molecular speeds, yields

$$\mathbf{F_D} = -\frac{4\pi}{3} \rho s^2 v_{th} \mathbf{v}. \tag{4.5}$$

The drag force, of course, acts in the opposite direction to the vector describing the relative velocity between the particle and the gas. Extensions to this formula to describe the case where the particle moves supersonically with respect to the gas ($v \gtrsim v_{th}$) can be found in Kwok (1975), though it is normally the subsonic regime that is relevant for protoplanetary disks.

4.1.2 Stokes drag

Once particles grow to a size much larger than the molecular mean-free path the interaction with the gas can be treated in classical fluid terms, without reference to the molecular nature of the gas. Drag in this regime is called Stokes drag. Naively, we might guess that the drag force would scale with the ram pressure experienced by the particle, and hence we write the force as

$$\mathbf{F_D} = -\frac{C_D}{2} \pi s^2 \rho v \mathbf{v}, \tag{4.6}$$

where C_D, the *drag coefficient*, describes how aerodynamic the particle is. In general, C_D will depend upon the shape of the particle, but for spherical particles it depends only upon the fluid Reynolds number (Eq. 3.44), which we define here on the scale of the particle as

$$\mathrm{Re} = \frac{2sv}{\nu_m}. \tag{4.7}$$

Note that the viscosity here is the (small) molecular viscosity of the gas, rather than any turbulent viscosity within the disk. In terms of the Reynolds number, Weidenschilling (1977b) quotes a piecewise expression for scaling of the drag coefficient

$$C_D \simeq 24\text{Re}^{-1}, \quad \text{Re} < 1 \tag{4.8}$$

$$C_D \simeq 24\text{Re}^{-0.6}, \quad 1 < \text{Re} < 800 \tag{4.9}$$

$$C_D \simeq 0.44, \quad \text{Re} > 800. \tag{4.10}$$

Comparison of the expressions for Epstein and Stokes drag shows that they are equal for a particle of size $s = 9\lambda/4$, and this can be taken as the transition size when constructing a smooth drag law that encompasses both regimes.

4.2 Dust settling

Aerodynamic drag on particles is important for understanding both the vertical distribution and radial motion of dust and larger bodies within the protoplanetary disk. More subtle issues concern the interaction between aerodynamic forces and turbulence, which we have already noted is likely to be a ubiquitous feature (albeit with poorly determined properties) of the disk. To begin with, we ignore turbulence and consider the vertical settling and growth of dust particles suspended in a laminar disk. We quantify the coupling between the solid and gas components of the disk by defining the *friction time scale* for a particle of mass m as

$$t_{\text{fric}} = \frac{mv}{|F_D|}, \tag{4.11}$$

where, as before, v is the relative velocity between the particle and the gas. The friction time scale measures the time in which drag modifies the relative velocity significantly. Writing the particle mass $m = (4/3)\pi s^3 \rho_m$ in terms of the material density ρ_m, t_{fric} takes on a simple form in the Epstein drag regime

$$t_{\text{fric}} = \frac{\rho_m}{\rho} \frac{s}{v_{\text{th}}}. \tag{4.12}$$

Adopting conditions appropriate for a particle at the mid-plane of the protoplanetary disk at $r = 1$ AU ($\rho = 10^{-9}$ g cm^{-3}, $\rho_m = 3$ g cm^{-3}, $v_{\text{th}} = 10^5$ cm s^{-1}) we obtain an estimate for the friction time scale for a particle of size $s = 1$ μm

$$t_{\text{fric}} \approx 3 \text{ s.} \tag{4.13}$$

Small dust particles are thus very tightly coupled to the gas.

We now consider the forces acting on a small dust particle at height z above the mid-plane of a laminar disk. Concentrating for now just on the vertical forces, the z component of the stellar gravity yields a downward force

$$|F_{\text{grav}}| = m\Omega^2 z, \tag{4.14}$$

where $\Omega = \sqrt{GM_*/r^3}$ is the local Keplerian angular velocity (cf. Section 2.2). The *gas* in the disk is supported against this force by an upwardly directed pressure gradient, but no such force acts on a solid particle. If started at rest a particle will therefore accelerate downward until the gravitational force is balanced by aerodynamic drag. In the Epstein regime we have

$$|F_D| = \frac{4\pi}{3}\rho s^2 v_{th} v. \tag{4.15}$$

In practice – given the very short friction time – force balance is attained almost instantaneously and the particle drifts toward the disk mid-plane with a terminal velocity given by equating $|F_D|$ and $|F_{grav}|$

$$v_{settle} = \frac{\rho_m}{\rho} \frac{s}{v_{th}} \Omega^2 z. \tag{4.16}$$

Inserting numerical values roughly appropriate for a 1 μm particle at $z \sim h$ at 1 AU from a Solar mass star ($\rho = 6 \times 10^{-10}\,\mathrm{g\,cm^{-3}}$, $z = 3 \times 10^{11}$ cm, $v_{th} = 10^5\,\mathrm{cm\,s^{-1}}$) one finds that $v_{settle} \approx 0.06\,\mathrm{cm\,s^{-1}}$ and that the settling time, defined as

$$t_{settle} = \frac{z}{|v_{settle}|}, \tag{4.17}$$

is about 1.5×10^5 yr. In the absence of turbulence we would therefore expect that micron-sized particles ought to sediment out of the upper layers of the disk on a time scale that is short compared to the disk lifetime.

Returning to Eq. (4.16), we note that the terminal velocity of a dust particle is inversely proportional to the gas density. Settling will therefore be faster at high z where the gas is tenuous. Using the Gaussian density profile appropriate for a vertically isothermal disk (Eq. 2.8), and noting that the mean thermal speed differs from the sound speed that determines the vertical scale-height h only by a numerical factor, we obtain a general expression for the settling time as a function of z

$$t_{settle} = \frac{2}{\pi} \frac{\Sigma}{\rho_m s \Omega} \exp\left[-\frac{z^2}{2h^2}\right]. \tag{4.18}$$

The strong z dependence implied by this formula means that in the absence of turbulence dust particles would be expected to settle out of the uppermost disk layers rather quickly.

4.2.1 Single particle settling with coagulation

Even if the neglect of turbulence were justified – and it is not – the estimate of the dust settling time given above would be incomplete because it ignores the likelihood that dust particles will collide with one another and grow during the settling process. The settling velocity increases with particle size, so any such coagulation hastens the collapse of the dust toward the disk mid-plane.

To estimate how fast particles could grow during sedimentation we appeal to a simple single particle growth model due to Safronov (1969) (see also Dullemond & Dominik, 2005). Imagine that a single "large" particle of radius s and mass $m = (4/3)\pi s^3 \rho_m$ is settling toward the disk mid-plane at velocity v_{settle} through a background of much smaller solid particles. By virtue of their small size the settling of the small particles can be neglected. If every collision leads to coagulation the large particle grows in mass at a rate that reflects the amount of solid material in the volume swept out by its geometric cross-section

$$\frac{dm}{dt} = \pi s^2 |v_{settle}| f \rho(z), \qquad (4.19)$$

where f is the dust to gas ratio in the disk. Substituting for the settling velocity one finds

$$\frac{dm}{dt} = \frac{3}{4} \frac{\Omega^2 f}{v_{th}} z m. \qquad (4.20)$$

Since $z = z(t)$ this equation cannot generally be integrated immediately,[1] but rather must be solved in concert with the equation for the height of the particle above the mid-plane

$$\frac{dz}{dt} = -\frac{\rho_m}{\rho} \frac{s}{v_{th}} \Omega^2 z. \qquad (4.21)$$

Solutions to these coupled equations provide a very simple model for particle growth and sedimentation in a nonturbulent disk.

Figure 4.1 shows numerical solutions to Eq. (4.20) and (4.21) for initial particle sizes of 0.01 μm, 0.1 μm, and 1 μm. The particles settle from an initial height $z_0 = 5h$ through a disk whose parameters are chosen to be roughly appropriate to a (laminar) Solar Nebula model at 1 AU from the Sun. Both particle growth and vertical settling are found to be extremely rapid. With the inclusion of coagulation, particles settle to the disk mid-plane on a time scale of the order of 10^3 yr – more than two orders of magnitude faster than the equivalent time scale in the absence of particle growth. By the time that the particles reach the mid-plane they have grown to a final size of a few mm, irrespective of their initial radius.

The single particle model described above is very simple, both in its neglect of turbulence and because it assumes that the only reason that particle–particle collisions occur is because the particles have different vertical settling velocities.

---

[1] Note, however, that if the particle grows rapidly (i.e. more rapidly than it sediments) then the form of the equation implies exponential growth of m with time.

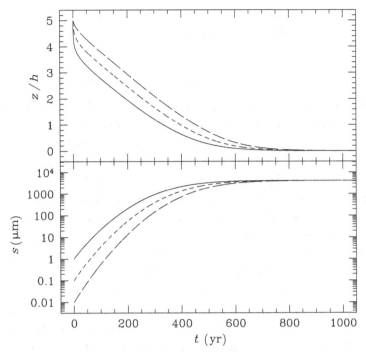

Fig. 4.1. The settling and growth of a single particle in a laminar (nonturbulent) protoplanetary disk. The model assumes that a single particle (with initial size $s = 1\,\mu$m (solid line), 0.1 μm (dashed line), or 0.01 μm (long dashed line)), accretes all smaller particles it encounters as it settles toward the disk mid-plane. The smaller particles are assumed to be at rest. The upper panel shows the height above the mid-plane as a function of time, the lower panel the particle radius s. For this example the disk parameters adopted are: orbital radius $r = 1$ AU, scale-height $h = 3 \times 10^{11}$ cm, surface density $\Sigma = 10^3$ g cm^{-2}, dust to gas ratio $f = 10^{-2}$, and mean thermal speed $v_{\rm th} = 10^5$ cm s^{-1}. The dust particle is taken to have a material density $\rho_{\rm m} = 3$ g cm^{-3} and to start settling from a height $z_0 = 5h$.

Other drivers of collisions include Brownian motion, turbulence, and differential *radial* velocities. The basic result, however, is confirmed by more sophisticated models (e.g. Dullemond & Dominik, 2005), which show that, if collisions lead to particle adhesion, growth from sub-micron scales up to small macroscopic scales (of the order of a mm) occurs rapidly. This means that there is no time scale problem associated with the very earliest phases of particle growth. Indeed, what is more problematic is to understand how the population of small grains – which are unquestionably present given the IR excesses characteristic of classical T Tauri stars – survive to late times. The likely solution to this quandary involves the inclusion of particle *fragmentation* in sufficiently energetic collisions, which allows a broad distribution of particle sizes to survive out to late times. Fragmentation is not likely given collisions at relative velocities of the order of a cm s^{-1} – values typical of settling for micron-sized particles – but becomes more probable for collisions at velocities of a m s^{-1} or higher.

4.2.2 Settling in the presence of turbulence

It is a matter of everyday experience that dust, which settles out of the air readily in an unused room, can remain suspended in the presence of vigorous air currents. The same physics applies within protoplanetary disks. Turbulence – probably but not necessarily the same as that responsible for angular momentum transport – acts to stir up small solid particles and this prevents them from settling into a thin layer at the disk mid-plane. The easiest regime to treat is that in which the particles are small enough to be well-coupled to the gas (mathematically we require that a dimensionless version of the friction time $\Omega t_{\mathrm{fric}} \ll 1$) and represent a negligible fraction of the total disk mass. These conditions are met for dust particles in protoplanetary disks with typical dust to gas ratios.

The conditions necessary for turbulence to stir up the dust enough to oppose vertical settling can be estimated by comparing the settling time (Eq. 4.18) with the time scale on which diffusion will erase spatial gradients in the particle concentration. To diffuse vertically across a scale z requires a time scale

$$t_{\mathrm{diffuse}} = \frac{z^2}{D},\tag{4.22}$$

where D is an anomalous (i.e. turbulent) diffusion coefficient whose magnitude we will discuss later. Equating the settling and diffusion time scales at $z = h$ we find that turbulence will inhibit the formation of a particle layer with a thickness less than h provided that

$$D \gtrsim \frac{\pi e^{1/2}}{2} \frac{\rho_{\mathrm{m}} s h^2 \Omega}{\Sigma}.\tag{4.23}$$

This result is not terribly transparent. We can cast it into a more interesting form if we assume that the turbulence stirring up the particles is the same turbulence responsible for angular momentum transport within the disk. In that case it is plausible[2] that the anomalous diffusion coefficient has the same magnitude and scaling as the anomalous viscosity, which motivates us to write

$$D \sim \nu = \frac{\alpha c_s^2}{\Omega}.\tag{4.24}$$

---

[2] A great deal of interesting complexity is being swept under the carpet here. Although D and ν have the same dimensions (cm^2 s^{-1}), and in the broadest sense arise "from the same turbulent processes," there is no detailed reason why the *vertical* diffusion of a trace scalar contaminant should be equivalent to the *radial* diffusion of angular momentum. For example, in magnetized disks angular momentum transport is dominated by Maxwell rather than fluid stresses. Nonetheless, numerical simulations suggest that taking $D = \nu$ is typically reasonable to within a numerical factor of a few.

With this form for D, the minimum value of α required for turbulence to oppose settling becomes

$$\alpha \gtrsim \frac{\pi e^{1/2}}{2} \frac{\rho_m s}{\Sigma}, \tag{4.25}$$

which is roughly the ratio between the column density through a single solid particle and that of the whole gas disk. For small particles this critical value of α is extremely small. If we take $\Sigma = 10^2 \, \mathrm{g \, cm^{-2}}$, $\rho_m = 3 \, \mathrm{g \, cm^{-3}}$, and $s = 1 \, \mu\mathrm{m}$, for example, we obtain $\alpha \gtrsim 10^{-5}$. This implies that small particles of dust will remain suspended throughout much of the vertical extent of the disk in the presence of turbulence with any plausible strength. For larger particles the result is different. If we consider particles of radius 1 mm – a size that we argued above might form very rapidly – we find that the critical value of $\alpha \sim 10^{-2}$. This value is comparable to most large scale estimates of α for protoplanetary disks. Particles of this size and above will therefore not have the same vertical distribution as the gas in the disk.

To proceed more formally, we can consider the solid particles as a separate fluid that is subject to the competing influence of settling and turbulent diffusion. If the "dust" fluid with density[3] ρ_d can be treated as a trace species within the disk (i.e. that $\rho_d/\rho \ll 1$) then it evolves according to an advection–diffusion equation of the form (Dubrulle *et al.*, 1995; Fromang & Papaloizou, 2006)

$$\frac{\partial \rho_d}{\partial t} = D \frac{\partial}{\partial z} \left[\rho \frac{\partial}{\partial z} \left(\frac{\rho_d}{\rho} \right) \right] + \frac{\partial}{\partial z} \left(\Omega^2 t_{\mathrm{fric}} \rho_d z \right). \tag{4.26}$$

Simple steady-state solutions to this equation can be found in the case where the dust layer is thin enough that the *gas* density varies little across the dust scale-height. In that limit, the dimensionless friction time Ωt_{fric} is independent of z and we obtain

$$\frac{\rho_d}{\rho} = \left(\frac{\rho_d}{\rho} \right)_{z=0} \exp \left[-\frac{z^2}{2h_d^2} \right], \tag{4.27}$$

where h_d, the scale-height describing the vertical distribution of the particle concentration ρ_d/ρ, is

$$h_d = \sqrt{\frac{D}{\Omega^2 t_{\mathrm{fric}}}}. \tag{4.28}$$

---

[3] The dust density ρ_d is the mass of solid particles per unit volume within the disk. It should not be confused with either the gas density ρ or the material density ρ_m which expresses the density of the matter that makes up the particles.

If, as previously, we assume that $D \sim \nu$, we can write a compact expression for the ratio of the concentration scale-height to the usual gas scale-height

$$\frac{h_{\rm d}}{h} \simeq \sqrt{\frac{\alpha}{\Omega t_{\rm fric}}}. \tag{4.29}$$

The condition for solid particles to become strongly concentrated toward the disk mid-plane is then that the dimensionless friction time is substantially greater than α. For any reasonable value of α this implies that substantial particle growth is required before settling takes place.

4.3 Radial drift of solid particles

The fact that solid particles do not experience the same pressure forces as the gas has even more important consequences for the *radial* dynamics of solids (Weidenschilling, 1977b). As we showed previously in Section 2.3, the gas in the disk is normally partially supported against gravity by an outward pressure gradient, and as a result orbits the star at a slightly sub-Keplerian velocity. If locally the mid-plane pressure can be written as a power-law in radius, $P \propto r^{-n}$, the actual gas orbital velocity can be written in terms of the Keplerian velocity $v_{\rm K} = \sqrt{GM_*/r}$ as

$$v_{\phi,{\rm gas}} = v_{\rm K} (1 - \eta)^{1/2}, \tag{4.30}$$

where $\eta = nc_{\rm s}^2/v_{\rm K}^2$ (Eq. 2.20). Let us now consider the implications of this sub-Keplerian rotation for solid particles of different sizes embedded within the gas. For a small dust particle, aerodynamic coupling to the gas is very strong ($\Omega t_{\rm fric} \ll 1$). To a good approximation the dust will be swept along with the gas, and its azimuthal velocity will equal that of the disk gas. Since this is sub-Keplerian, the centrifugal force will be insufficient to balance gravity, and the particle will spiral inward at its radial terminal velocity. Inward radial drift also occurs for large rocks that are poorly coupled to the gas ($\Omega t_{\rm fric} \gg 1$). In this case the aerodynamic forces can be regarded as perturbations to the orbital motion of the body, which orbits the star with an azimuthal velocity that is close to the Keplerian speed. This is faster than the motion of the disk gas, and as a result the rock experiences a headwind that tends to remove angular momentum from the orbit. The loss of angular momentum again results in inward drift.

With this physical understanding in mind we proceed to calculate the rate of radial drift as a function of the friction time of particles located at the disk mid-plane (the analogous calculation for $z \neq 0$ can be found in Takeuchi & Lin, 2002). If the radial and azimuthal velocities of the particle are v_r and v_ϕ respectively, the equations of motion including the aerodynamic drag forces can be written as

$$\frac{dv_r}{dt} = \frac{v_\phi^2}{r} - \Omega_K^2 r - \frac{1}{t_{\text{fric}}}\left(v_r - v_{r,\text{gas}}\right), \tag{4.31}$$

$$\frac{d}{dt}\left(rv_\phi\right) = -\frac{r}{t_{\text{fric}}}\left(v_\phi - v_{\phi,\text{gas}}\right). \tag{4.32}$$

We simplify the azimuthal equation by noting that the specific angular momentum always remains close to Keplerian (i.e. the particle spirals in through a succession of almost circular, almost Keplerian orbits)

$$\frac{d}{dt}\left(rv_\phi\right) \simeq v_r \frac{d}{dr}\left(rv_K\right) = \frac{1}{2}v_r v_K. \tag{4.33}$$

This yields

$$v_\phi - v_{\phi,\text{gas}} \simeq -\frac{1}{2}\frac{t_{\text{fric}} v_r v_K}{r}. \tag{4.34}$$

Turning now to the radial equation, we substitute for Ω_K using Eq. (4.30). Retaining only the lowest order terms

$$\frac{dv_r}{dt} = -\eta\frac{v_K^2}{r} + \frac{2v_K}{r}\left(v_\phi - v_{\phi,\text{gas}}\right) - \frac{1}{t_{\text{fric}}}\left(v_r - v_{r,\text{gas}}\right). \tag{4.35}$$

The dv_r/dt term is negligible. Dropping that term, and eliminating $(v_\phi - v_{\phi,\text{gas}})$ between Eq. (4.34) and (4.35), we obtain

$$v_r = \frac{(r/v_K)t_{\text{fric}}^{-1} v_{r,\text{gas}} - \eta v_K}{(v_K/r)t_{\text{fric}} + (r/v_K)t_{\text{fric}}^{-1}}. \tag{4.36}$$

This result can be cast into a more intuitive form by defining a dimensionless stopping time

$$\tau_{\text{fric}} \equiv t_{\text{fric}}\Omega_K, \tag{4.37}$$

in terms of which the particle radial velocity is

$$v_r = \frac{\tau_{\text{fric}}^{-1} v_{r,\text{gas}} - \eta v_K}{\tau_{\text{fric}} + \tau_{\text{fric}}^{-1}}. \tag{4.38}$$

Let us note some special cases of this general result. For small particles that are tightly coupled to the gas ($\tau_{\text{fric}} \ll 1$), radial drift occurs at a speed

$$v_r \simeq v_{r,\text{gas}} - \eta\tau_{\text{fric}} v_K. \tag{4.39}$$

Such particles are dragged in with the gas, on top of which they experience a radial drift relative to the gas at a rate which is linear in the dimensionless stopping time. For very large particles, conversely, the radial drift velocity decreases linearly with the stopping time.

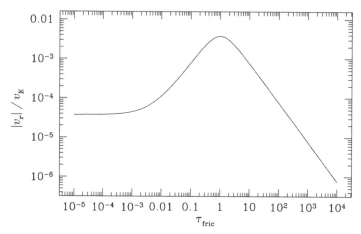

Fig. 4.2. The radial drift velocity of particles at the mid-plane of the protoplanetary disk is plotted as a function of the dimensionless stopping time $\tau_{\text{fric}} = t_{\text{fric}}\Omega_{\text{K}}$. The model plotted assumes that $\eta = 7.5 \times 10^{-3}$ and that $v_{r,\text{gas}}/v_{\text{K}} = -3.75 \times 10^{-5}$. These values are approximately appropriate for a disk with $h/r = 0.05$ and $\alpha = 10^{-2}$ at 5 AU.

Figure 4.2 shows the radial drift velocity as a function of the dimensionless stopping time for parameters $(\eta,\ v_{r,\text{gas}}/v_{\text{K}})$ that are approximately appropriate for the protoplanetary disk at a radius of 5 AU. The drift velocity peaks when $\tau_{\text{fric}} \simeq 1$ at a value

$$v_{r,\text{peak}} \simeq -\frac{1}{2}\eta v_{\text{K}}, \tag{4.40}$$

that depends only upon the pressure gradient in the disk via the dependence of η on the sound speed and surface density gradients. The particle size that corresponds to $\tau_{\text{fric}} \simeq 1$ can be computed based on the formulae for the friction time in the appropriate drag regime (either Epstein or Stokes). In the Epstein regime, for example, a dimensionless stopping time of unity occurs for a particle size

$$s(\tau = 1) = \frac{\rho v_{\text{th}}}{\rho_{\text{m}}\Omega_{\text{K}}}. \tag{4.41}$$

At 5 AU in a disk with $\Sigma = 10^2\ \text{g cm}^{-2}$ and $h/r = 0.05$ the fastest drift occurs for a particle size $s \simeq 20\ \text{cm}$.[4] This is typical – in the inner disk (1–10 AU) the most rapid radial drift coincides with particle sizes in the 10 cm to a few m range.

Figure 4.3 plots the minimum radial drift time scale (i.e. the time scale evaluated at the peak of the curve in Figure 4.2) $t_{\text{drift}} = r/|v_{r,\text{peak}}|$ as a function of radius in the disk. Throughout the main planet-forming regions of the disk this time scale

---

[4] With these parameters the mean free path is of the order of a meter, so it is consistent to use the Epstein formula.

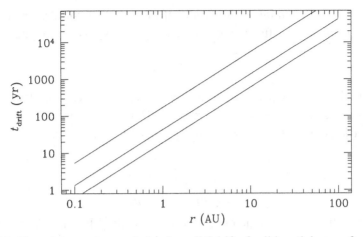

Fig. 4.3. The *minimum* time scale for the radial drift of solid particles as a function of radius, for disk models in which $\Sigma \propto r^{-1}$ and $h/r = 0.025$ (uppermost line), $h/r = 0.05$, or $h/r = 0.075$ (bottom line).

is extremely short – of the order of 10^3 yr or less for reasonable disk parameters. This is an important result, from which flow two robust conclusions:

- **Planetesimal formation must be rapid**, at least if it occurs via a cascade of pairwise collisions that lead to growth. This is an inescapable conclusion – if growth through the cm–m size scale were not very rapid the vast majority of the solid material in the disk would drift toward the star to be evaporated in the hot inner regions of the protoplanetary disk.
- **Radial redistribution of solids is very likely to occur**. Radial flow of solid particles on a time scale shorter than the disk lifetime occurs not just at the peak of the radial velocity drift curve, but also for substantially smaller and larger particles. Local enhancements or depletions of solids (relative to the gas surface density) will occur as a result.

It is worth noting that these inferences follow from rather simple and well-understood physics, namely the action of aerodynamic drag on particles orbiting within a sub-Keplerian gas disk.

4.3.1 Radial drift with coagulation

The size dependence of the radial drift velocity introduces a relative velocity between particles of different sizes, which can promote collisions and (possibly) growth via coagulation. We can use the same arguments that we employed to study coagulation during vertical settling to assess whether this potentially beneficial aspect of radial drift outweighs the deleterious effects of radial drift in depleting the solid surface density. As previously, we note that in the limit where all collisions are adhesive, the growth rate of a particle of radius s that collides primarily with smaller particles[5] as it drifts inward at the disk mid-plane is approximately

$$\frac{dm}{dt} = \pi s^2 |v_r| f \rho_0, \qquad (4.42)$$

where f is the ratio of particle to gas density at $z = 0$, and ρ_0, the mid-plane gas density, is given by $\rho_0 = (1/\sqrt{2\pi}) \Sigma / h$. Comparing the growth time scale $t_{\text{grow}} = m/(dm/dt)$ to the drift time scale $t_{\text{drift}} = r/|v_r|$ we find that $t_{\text{grow}} < t_{\text{drift}}$ for particles of size

$$s \lesssim \frac{3f}{4\sqrt{2\pi}} \left(\frac{h}{r} \right)^{-1} \frac{\Sigma}{\rho_m}. \qquad (4.43)$$

If we assume that a modest amount of vertical settling has already taken place, appropriate values for the parameters in the inner disk might be $f = 0.1$, $\Sigma = 10^3 \, \text{g cm}^{-2}$, $\rho_m = 3 \, \text{g cm}^{-3}$, and $h/r = 0.05$. With these values we find that particles with a size up to $s \simeq 2$ m would collide with at least their own mass of other particles during their inward drift. This implies that the high peak value of the radial drift speed is not, in and of itself, an insurmountable barrier to ongoing growth, provided that vertical settling has taken place. What *is* a serious problem is the fact that the resultant high relative collision velocities between large rocks will very likely invalidate the assumption that collisions will lead to growth. Indeed, at the peak of the radial velocity drift curve it is possible that collisions will actually break up particles.

4.3.2 Particle concentration at pressure maxima

Since both the mid-plane density and the mid-plane sound speed typically decrease with radius, it is normally the case that the disk pressure has a maximum at or close to the star. This results in sub-Keplerian gas velocities across most of the disk, and inward radial drift. The more general rule, however, is that particles drift in the direction of the pressure gradient. Outward drift is therefore possible if the disk possesses a local pressure maximum. This possibility may be of interest if, for example, turbulence in the disk is strong enough to create local pressure maxima. In that situation, which is illustrated in Fig. 4.4, particles would be expected to flow towards the maximum from both smaller and larger radii. The time scale for this process is even faster than the global drift time scale. If the pressure maximum has some radial scale Δr, the *local* pressure gradient $\sim P/\Delta r$ exceeds the global gradient $\sim P/r$. The time scale for solids to pile-up at the pressure maximum

---

[5] We assume that the particles in question have settled to $z \approx 0$, and that their size is such that they lie on the left-hand-side of the peak in the radial drift velocity curve plotted in Fig. 4.2.

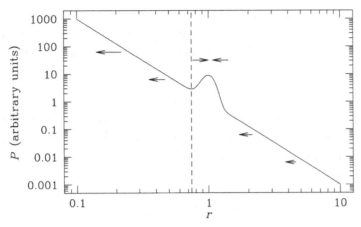

Fig. 4.4. How a nonmonotonic pressure profile in the disk can result in particle pile-up. If the pressure at the disk mid-plane has a local maximum, particles drift towards the location of the maximum from both sides, resulting in a rapid concentration of solids at that point.

is then shorter than the global drift time scale by a factor of $(r/\Delta r)^2$. Another relevant circumstance in which inward drift can be halted occurs in the presence of a massive planet, which as we will show later tends to create an annular gap in the disk surface density near the location of its orbit. A disk with such a gap possesses a pressure maximum at the inner edge of the disk that is exterior to the planet. Solids with some range of sizes would be expected to accumulate at that location.

4.3.3 Turbulent radial diffusion

The presence of turbulence within the disk modifies the radial transport of solid particles, though the differences between the laminar and turbulent cases are less dramatic for radial drift than for vertical settling. Whereas turbulence essentially precludes settling for all but the largest particles, it does not alter the *mean* sub-Keplerian flow that is responsible for radial drift. Substantial rocks will still drift inwards rapidly, unless, as discussed above, the turbulence is so strong as to create local pressure maxima. Where turbulence matters most is for small particles that are well-coupled to the gas. Such particles can diffuse in a turbulent flow as well as being advected with the mean gas motion.

In general there are three processes to consider when modeling the radial transport of solids within a turbulent disk: advection with the mean flow, radial drift relative to the gas due to aerodynamic drag, and turbulent diffusion. Let us consider the limit where advection and diffusion are the dominant processes. This limit is valid for trace gas species and (approximately) for small particles that are sufficiently well-coupled to the gas that advection dominates over aerodynamic drag in Eq. (4.38). Writing the surface density of the gas or dust (generically the "contaminant") as Σ_d, we define the concentration of the contaminant as

$$f = \frac{\Sigma_d}{\Sigma}. \tag{4.44}$$

This is the dust to gas ratio that we have discussed before, except that now we seek to determine how f evolves with radius and time in the disk. If the contaminant is neither created nor destroyed within the region of the disk under consideration, continuity demands that

$$\frac{\partial \Sigma_d}{\partial t} + \nabla \cdot \mathbf{F}_d = 0, \tag{4.45}$$

where $\mathbf{F}_d$, the flux, can be decomposed into two parts: an advective part describing transport of the dust or gas with the mean disk flow, and a diffusive part describing the tendency of turbulence to equalize the concentration of the contaminant across the disk. For $f \ll 1$ we can reasonably assume that the diffusive properties of the disk depend only on the *gas* surface density, in which case the flux can be written as

$$\mathbf{F}_d = \Sigma_d \mathbf{v} - D\Sigma \nabla \left(\frac{\Sigma_d}{\Sigma} \right). \tag{4.46}$$

Here $\mathbf{v}$ is the mean velocity of gas in the disk and D is the usual turbulent diffusion coefficient. We note that the diffusive term vanishes if f is constant. Combining this equation with the continuity equation for the gaseous component, we obtain an evolution equation for f in an axisymmetric disk. In cylindrical polar coordinates

$$\frac{\partial f}{\partial t} = \frac{1}{r\Sigma} \frac{\partial}{\partial r} \left(Dr\Sigma \frac{\partial f}{\partial r} \right) - v_r \frac{\partial f}{\partial r}. \tag{4.47}$$

In common with the equation describing the settling of solid particles in a turbulent disk (Eq. 4.26), this is an advection–diffusion equation, though here the advective component is due to the radial flow of the disk gas rather than settling. It is straightforward to generalize this equation to account for the radial drift of larger particles that are imperfectly coupled to the gas – all that is required is to add an additional flux to Eq. (4.46).

Equation (4.47) expresses a competition between diffusion, whose strength depends upon the turbulent diffusion coefficient D, and radial advection at a velocity v_r. For a steady disk away from the boundaries, the radial velocity can be written in terms of the viscosity as

$$v_r = -\frac{3\nu}{2r}, \tag{4.48}$$

so, as was the case with vertical settling, it is the ratio of the two transport coefficients that is critical. This ratio is called the Schmidt number

$$Sc \equiv \frac{\nu}{D}. \tag{4.49}$$

Diffusion becomes increasingly more important for low values of the Schmidt number.

In most cases of interest the contaminant equation (Eq. 4.47), where necessary modified to allow for radial drift of particles, needs to be solved numerically along with the evolution equation for the gas surface density (Eq. 3.124). We can gain considerable insight into the general behavior, however, by examining the properties of analytic solutions available for some special cases (Clarke & Pringle, 1988). Let us imagine a steady disk in which the surface density profile $\Sigma \propto r^{-\gamma}$ (correspondingly, the viscosity scales with radius as $\nu \propto r^{\gamma}$). At some instant a ring of contaminant is injected into the disk at $r = r_0$. Writing $x = r/r_0$, we define $P(> x)$ to be the maximum fraction of contaminant that is ever at a disk radius of x or larger. For $\gamma = 2$ this quantity can be expressed analytically in terms of the complementary error function[6]

$$P(> x) = \frac{1}{2} \, \text{erfc} \left[\left(\frac{3}{2} Sc \ln x \right)^{1/2} \right]. \tag{4.50}$$

The solution is plotted for a variety of Schmidt numbers in Fig. 4.5. The extent to which contaminant can diffuse "upstream" (radially outward for a steady disk in which $v_r < 0$) is a rather sensitive function of Sc. If the Schmidt number is significantly greater than unity (i.e. the anomalous viscosity is greater than the turbulent diffusivity) almost no upstream diffusion occurs. Values of $Sc = 0.33$ or lower, on the other hand, result in significant amounts of dust or gas diffusing outward against the mean flow to large distances. Although the specific disk model for which the analytic solution applies is not very realistic ($\gamma = 1$ or $\gamma = 3/2$ are more typical values than $\gamma = 2$) the basic conclusion holds more generally – the extent of outward diffusion depends on the Schmidt number.

4.4 Diffusion of large particles

From the preceding discussion it ought to be clear that determining the value of the Schmidt number is critical if we want to quantify either the vertical distribution of particles or their radial transport through the disk. If we restrict ourselves to

---

[6] The complementary error function $\text{erfc}(x) \equiv 1 - \text{erf}(x)$, where, $\text{erf}(x) \equiv \frac{2}{\sqrt{\pi}} \int_0^x e^{-t^2} \, dt$ is the error function.

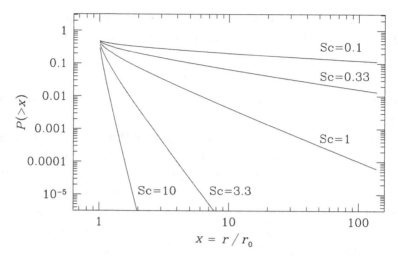

Fig. 4.5. Demonstration of how the Schmidt number ($Sc \equiv \nu/D$, the ratio of the turbulent viscosity to the turbulent diffusivity) affects the ability of a trace contaminant (a minor gas species, or tightly coupled dust particles) to diffuse upstream against the inward flow of a steady accretion disk. The quantity $P(> x)$ is the maximum fraction of contaminant, released at $x = r/r_0 = 1$ in a steady disk with $\Sigma \propto r^{-2}$, that ever reaches radius x or larger. Upstream diffusion is significant only for $Sc \leq 1$.

considering gaseous species, or very small dust particles for which the dimensionless friction time $\tau_{\mathrm{fric}} \ll 1$, the Schmidt number is best thought of as an intrinsic property of the turbulence present within the disk. To an order of magnitude it is normally reasonable to assume that $Sc \sim 1$, but deviations from a Schmidt number of unity can be a factor of several in either direction and may differ between the vertical and radial directions. Numerical simulations of the relevant turbulent processes are needed in order to quantify such departures.

For larger particles, the diffusion coefficient D_p is predicted to vary with the particle size[7] according to the value of the dimensionless friction time τ_{fric}. This is easiest to visualize if we consider a large rock for which $\tau_{\mathrm{fric}} \gg 1$. A rock is not swept up by turbulence in the relatively diffuse gas of the disk. Instead one can envisage the rock as orbiting the star with a well-defined semi-major axis, eccentricity, and inclination. The rock is buffeted by random aerodynamic forces due to the presence of turbulence, and these cause a slow drift in the orbital elements. Particles cannot plausibly diffuse *faster* than gas molecules, so we anticipate that

[7] Confusingly, different authors use the term "Schmidt number" to refer to either ν/D or D/D_p. These are different physical quantities. Here we define Sc to be the ratio of the kinematic viscosity in the disk to the gaseous diffusion coefficient. In this section we discuss the relation between the gas and particle diffusivities, but refrain from giving that ratio a name.

the ratio D/D_p is bounded from below at unity. As particles become larger and the coupling becomes weaker, the ratio will increase.

We can illustrate the dependence of D_p on τ_{fric} using simple arguments borrowed from Youdin & Lithwick (2007). We assume that the disk is turbulent, and that the turbulence can be described using the hydrodynamic picture of interacting eddies developed by Kolmogorov. The characteristic turnover time scale of the eddies is written in dimensionless form as

$$\tau_{eddy} \equiv \Omega t_{eddy}, \tag{4.51}$$

exactly analogous to the definition of the dimensionless friction time $\tau_{fric} \equiv \Omega t_{fric}$. The characteristic velocity of the turbulent fluctuations δv_g is defined such that the *gaseous* diffusion coefficient

$$D = \delta v_g^2 t_{eddy}. \tag{4.52}$$

For simplicity we focus on the limit in which $\tau_{fric} \gg 1$ and $\tau_{eddy} \ll 1$. In this regime of friction time a particle on a noncircular or noncoplanar orbit executes (to leading order) epicyclic oscillations with a frequency $\kappa = \Omega$. During each oscillation of duration Ω^{-1} the particle receives $N \sim \tau_{eddy}^{-1}$ independent impulses, each imparting a velocity kick

$$\delta v_p \sim \frac{\tau_{eddy}}{\tau_{fric}} \delta v_g. \tag{4.53}$$

The kicks are independent, so they accumulate as a random walk. After a time Ω^{-1}

$$\delta v_p \sim \frac{\tau_{eddy}}{\tau_{fric}} \delta v_g \sqrt{N} \sim \frac{\sqrt{\tau_{eddy}}}{\tau_{fric}} \delta v_g, \tag{4.54}$$

and the particle has drifted a distance $\delta l \sim \delta v_p \Omega^{-1}$. Noting that the diffusion coefficient (Eq. 4.52) can be generically rewritten in the form $D_p \sim \delta l^2 \Omega$ we obtain, in the limit of large τ_{fric}, that $D_p \sim D/\tau_{fric}^2$. Since we have argued that $D_p \to D$ for $\tau_{fric} \ll 1$ we can guess a general expression for the ratio of the gas to the particle diffusion coefficients

$$\frac{D}{D_p} \sim 1 + \tau_{fric}^2. \tag{4.55}$$

This agrees to within order unity factors with a more formal analysis of radial diffusivities (Youdin & Lithwick, 2007).

The suppression of particle diffusivity for $\tau_{fric} \gg 1$ means that radial diffusion is an important effect only for relatively small particles ($s \lesssim 1$ mm) – and even then only in the case where the Schmidt number is relatively low. For larger particles, radial drift due to the mean aerodynamic drag dominates and can result in large scale redistribution of solid material.

4.5 Planetesimal formation via coagulation

Although the forces acting on individual solid particles within the disk are reasonably well understood, no consensus has been reached as to how this leads to the formation of km-scale planetesimals. The simplest model is based on the idea that growth to km-scale occurs via a succession of pairwise particle–particle collisions that lead, on average, to growth. We have already noted – in the context of an admittedly over-simplified model of growth during settling – that adhesive collisions between small particles result in particle growth on very short time scales, and in general coagulation appears unavoidable for particles less than about a mm in size. It is harder to envisage large boulders of a meter or so in size sticking together – particularly given that particles in this size range have substantial relative velocities due to aerodynamic effects – and this size regime is also inaccessible to direct experiments. It is this gap in our knowledge that poses a challenge for coagulation models of planetesimal formation.

Whether planetesimals can form via coagulation depends upon two factors: how frequently do particles collide, and what is the outcome of those collisions? For a single population of particles the collision time scale is

$$t_{\text{collide}} = \frac{1}{n\sigma \, \Delta v}, \tag{4.56}$$

where n is the particle number density, σ the cross-section for collisions, and Δv the relative velocity. If the particles are spheres of radius s with mass $m = (4/3)\pi s^3 \rho_{\text{m}}$ the cross-section is given by

$$\sigma = \pi(2s)^2, \tag{4.57}$$

while the number density

$$n = \frac{f\rho}{m}. \tag{4.58}$$

Here ρ is the gas density and f, the ratio of the density of solid particles to the gas density, incorporates any enhancement or depletion of solids due to vertical settling or radial drift. Determining the appropriate value of Δv is trickier. For small dust particles the dominant effect will be due to Brownian motion, which, for a particle in thermal equilibrium with a gas at temperature T, introduces a random velocity of the order of $(1/2)mv^2 \sim k_B T$. For particles of mass m_1 and m_2 the collision velocity in the Brownian motion regime is

$$\Delta v = \sqrt{\frac{8k_B T(m_1 + m_2)}{\pi m_1 m_2}}. \tag{4.59}$$

In the simple case where all the particles have the same mass $m_1 = m_2 = m$, Eq. (4.56) through (4.59) yield as an estimate of the collision time scale

$$t_{collide} = \frac{\pi \rho_m^{3/2} s^{5/2}}{6\sqrt{3k_B T} f \rho}.$$ (4.60)

Substituting typical numerical values for conditions in the inner protoplanetary disk (a gas density $\rho = 10^{-10}\,\mathrm{g\,cm^{-3}}$, a dust to gas ratio $f = 10^{-2}$, a material density $\rho_m = 3\,\mathrm{g\,cm^{-3}}$, and a temperature $T = 300\,\mathrm{K}$) we find

$$t_{collide} \simeq 24 \left(\frac{s}{1\,\mu m}\right)^{5/2} \mathrm{yr}.$$ (4.61)

This is a very short time scale. Small dust particles will inevitably collide in the environment of the protoplanetary disk even if Brownian motion is the only process that yields a relative velocity between particles. If the collisions are adhesive, particle growth will be rapid.

For larger particles the relative velocity is determined by some combination of differential settling, differential radial drift, and velocity induced by imperfect coupling to turbulence within the disk. To get a very crude idea of when these processes start to dominate over Brownian motion, we can compare the relative velocity induced by Brownian motion (Eq. 4.59) with the settling velocity[8] in a laminar disk (Eq. 4.16). Equating these expressions one finds that Brownian motion dominates for

$$s < \left(\frac{4}{\pi^{3/2}} \sqrt{\frac{6}{\mu m_H}} \frac{\rho}{\rho_m^{3/2}} \frac{kT}{\Omega^2 z}\right)^{2/5}.$$ (4.62)

This messy expression is rather unenlightening, but because the relative velocity due to Brownian motion *decreases* with particle mass while that due to settling *increases*, the result is only weakly dependent on the value of the parameters. At a height of $z = 3 \times 10^{11}\,\mathrm{cm}$ at 1 AU, in a disk with $T = 300\,\mathrm{K}$ and $\rho = 10^{-10}\,\mathrm{g\,cm^{-3}}$ the critical size is

$$s < 1\,\mu m.$$ (4.63)

Above this size the estimate of the collision time scale derived for Brownian motion no longer applies. Rapid particle growth, however, will persist as long as collisions remain adhesive. For the case of settling in a laminar disk, in fact, the dependence of the settling velocity on particle size, $\Delta v \propto s$, exactly compensates for the increase in the collision time scale that would otherwise occur because of the variation

---

[8] The settling velocity is an absolute rather than a relative velocity, so this comparison implicitly assumes that there is a range of particle sizes such that the relative velocities are comparable to the absolute velocities.

of the remaining terms $n\sigma \propto s^{-1}$. Indeed, as we showed in Section 4.2.1 using a simple model of particle growth during settling in a laminar disk, growth up to mm or cm size scales appears possible on time scales that are of the order of 10^3 yr in the inner disk. Although the details of how particles grow will of course vary according to the nature and strength of disk turbulence, it is generally true that the collision frequency is high enough that growth up to the cm size regime presents no time scale problems within protoplanetary disks.

Can larger objects also grow via coagulation? For an estimate, let us consider a population of boulders of size $s = 1$ m orbiting at 1 AU. The collision time scale is still given by Eq. (4.56), but in this size regime the relative velocity is dominated by radial drift due to aerodynamic forces against the disk gas. If we assume that a boulder of this size is close to the peak of the radial drift curve (shown as Fig. 4.2) then a plausible value for the relative velocity is,

$$\Delta v \sim 10^{-3} v_K \sim 3 \times 10^3 \, \text{cm} \, \text{s}^{-1}. \tag{4.64}$$

One may legitimately worry about whether collisions at $30 \, \text{m} \, \text{s}^{-1}$ will result in growth as opposed to fragmentation.[9] Finessing that question for the time being, however, the resulting collision time scale is

$$t_{\text{collide}} = \frac{1}{3} \frac{\rho_m}{f\rho} \frac{s}{\Delta v}. \tag{4.65}$$

Adopting the same disk parameters as those given above, we find that for $f = 10^{-2}$

$$t_{\text{collide}} \sim 10^3 \, \text{yr}. \tag{4.66}$$

This time scale is short, but it still exceeds the time scale on which radial drift would cause boulders of this size to migrate into the star. For these specific parameters a coagulation model of growth fails. However, as we showed in Section 4.3.1 via a slightly different argument, we only need to increase f by a relatively modest factor (to $f \sim 0.1$) in order to reach a regime where growth outstrips loss due to radial drift. Since vertical settling toward the mid-plane is very likely to occur for particles whose size exceeds 1 mm there can be in principle enough collisions to continue growth via coagulation despite the rapid inward drift rate. Nonetheless, it is clear from these simple arguments that we cannot tolerate too much inefficiency – in the form of collisions that do not result in adhesion – if we want to grow toward planetesimals via coagulation of meter-sized objects.

---

[9] The noted astrophysicist Doug Lin has been known to challenge advocates of coagulation theories to retire to the desert and return only once they have gotten such rocks to stick together upon impact!

4.5.1 Coagulation equation

Thus far we have been proceeding as if growth occurs for all particles at the same rate, so that the typical collisions are between particles of comparable size. Obviously this need not be true, and indeed the observation that infrared excesses due to disks persist for millions of years implies that small particles must survive (or, be regenerated via erosive collisions) even as others grow. The mathematical treatment of the more realistic situation – where the disk simultaneously contains particles of all sizes colliding with each other – is based upon the *coagulation equation* developed by Smoluchowski (1916). The formalism that results is easy to state but difficult to solve except by numerical means.

Suppose that at some time the number of solid particles per unit volume with masses in the range between m and $(m + dm)$ is $n(m)dm$. As time proceeds, the number of particles of mass m increases whenever there is an adhesive collision between any two particles whose masses *sum* to m, and decreases whenever a particle of mass m coagulates with any other particle. Mathematically

$$\frac{\partial n(m)}{\partial t} = \frac{1}{2} \int_0^m A(m', m - m')n(m')n(m - m')dm'$$

$$- n(m) \int_0^\infty A(m', m)n(m')dm', \qquad (4.67)$$

where the factor of one half eliminates double counting of the collisions that increase the number of particles of mass m. As written this equation is extremely general, and it finds applications in physical chemistry, biology, and cosmology as well as in planet formation. The physics of any specific application is encoded in the *reaction kernel A*, which can be written in our case as

$$A(m_1, m_2) = P(m_1, m_2, \Delta v)\Delta v(m_1, m_2)\sigma(m_1, m_2). \qquad (4.68)$$

Here $P(m_1, m_2, \Delta v)$ is the probability that a collision between two particles of mass m_1 and m_2 leads to adhesion, Δv is the relative velocity at collision, and σ is the collision cross-section. It is straightforward to further generalize the coagulation equation to include fragmentation and/or populations of particles with different physical properties (e.g. some particles may be compact spheres, while others are porous or fractal aggregates).

The coagulation equation (4.67) is also important in the theory of terrestrial planet formation, and we defer detailed discussion of analytic solutions to the equation until Section 5.6. Suffice to say that it has the form of an integro–differential equation for the time evolution of the particle mass distribution, and that it is difficult to solve. A handful of analytic solutions are known for particular choices of the reaction kernel, but the main use of these solutions is to verify numerical solution techniques[10] as the tractable reaction kernels are not particularly realistic for most applications. Two general points, however, are worth noting now. First, growth of particles of mass m is described by a weighted sum over all possible collisions that yield the correct total mass. This means that even if the kernel is near zero for some combinations of masses (perhaps for near-equal mass collisions of meter-scale bodies), growth is still possible if there is a high probability that other types of collision lead to adhesion. Second, the sticking probability P is evidently of central importance when assessing whether growth can occur and how rapid the process is. In general P will be a function of the masses of the particles involved, their collision velocities, and additional parameters describing the shape and strength of the objects.

4.5.2 Sticking efficiencies

Determining the sticking efficiency for collisions as a function of particle size, composition, and relative velocity, is one of the few aspects of planet formation where laboratory experiments – rather than theoretical calculations or astronomical observations – have primacy (a recent review of such experiments has been given by Dominik *et al.*, 2007). The parameter space of interest is large. As we have noted, the relative velocities of bodies upon collision range from $\Delta v \sim 0.1 \, \text{cm s}^{-1}$ (for Brownian motion of micron-sized dust particles) up to $\Delta v \sim 10-100 \, \text{m s}^{-1}$ (for the relative velocity of meter-sized rocks subject to radial drift). Of even greater importance is the composition of the colliding objects – which here means not just their chemical make-up (silicates, water ice, etc), but also their internal structure (solid particles, or aggregates with varying shapes and strengths).

Although a detailed theoretical understanding of laboratory results on sticking efficiencies remains elusive, the basic considerations that determine whether two particles stick upon collision are readily understood (e.g. Youdin, 2008). We consider two identical particles of mass m, which have a relative velocity (when they are well separated) of Δv. Upon collision we assume that short range adhesive forces act to bind the particles together with an energy ΔE_s, and that a fraction f of the impact energy is dissipated. Requiring that the particles remain bound (total energy less than zero) upon rebounding from each other, then yields a simple expression for the maximum impact velocity that results in sticking

$$\frac{1}{4}m\Delta v^2 \leq \frac{f}{(1-f)}|\Delta E_s|. \tag{4.69}$$

---

[10] The coagulation equation is rather different from most equations encountered in astrophysics, and there are a number of subtleties involved in developing an accurate and efficient numerical scheme for its solution. Study of existing algorithms (e.g. Lee, 2000) is *highly* recommended before starting from scratch on your own scheme.

For a given impact velocity, particles can therefore adhere to each other either as a consequence of strong surface forces (i.e. the particles are physically "sticky"), or if their internal structure can absorb the energy of the impact efficiently.

For the smallest particles the role of surface forces is of paramount importance. Dust grains that are nonmagnetic and electrically neutral,[11] adhere as a consequence of induced dielectric forces if the grains collide gently enough. For spherical grains with $s \approx 0.5$ μm a sharp transition from adhesion ($P = 1$) to bouncing ($P = 0$) occurs at a threshold relative velocity that is in the range of $1–2 \, \mathrm{m \, s^{-1}}$. No such transition exists for grains of irregular shape, and although the sticking efficiency declines as Δv increases there is a nonzero probability of adhesion up to collision velocities as large as $100 \, \mathrm{m \, s^{-1}}$. Taken together with our previous results on the collision frequency, these results suggest that there is no impediment to the rapid growth of micron-sized particles via agglomeration, perhaps up to mm-size dimensions.

For larger particles, the declining ratio of surface area to mass means that surface forces become less important as a determining factor in collision outcomes, which are instead controlled increasingly by the ability of particles to dissipate energy upon collision. Solid bodies fail to dissipate sufficient energy and tend to rebound rather than stick at typical collision velocities. For example the coefficient of restitution for icy spheres, defined as the ratio of the rebound velocity to the impact velocity

$$\epsilon \equiv \frac{v_{\mathrm{r}}}{v_{\mathrm{i}}}, \tag{4.70}$$

has been measured to be $\epsilon \approx 1$ for essentially all collisions that are not energetic enough to induce fracture in the targets (Supulver *et al.*, 1995; Higa *et al.*, 1998). A layer of frost on the surface of the ice results in substantially stickier particles, but even this is not sufficient to result in adhesion in the violent collisions with $\Delta v \sim 10 \, \mathrm{m \, s^{-1}}$ that are predicted to occur for meter-scale bodies.

Given that collision velocities within the disk are probably high enough to fracture initially solid bodies (for ice at low temperature, the onset of fracture occurs at $\Delta v \simeq 1–2 \, \mathrm{m \, s^{-1}}$ for cm-sized particles), it is reasonable to assume that the typical structure of cm-sized particles may instead resemble a loosely bound aggregate of dusty or icy particles. Such aggregates can dissipate a greater fraction of the kinetic energy of impact internally, and experiments confirm that this can result in sticking probabilities that are substantially more favorable for ongoing particle growth. In particular, collisions between mm- and cm-sized aggregates of

---

[11] Small grains that rebound upon collision can develop net charges via charge exchange, a phenomenon known as the triboelectric effect. The resulting electrostatic forces exceed the dielectric forces between uncharged particles by a factor of the order of 10^3, promoting faster aggregation.

SiO_2 particles result in sticking at cm s^{-1} impact velocities, with a transition to rebound only for velocities of the order of a m s^{-1}. Even more intriguingly, violent impacts with $\Delta v \gtrsim 10$ m s^{-1} may still yield net growth of the target, although this is accompanied by a spray of small debris (Wurm *et al.*, 2005). These results imply that growth of aggregate particles via collisions is plausible for $\Delta v \lesssim 1$ m s^{-1}, and may be possible even at substantially higher impact velocities of 10–20 m s^{-1}.

4.6 Goldreich–Ward mechanism

Thus far we have assumed that the only important interactions between particles are physical collisions, and that those particles are dynamically unimportant for the evolution of the gas disk. These assumptions are valid for small dust particles distributed uniformly throughout the gas disk – since the total mass of solids is negligible compared to the mass of gas at the epoch when the disk forms – but they may be locally violated due to some combination of vertical settling, radial drift, and photoevaporation. If the solid particles start to play a dynamical role a number of new physical effects may occur:

- Gravitational instability within a dense layer of particles located close to the disk mid-plane. Safronov (1969) and, independently, Goldreich & Ward (1973) proposed that such an instability might result in the prompt formation of planetesimals. Although it is now known that the simplest version of their theory, known as the *Goldreich–Ward mechanism*, fails, closely related physical ideas remain important.
- Modification of the properties of turbulence within the gas due to feedback from the solid component.
- The existence of new two-fluid instabilities that arise because of the coupling between the solid and gaseous components. These might result in clumping of the solid particles (over and above overdensities that would occur if the solids were passive tracers within the turbulent gas flow), and promote planetesimal formation either via direct collisions or via gravitational collapse.

We begin by analyzing the stability of a thin particle layer to gravitational collapse. The physical considerations are identical to those discussed heuristically in Section 3.4.4 for the stability of a *gaseous* disk, with the particle velocity dispersion σ taking the place of the gas sound speed, and the control parameter is once again Toomre's Q (Eq. 3.92). Let us derive that result more formally.

4.6.1 Gravitational stability of a particle layer

We consider a razor-thin disk of particles with uniform surface density Σ_0 and constant velocity dispersion σ orbiting the star in circular orbits in the $z = 0$ plane.

In cylindrical polar coordinates (r, ϕ, z) the density of the disk is given by

$$\rho_0(r, \phi, z) = \Sigma_0 \delta(z), \tag{4.71}$$

where $\delta(z)$ is a Dirac delta-function, while the velocity field is

$$\mathbf{v}_0(r, \phi, z) = (0, r\Omega, 0). \tag{4.72}$$

Here $\Omega = \Omega(r)$ is the angular velocity of the particles. The angular velocity need not be Keplerian, but for circular orbits we require that the centrifugal force balance gravity. If the gravitational potential is Φ_0,

$$\Omega^2 r = -\frac{d\Phi_0}{dr}. \tag{4.73}$$

Note that because of the assumptions of constant density and constant velocity dispersion, the equivalent of a pressure gradient force does not enter into the problem.

To proceed we now make a number of simplifications. First, we assume that coupling of the particles to any gas present can be ignored. Second, we assume that the particle layer can be treated as a *fluid* in which the two dimensional sound speed, defined in terms of the pressure p and surface density Σ in the usual way via

$$c_s^2 \equiv \frac{dp}{d\Sigma}, \tag{4.74}$$

is equivalent to the particle velocity dispersion σ. This is not formally correct, and it is not obvious that it is correct at all. If the particles are collisionless, their "pressure" need not be isotropic and their dynamics should be described not by the fluid equations but by the collisionless Boltzmann equation. It turns out, however, that the fluid and collisionless stability criteria are practically indistinguishable (see, e.g. Binney & Tremaine, 1987), with the fluid version being much easier to derive. Adopting the fluid description, the basic equations describing the dynamics of the particle disk are the continuity and momentum equations, together with Poisson's equation for the gravitational field

$$\frac{\partial \Sigma}{\partial t} + \nabla \cdot (\Sigma \mathbf{v}) = 0, \tag{4.75}$$

$$\frac{\partial \mathbf{v}}{\partial t} + (\mathbf{v} \cdot \nabla)\mathbf{v} = -\frac{\nabla p}{\Sigma} - \nabla \Phi, \tag{4.76}$$

$$\nabla^2 \Phi = 4\pi G \rho. \tag{4.77}$$

Our task is to determine, given these equations, the conditions under which the initial equilibrium state of the disk (defined by Eq. 4.71, 4.72, and 4.73) is stable against the effect of disk self-gravity, which unopposed would tend to result in the particles clumping into dense clumps or rings. This can be accomplished with a standard linear stability analysis. We consider infinitesimal axisymmetric perturbations to the equilibrium state

$$\Sigma = \Sigma_0 + \Sigma_1(r, t), \tag{4.78}$$

$$p = p_0 + p_1(r, t), \tag{4.79}$$

$$\Phi = \Phi_0 + \Phi_1(r, t), \tag{4.80}$$

$$\mathbf{v} = \mathbf{v}_0 + \left[v_r(r, t), \delta v_\phi(r, t), 0 \right], \tag{4.81}$$

that have a spatial and temporal dependence given by (using the surface density as an example)

$$\Sigma_1(r, t) \propto \exp[i(kr - \omega t)]. \tag{4.82}$$

Here k is the spatial wavenumber of the perturbation (related to the wavelength via $\lambda = 2\pi/k$) and ω is the temporal frequency. Ultimately, what we need to determine is whether there are values of k (i.e. spatial scales) for which ω is imaginary, since imaginary values of the frequency will result in exponentially growing perturbations. Making one further approximation, we assume that for the perturbations of interest

$$kr \gg 1. \tag{4.83}$$

This amounts to considering disturbances that are small compared to the radial extent of the disk.

We now substitute the expressions for the surface density, pressure, gravitational potential and velocity into the fluid equations, discarding any terms we encounter that are quadratic in the perturbed quantities (this is a *linear* stability analysis). For the continuity equation this yields,

$$-i\omega \Sigma_1 + v_r \Sigma_0 \left(\frac{1}{r} + ik \right) = 0, \tag{4.84}$$

which simplifies further in the local limit ($kr \gg 1$) to

$$-\omega \Sigma_1 + k v_r \Sigma_0 = 0. \tag{4.85}$$

Deriving the analogous algebraic equations from the momentum equation requires us to express the convective operator $(\mathbf{v} \cdot \nabla)\mathbf{v}$ in cylindrical coordinates. This takes the rather unwieldy form

$$(\mathbf{v} \cdot \nabla)\mathbf{v} = \left[v_r \frac{\partial v_r}{\partial r} + \frac{v_\phi}{r} \frac{\partial v_r}{\partial \phi} + v_z \frac{\partial v_r}{\partial z} - \frac{v_\phi^2}{r}, \right.$$
$$\left. v_r \frac{\partial v_\phi}{\partial r} + \frac{v_\phi}{r} \frac{\partial v_\phi}{\partial \phi} + v_z \frac{\partial v_\phi}{\partial z} + \frac{v_r v_\phi}{r}, \; v_r \frac{\partial v_z}{\partial r} + \frac{v_\phi}{r} \frac{\partial v_z}{\partial \phi} + v_z \frac{\partial v_z}{\partial z} \right]. \tag{4.86}$$

With this in hand, the momentum equation reduces immediately to

$$-i\omega v_r - 2\Omega \delta v_\phi = -\frac{1}{\Sigma_0}\frac{dp_1}{dr} - \frac{d\Phi_1}{dr}, \tag{4.87}$$

$$-i\omega \delta v_\phi + v_r\left[\Omega + \frac{d}{dr}(r\Omega)\right] = 0, \tag{4.88}$$

where the two equations come from the radial and azimuthal components respectively.

The next step is to relate the perturbations in pressure and gravitational potential expressed on the right-hand-side of Eq. (4.87) to perturbations in the surface density. For the pressure term this is straightforward. Equation (4.74) implies that,

$$\frac{1}{\Sigma_0}\frac{dp_1}{dr} = \frac{1}{\Sigma_0}c_s^2 ik\Sigma_1. \tag{4.89}$$

Dealing with the potential perturbations requires more work. Starting from the linearized Poisson equation

$$\nabla^2\Phi_1 = 4\pi G\Sigma_1\delta(z), \tag{4.90}$$

we write out the Laplacian explicitly and simplify making use of the fact that for short wavelength perturbations $kr \gg 1$. This yields a relation between the density and potential fluctuations

$$\frac{d^2\Phi_1}{dz^2} = k^2\Phi_1 + 4\pi G\Sigma_1\delta(z). \tag{4.91}$$

For $z \neq 0$ the only solution to this equation that remains finite for large $|z|$ has the form

$$\Phi_1 = C\exp[-|kz|], \tag{4.92}$$

where C remains to be determined. To do so we integrate the Poisson equation vertically between $z = -\epsilon$ and $z = +\epsilon$

$$\int_{-\epsilon}^{+\epsilon}\nabla^2\Phi_1 dz = \int_{-\epsilon}^{+\epsilon}4\pi G\Sigma_1\delta(z)dz. \tag{4.93}$$

Noting that both $\partial^2\Phi_1/\partial x^2$ and $\partial^2\Phi_1/\partial y^2$ are continuous at $z = 0$, whereas $\partial^2\Phi_1/\partial z^2$ is *not*, we obtain,

$$\frac{d\Phi_1}{dz}\bigg|_{-\epsilon}^{+\epsilon} = 4\pi G\Sigma_1. \tag{4.94}$$

Taking the limit $\epsilon \to 0$ we find that $C = -2\pi G\Sigma_1/|k|$, and hence that the general relation between potential and surface density fluctuations on the $z = 0$ plane is

$$\Phi_1 = -\frac{2\pi G\Sigma_1}{|k|}. \tag{4.95}$$

Taking the radial derivative

$$\frac{d\Phi_1}{dr} = -\frac{2\pi ik G\Sigma_1}{|k|}, \tag{4.96}$$

which allows us to eliminate the potential from the right-hand-side of Eq. (4.87) in favor of the surface density. The result is,

$$-i\omega v_r - 2\Omega\delta v_\phi = -\frac{1}{\Sigma_0}c_s^2 ik\Sigma_1 + \frac{2\pi ik G\Sigma_1}{|k|}. \tag{4.97}$$

Finally, we are ready to derive the functional relationship between ω and k, known as the dispersion relation. Eliminating v_r and δv_ϕ between Eq. (4.85), (4.88), and (4.97) we find that

$$\omega^2 = \kappa^2 + c_s^2 k^2 - 2\pi G\Sigma_0|k|, \tag{4.98}$$

where the *epicyclic frequency* κ is defined as

$$\kappa^2 \equiv 4\Omega^2 + 2r\Omega\frac{d\Omega}{dr}. \tag{4.99}$$

In a Keplerian potential $\kappa^2 = \Omega^2$.

The basic properties of the dispersion relation for a self-gravitating particle (or gas) disk (Eq. 4.98) are readily apparent. Recall that for instability to occur, we require that $\omega^2 < 0$ so that ω itself is imaginary and perturbations grow exponentially with time.[12] Both rotation and pressure – the first and second terms on the right-hand-side of the dispersion relation – are unconditionally positive and act as stabilizing influences, with rotation stabilizing all scales equally and pressure preferentially stabilizing short wavelength (large k) perturbations. Self-gravity, the third term, is destabilizing with a spatial dependence (linear in k) that lies intermediate between that of rotation and pressure. These dependencies are shown in Fig. 4.6.

Setting $d\omega^2/dk = 0$, we find that the wavenumber of the minimum in $\omega^2(k)$ is given by

$$k_{min} = \frac{\pi G\Sigma_0}{c_s^2}. \tag{4.100}$$

---

[12] There will also be exponentially *decaying* solutions, but the growing modes will rapidly dominate.

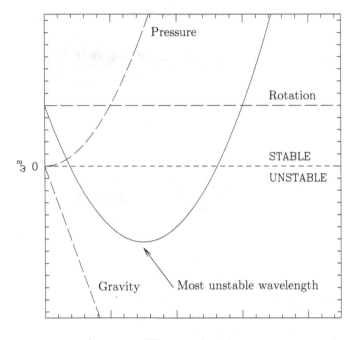

Fig. 4.6. The dispersion relation (solid curve) for axisymmetric perturbations to a self-gravitating razor-thin gaseous disk. The short-dashed line shows the boundary between instability ($\omega^2 < 0$) and stability ($\omega^2 > 0$). The long-dashed lines show the individual contributions from rotation (stabilizing at all wavenumbers), pressure (stabilizing at large wavenumber/short wavelength) and self-gravity (destabilizing).

If the minimum in the $\omega^2(k)$ curve falls into the unstable region, modes with $k \simeq k_{\min}$ will display the fastest exponential growth. The condition for the disk to be marginally unstable to gravitational instability is then obtained by requiring that $\omega^2(k_{\min}) = 0$. For a Keplerian disk instability requires that

$$Q \equiv \frac{c_s \Omega}{\pi G \Sigma_0} < 1, \tag{4.101}$$

where Q is customarily referred to as the "Toomre Q" parameter after Alar Toomre's 1964 paper. Up to numerical factors, the result matches that deduced using time scale arguments in Section 3.4.4. Additional numerical factors arise when the analysis is generalized to allow for the possibility of *nonaxisymmetric* instability within the disk. For nonaxisymmetric modes the control parameter is still Q, but instability sets in more easily (i.e. at higher Q) and results in the development of spiral arms within the disk rather than the rings that would be the endpoint of the purely axisymmetric instability. In most circumstances the subtleties introduced by

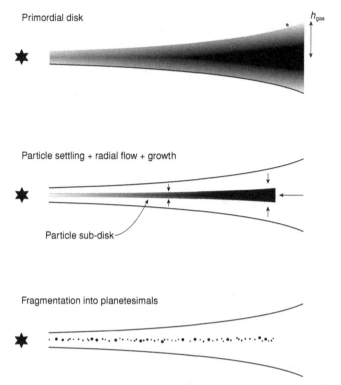

Fig. 4.7. How planetesimals form within the classical Goldreich–Ward scenario.

the presence of nonaxisymmetric modes are of only moderate importance, and it suffices to note that the disk will become unstable to its own self-gravity when

$$Q \overset{<}{\sim} Q_{\text{crit}}, \tag{4.102}$$

with Q_{crit} being of the order of, but slightly larger than, unity.

4.6.2 Application to planetesimal formation

With the mathematical formalities of disk instability in place we can proceed to consider how this process might play a role in planetesimal formation. The basic idea, illustrated schematically in Fig. 4.7 involves three stages

(1) Initially the solid component of the disk is well-mixed with gas. The Q of the solid component is very large and the effects of self-gravity play no role in the evolution.
(2) Over time the dust settles vertically to form a thin sub-disk of particles around the $z = 0$ plane. As we have already noted, even the slightest breath of turbulence suffices to stir up dust particles, so substantial settling requires at least some collisional growth

to have occurred. *Radial* drift can also in principle contribute to an increase in the mid-plane particle density in the inner disk.

(3) Due to some combination of high surface density and/or low velocity dispersion, the particle sub-disk becomes unstable according to the Q criterion. This may lead to the formation of bound clumps of particles, which rapidly agglomerate to form planetesimals.

Deferring for the moment the question of whether this sequence of events can plausibly occur, we can ask what would be the properties of planetesimals formed via gravitational instability? If the particle layer has velocity dispersion σ and surface density Σ_s its vertical scale-height will be

$$h_d = \frac{\sigma}{\Omega},$$ (4.103)

while the most unstable wavelength

$$\lambda \sim \frac{2\pi}{k_{min}} = \frac{2\sigma^2}{G\Sigma_s}.$$ (4.104)

For $Q = Q_{crit}$ these scales are comparable, so instability, if it occurs, results in the collapse of relatively small patches of the particle layer into planetesimals.

The mass of clumps formed via gravitational instability is of the order of

$$m_p \sim \pi\lambda^2\Sigma_s \sim 4\pi^5 G^2 Q_{crit}^4 \frac{\Sigma_s^3}{\Omega^4}.$$ (4.105)

Adopting parameters that might be appropriate for the protoplanetary disk at 1 AU ($\Sigma_s = 10\,\mathrm{g\,cm^{-2}}$, $Q_{crit} = 1$) the estimated planetesimal mass is

$$m_p(1\ \mathrm{AU}) \sim 3 \times 10^{18}\ \mathrm{g},$$ (4.106)

which corresponds to a spherical body with a radius of 5–10 km. If we assume – rather unrealistically – that once gravitational instability of the particle layer sets in collapse occurs on the free-fall time scale, then the formation time is very short, being less than a year for the fiducial parameters above.

In principle a particle layer that fragments to form planetesimals could be composed of objects of any size, provided that the conditions for instability are met. The case that attracts greatest interest in the literature, however, is that where the particles have sizes of the order of a cm or smaller. Such particles are small enough not to suffer the potentially devastating rapid radial drift of larger bodies, and there does not appear to be any particular obstacle to forming them rapidly via pairwise collisions. Subsequent instability of a layer made of small particles then has the feature – understandably attractive to many authors – of rapidly forming planetesimals in a way that *bypasses* all of the potential hurdles involved in particle growth through the meter-scale regime.

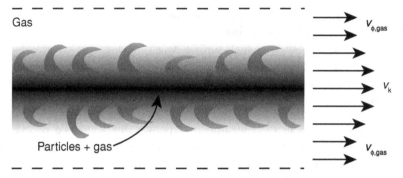

Fig. 4.8. How the presence of a dense particle disk results in excitation of turbulence within the flow.

4.6.3 Self-excited turbulence

Since the surface density in solid material is typically only of the order of 1% of the gas surface density, a very thin particle layer is required in order for gravitational instability to set in. For example, if the surface density in solids at 1 AU is $\Sigma_s = 10 \, \mathrm{g \, cm^{-2}}$, we require a velocity dispersion of $\sigma \simeq 10 \, \mathrm{cm \, s^{-1}}$ in order to attain $Q = 1$. If the *gas* disk has $h/r = 0.05$ at this radius, the relative thickness of the particle and gaseous disks is then

$$\frac{h_d}{h} \sim 10^{-4}. \tag{4.107}$$

In the absence of gas, attaining such a razor-thin particle disk is not implausible – the vertical thickness of Saturn's rings, for example, is of the order of only 10 m. In the gas-rich environment of the protoplanetary disk, however, we need to consider carefully whether turbulence will preclude the particle layer from ever becoming thin enough to become unstable.

Intrinsic disk turbulence provides one barrier to the development of very thin particle layers. Equation (4.29) yields an estimate of the ratio of the thicknesses of the particle and gas disks as a function of α and the dimensionless friction time Ωt_{fric}. From that analysis it is obvious that the classical Goldreich–Ward instability cannot work for small particles (those with $\Omega t_{\mathrm{fric}} \ll 1$ that have not yet grown to the size where radial drift becomes rapid) unless α is very small. Fully turbulent regions of the disk are unpromising sites for planetesimal formation via gravitational instability.

One might argue, of course, that the mid-plane of the disk is *not* necessarily intrinsically turbulent (e.g. the discussion in Section 3.5.1). Unfortunately, even if the gas disk on its own were laminar the presence of the particle layer tends to excite turbulence, via the mechanism depicted in Fig. 4.8. The crucial physical point is that for gravitational instability to occur we require that the local particle density *exceed* the local gas density (by as much as two orders of magnitude if we adopt the parameters given above). Within the particle layer the gas is therefore subdominant to the particles, and the orbital velocity of the flow will be comparable to the Keplerian velocity. Just above the particle layer, on the other hand, the gas-dominated disk orbits at slightly less than the Keplerian speed due to the influence of radial pressure gradients. As a consequence there will exist a vertical shear which, if it is too large, will be unstable to the development of Kelvin–Helmholtz instabilities. This self-excited turbulence can prevent the particle layer from ever settling to the point where self-gravity could set in (Cuzzi *et al.*, 1993).

To estimate the seriousness of this impediment to planetesimal formation we need to consider the hydrodynamic stability of a stratified shear flow. The simplest case to analyze[13] is that of a nonrotating flow (i.e. the Coriolis force is neglected) with a density profile $\rho(z)$ and a velocity profile $v_\phi(z)$. The flow is assumed to be in hydrostatic equilibrium with a vertical pressure gradient being balanced by the vertical component of gravity g_z. Stability against shear instabilities in this situation is measured by the *Richardson number*

$$\mathrm{Ri} \equiv -\frac{g_z \mathrm{d} \ln \rho/\mathrm{d}z}{(\mathrm{d}v_\phi/\mathrm{d}z)^2}.$$
(4.108)

A necessary condition for instability is that $\mathrm{Ri} < 0.25$ somewhere within the flow.

Let us evaluate the Richardson number across a particle layer of vertical height h_d, at radius r within a gas disk of scale-height h and in which the mid-plane pressure varies as $P \propto r^{-n}$. We assume that g_z is dominated by the vertical component of the star's gravity, that the density in $\mathrm{d} \ln \rho/\mathrm{d}z$ is the total density (gas plus particles), and that the velocity shear reflects the difference between the Keplerian velocity and the gas disk velocity across the height of the layer. Collecting together previously derived results we then estimate these terms as

$$g_z = \Omega^2 h_\mathrm{d},$$
(4.109)

$$\frac{\mathrm{d} \ln \rho}{\mathrm{d}z} = -\frac{1}{h_\mathrm{d}},$$
(4.110)

$$\left(\frac{\mathrm{d}v_\phi}{\mathrm{d}z}\right)^2 = \frac{n^2}{4}\left(\frac{h}{r}\right)^4 \frac{\Omega^2 r^2}{h_\mathrm{d}^2}.$$
(4.111)

---

[13] As several authors have recently emphasized, this analysis is considerably too simple to yield an accurate answer (Garaud & Lin, 2004; Gómez & Ostriker, 2005; Chiang, 2008). In more complete analyses the presence of the Coriolis force tends to further destabilize the disk flow, while the presence of *radial* shear is a stabilizing effect. The calculation presented here is intended only to illustrate the basic physical considerations at work.

The resulting Richardson number is (adopting $n = 3$)

$$\text{Ri} \simeq 0.25 \left(\frac{h/r}{0.05} \right)^{-2} \left(\frac{h_{\mathrm{d}}/h}{0.0375} \right)^2. \tag{4.112}$$

Clearly, attaining the very thin layers needed to induce gravitational instability is rather difficult. As the particle layer settles towards the mid-plane, self-excited turbulence is liable to set in long before $Q \sim 1$.

Notwithstanding these difficulties, self-gravity within a particle layer may still be viable as a route to planetesimal formation. To start with, the critical Richardson number below which turbulence sets in may be significantly smaller than 0.25 for a vertical shear flow within the protoplanetary disk, in which rotation and radial shear cannot be neglected. More important than this formal result, however, is the fact that the conditions for self-gravity to set in are greatly relaxed if the ratio of solids to gas – which we previously fixed at the nominal value of 10^{-2} – can be increased. Such an increase might occur due to radial drift of particles inward (Youdin & Shu, 2002), while potentially much larger *local* enhancements of the particle density might occur if other instabilities result in clumping of the solid material (into clumps, streams, spiral arms, etc). Numerical simulations, for example by Johansen *et al.* (2007), suggest that local particle overdensities may reach the large values required to initiate self-gravity, though additional work is needed to assess whether this results in a robust path to planetesimal formation.

4.7 Routes to planetesimal formation

The topics covered in this chapter include some of the most important, yet least well-understood, aspects of planet formation. To summarize the current state of knowledge, there appears to be no theoretical impediment to the rapid growth of dusty or icy particles up to small macroscopic dimensions ($s \sim 1$ mm). The growth mechanism at these scales is pairwise collisions that result in sticking, and the time scale in the inner disk can be surprisingly short – of the order of 10^3–10^4 yr within a few AU of the star. The fact that dust is clearly still present in the inner regions of disks with ages of several Myr strongly suggests that regeneration of dust, via erosive collisions, accompanies an overall trend toward growth.

Growth beyond the mm or cm size regime presents greater challenges. A strong argument can be made – based on the rapid radial drift of m-scale bodies due to aerodynamic forces – that growth all the way up to km-scale planetesimals must be rapid, with a time scale less than 10^5 yr across the entire radial extent of the disk. The mechanism that allows such rapid growth is unknown. One possibility is that rapid pairwise growth continues all the way from dust scales up to planetesimals. The open question is whether net growth is possible at intermediate size scales given realistic values for particle–particle collision velocities within turbulent disks. A second possibility is that gravitational instability forms planetesimals rapidly from much smaller objects, bypassing many of the scales for which collisional growth is most uncertain. The open question for this class of models is whether the local particle density can reach the large values needed to trigger collapse.

Our theoretical ignorance as to how planetesimals form should not obscure the empirical observation that the planetesimal formation process, at least in the Solar System, appears to have been rather robust. Solid bodies of varying composition, which are presumably descended from planetesimals, are found all the way from Mercury (at 0.4 AU) out to the edge of the classical Kuiper Belt (at about 47 AU) – a radial extent of two orders of magnitude. It therefore seems plausible to adopt as a working hypothesis that planetesimals typically form rapidly with a smooth radial distribution, and study how those planetesimals grow into larger bodies. This is the approach that we will adopt in the next chapter.

4.8 Further reading

The evolution of dust within protoplanetary disks is reviewed in the article "Growth of dust as the initial step toward planet formation," C. Dominik, J. Blum, J. N. Cuzzi, & G. Wurm (2007) in *Protostars and Planets V*, B. Reipurth, D. Jewitt, & K. Keil (eds.), Tucson: University of Arizona Press.

A different perspective on planetesimal formation is given in "From grains to planetesimals," A. Youdin (2010) published in the *European Astronomical Society Publications Series*.

5

Terrestrial planet formation

Once planetesimals have formed, the dominant physical process that controls further growth is their mutual gravitational interaction. Conventionally the only further role the gas disk plays in terrestrial planet formation is to provide a modest degree of aerodynamic damping of protoplanetary eccentricity and inclination. In this limit the physics involved – Newtonian gravity – is simple and the problem of terrestrial planet formation is well posed. It is not, however, easy to solve. It would take 4×10^9 planetesimals with a radius of 5 km to build the Solar System's terrestrial planets, and it is infeasible to directly simulate the N-body evolution of such a system for long enough (and with sufficient accuracy) to watch planets form. Instead a hybrid approach is employed. For the earliest phases of terrestrial planet formation a statistical approach, similar to that used in the kinetic theory of gases, is both accurate and efficient. When the number of dynamically significant bodies has dropped to a manageable number (of the order of hundreds or thousands), direct N-body simulations become feasible, and these are used to study the final assembly of the terrestrial planets. Using this two-step approach has known drawbacks (for example, it is difficult to treat the situation where a small number of protoplanets co-exist with a large sea of very small bodies), but nevertheless it provides a reasonably successful picture for how the terrestrial planets formed.

The main focus of this chapter is to outline the physical concepts that underlie statistical models of terrestrial planet formation. We will concentrate on two questions:

(1) Suppose that we have a population of bodies orbiting the star in approximately circular orbits. The population can be characterized by its mass distribution, and by the distributions of orbital eccentricity (e) and inclination (i) (note that these distributions will themselves be function of mass). Given these quantities, **what is the rate of collisions between bodies?**

(2) **What are the consistent distributions of** e **and** i given a population of bodies with a specified mass distribution? Gravitational perturbations among a population of bodies on almost circular orbits will tend to increase the random component of their velocity (or, equivalently, increase e and i), while gas drag and physical collisions can damp random motions.

In the inner Solar System, statistical models describe growth from planetesimals up to masses that approach that of the Moon or even Mars. The final assembly of the Earth and Venus from these building blocks can only be studied via N-body methods, and for this final stage analytic arguments provide only limited insight. A summary of some of the techniques employed for N-body simulations is given in Appendix 2, and a number of illustrative results from such calculations are reproduced here.

5.1 Physics of collisions

Terrestrial planets form from planetesimals as the endpoints of a cascade of pairwise collisions. For the most part the gravity of growing planets is strong enough that we can assume that most of the mass of the two colliding bodies ends up agglomerating into a single larger object. For masses and collision velocities for which this is true we can gloss over the detailed physics of the collisions, and the primary input to models of growth is the collision cross-section. The cross-section is enhanced by the gravity of the bodies ("gravitational focusing"), and modified as a result of the tidal gravitational field of the star. For small bodies and large impact velocities the assumption of perfect accretion can fail, and we need to consider the strength of the bodies explicitly to determine whether collisions lead to agglomeration or fragmentation.

5.1.1 Gravitational focusing

For sufficiently small bodies, the effects of gravity can be ignored for the purposes of determining whether they will physically collide. A massive planet, on the other hand, will deflect the trajectories of other bodies toward it, and as a result has a collision cross-section that is much larger than its physical cross-section.

To evaluate the magnitude of this *gravitational focusing*, consider two bodies of mass m, moving on a trajectory with impact parameter b, as shown in Fig. 5.1. The relative velocity at infinity is σ. At closest approach, the bodies have separation R_c and velocity v_{max}. Equating energy in the initial (widely separated) and final (closest approach) states we have

$$\frac{1}{4}m\sigma^2 = mv_{max}^2 - \frac{Gm^2}{R_c}.$$

(5.1)

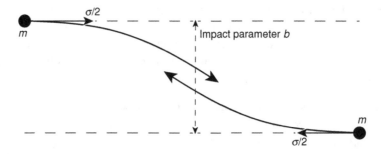

Fig. 5.1. Gravitational focusing. If the random velocity σ of bodies is smaller than the escape speed from their surface, the cross-section for physical collisions is boosted beyond the geometric cross-section by gravitational focusing.

Noting that there is no radial component to the velocity at the point of closest approach, angular momentum conservation gives

$$v_{max} = \frac{1}{2}\frac{b}{R_c}\sigma. \tag{5.2}$$

If the sum of the physical radii of the bodies is R_s, then for $R_c < R_s$ there will be a physical collision, while larger R_c will result in a harmless flyby. The *largest* value of the impact parameter that will lead to a physical collision is thus

$$b^2 = R_s^2 + \frac{4GmR_s}{\sigma^2}, \tag{5.3}$$

which can be expressed in terms of the escape velocity from the point of contact, $v_{esc}^2 = 4Gm/R_s$ as

$$b^2 = R_s^2\left(1 + \frac{v_{esc}^2}{\sigma^2}\right). \tag{5.4}$$

The cross-section for collisions is then

$$\Gamma = \pi R_s^2\left(1 + \frac{v_{esc}^2}{\sigma^2}\right), \tag{5.5}$$

where the term in brackets represents the enhancement to the physical cross-section due to gravitational focusing. Clearly a planet growing in a "cold" planetesimal disk for which $\sigma \ll v_{esc}$ will grow much more rapidly as a consequence of gravitational focusing.

5.1.2 Shear versus dispersion dominated encounters

Our derivation of the gravitational focusing term assumed that the only significant forces acting upon the colliding bodies were those due to their mutual gravity. This is a good approximation for studies, for example, of stellar collisions within a star cluster, but its legitimacy must always be evaluated carefully for planetary accretion where the gravity of the star is often important. What matters is not so much the total gravitational force due to the star, but rather the *difference* in the force that is experienced by two bodies on similar orbits (i.e. the tidal gravitational field). Three-body dynamics is more complex than the two-body case, and in general is not amenable to fully analytic treatments.

The condition for three-body dynamics to be important can be estimated from a time scale argument. We first estimate the radius within which the gravity of the protoplanet, with mass M_p and orbital radius a, dominates over the stellar tidal field. To do this we equate the orbital frequency of a protoplanet around the star ($\sqrt{GM_*/a^3}$) to that of a test particle orbiting the planet at radius r ($\sqrt{GM_p/r^3}$). This yields an estimate of the radius of the *Hill sphere*

$$r_H \sim \left(\frac{M_p}{M_*}\right)^{1/3} a. \tag{5.6}$$

Within r_H two-body effects provide an adequate description of the dynamics. Likewise, we can define a characteristic velocity (the Hill velocity) as the orbital velocity around the protoplanet at the distance of the Hill radius

$$v_H \sim \sqrt{\frac{GM_p}{r_H}}. \tag{5.7}$$

If the random velocity σ of small bodies is large compared to the Hill velocity ($\sigma > v_H$), then the rate with which they enter the Hill sphere and collide is determined by two-body dynamics. This regime is described as **dispersion dominated**. If, conversely, $\sigma < v_H$ then three-body effects must be considered and the system is said to be **shear dominated**.

The concept of the Hill sphere is important enough as to be worth deriving more rigorously. To do so, we consider the motion of a test particle (a body whose mass is negligible) in a binary system consisting of a star of mass M_* and a protoplanet of mass M_p. We assume that the protoplanet orbits the star in a circular orbit with angular frequency Ω, and work in a co-rotating coordinate system such that the line joining the star to the planet coincides with the x axis. The origin of the coordinates is taken to be the center of mass. This geometry is shown in Fig. 5.2. The motion of the test particle then obeys

$$\ddot{\mathbf{r}} = -\nabla\Phi - 2(\Omega \times \dot{\mathbf{r}}) - \Omega \times (\Omega \times \mathbf{r}), \tag{5.8}$$

$$\Phi = -\frac{GM_*}{r_*} - \frac{GM_p}{r_p}, \tag{5.9}$$

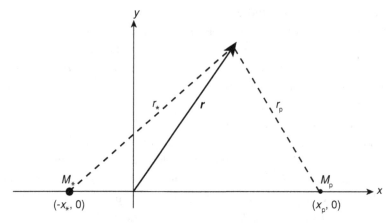

Fig. 5.2. Geometry for the derivation of Hill's equations.

where the dots denote time derivatives. Expressed in terms of components we have

$$\ddot{x} - 2\Omega\dot{y} - \Omega^2 x = -G\left[\frac{M_*(x + x_*)}{r_*^3} + \frac{M_p(x - x_p)}{r_p^3}\right], \tag{5.10}$$

$$\ddot{y} + 2\Omega\dot{x} - \Omega^2 y = -G\left[\frac{M_*}{r_*^3} + \frac{M_p}{r_p^3}\right]y, \tag{5.11}$$

$$\ddot{z} = -G\left[\frac{M_*}{r_*^3} + \frac{M_p}{r_p^3}\right]z. \tag{5.12}$$

These equations are valid for arbitrary masses M_* and M_p. We can simplify them considerably, however, by considering the limit $M_p \ll M_*$. In this limit $|x_*| \ll |x_p|$ and $\Omega^2 = GM_*/x_p^3$. We also shift coordinates so that $x = 0$ coincides with the position of the protoplanet, and consider motion at a distance $\Delta \equiv (x^2 + y^2)^{1/2}$ from the protoplanet that is small compared to the orbital radius. Simple algebraic manipulation of the equations then yields an approximate set of equations of motion that take the form

$$\ddot{x} - 2\Omega\dot{y} = \left(3\Omega^2 - \frac{GM_p}{\Delta^3}\right)x, \tag{5.13}$$

$$\ddot{y} + 2\Omega\dot{x} = -\frac{GM_p}{\Delta^3}y. \tag{5.14}$$

These are known as Hill's equations. They describe the motion of a small body (in our case a planetesimal) in the vicinity of a larger body (a protoplanet) that has a circular orbit around the star.

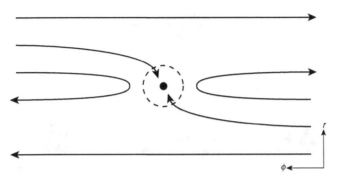

Fig. 5.3. Trajectories of test particles on almost circular orbits that encounter a protoplanet, as viewed in the frame co-rotating with the protoplanet. In the shear dominated regime particles whose orbits are *too close* to that of the protoplanet never enter the latter's Hill sphere and collide with the planet.

The left-hand-side of Hill's equations merely describes epicyclic motion of a test particle around the star. Looking at the right-hand-side, we note that the radial (in the x direction) force vanishes for $\Delta = r_{\mathrm{H}}$, where r_{H}, the Hill sphere radius, is given in terms of the orbital radius a as,

$$r_{\mathrm{H}} = \left(\frac{M_{\mathrm{p}}}{3M_*} \right)^{1/3} a. \qquad (5.15)$$

This is a more formal derivation of the distance from the protoplanet within which the gravitational attraction of the protoplanet dominates over the tidal gravitational field of the star.[1]

Hill's equations can be used to compute the trajectories of test particles that move in the vicinity of a larger body. The resulting trajectories are shown schematically in Fig. 5.3 for the case of test particles with almost circular orbits. As one might expect, particles on near circular orbits that pass more than a few r_{H} from the protoplanet are essentially unperturbed by the presence of the protoplanet, and will not collide. More interestingly, particles that are on orbits that are *too close* in radius to the protoplanet follow what are referred to as horseshoe or tadpole orbits, which also fail to enter the Hill sphere and do not contribute to the collision cross-section. The dynamics of these orbits is the same as that of the Trojan asteroids in the Solar System, of which the largest population is that in 1:1 resonance with Jupiter. Only for a range of intermediate separations is the perturbation from the protoplanet able

---

[1] In the astrophysical literature you will sometimes find reference to the *Roche lobe* of the protoplanet, which is essentially the same concept as the Hill sphere. Less frequently you may find discussion of the *Tisserand sphere of influence*, which is the region within which motion of test particles is better described by the two-body dynamics of the planet plus the test particle than by the two-body dynamics of the star plus the test particle. The radius of the Tisserand sphere of influence scales slightly differently with planet mass than the Hill sphere radius, but the distinction is immaterial for our purposes.

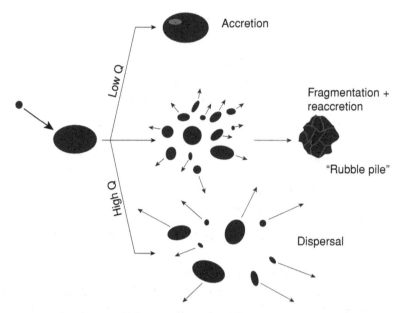

Fig. 5.4. Possible outcomes of collisions between bodies.

to overcome the tidal gravitational force and bring the test particle into the region where a collision can occur. We will discuss the resulting collision rate later, but for now the main point to note is simply that the dynamics of collisions is qualitatively different in the shear and dispersion dominated cases.

5.1.3 Accretion versus disruption

When two initially solid bodies physically collide the outcomes can be divided broadly into three categories:

- **Accretion**. All or most of the mass of the impactor becomes part of the mass of the final body, which remains solid. Small fragments may be ejected, but overall there is net growth.
- **Shattering**. The impact breaks up the target body into a number of pieces, but these pieces remain part of a single body (perhaps after reaccumulating gravitationally). The structure of the shattered object resembles that of a *rubble pile*.
- **Dispersal**. The impact fragments the target into two or more pieces that do not remain bound.

These possibilities are illustrated in Fig. 5.4. If the target is itself a rubble pile then the first possibility – ending up with a solid body – is unlikely, but the collision could still either disperse the pieces or merely rearrange them into a larger but still shattered object.

To delineate the boundaries between these regimes quantitatively, we consider an impactor of mass m colliding with a larger body of mass M at velocity v. We define the specific energy Q of the impact via

$$Q \equiv \frac{mv^2}{2M},$$ (5.16)

and postulate, plausibly, that this parameter largely controls the result. The thresholds for the various collision outcomes can then be expressed in terms of Q. Conventionally, we define the threshold for catastrophic disruption Q_D^* as the minimum specific energy needed to disperse the target in two or more pieces, with the largest one having a mass $M/2$. Similarly Q_S^* is the threshold for shattering the body. More work is required to disperse a body than to shatter it, so evidently $Q_D^* > Q_S^*$. It is worth keeping in mind that in detail the outcome of a particular collision will depend upon many factors, including the mass ratio between the target and the impactor, the angle of impact, and the shape and rotation rate of the bodies involved. Quoted values of Q_D^* are often averaged over impact angles, but even when this is done the parameterization of collision outcomes in terms of Q is only an approximation.

The estimated values of Q_D^* for a target of a particular size vary by more than an order of magnitude depending upon the composition of the body, which can broadly be categorized into solid or shattered rock, and solid or porous ice. For any particular type of body, however, two distinct regimes can be identified:

- **Strength dominated regime.** The ability of small bodies to withstand impact without being disrupted depends upon the material strength of the object. In general the material strength of bodies declines with increasing size, owing to the greater prevalence of defects that lead to cracks. In the strength dominated regime Q_D^* decreases with increasing size.
- **Gravity dominated regime.** Large bodies are held together primarily by gravitational forces. In this regime Q_D^* must at the very least exceed the specific binding energy of the target, which scales with mass M and radius s as $Q_B \propto GM/s \propto \rho_m s^2$. In practice it requires a great deal more than this minimum amount of energy to disrupt the target – so Q_B is *not* a good estimate of Q_D^* – but nonetheless Q_D^* does increase with increasing size.

Although the transition between these regimes is reasonably sharp there is *some* influence of the material properties (in particular the shear strength) on the catastrophic disruption threshold for smaller bodies within the gravity dominated regime.

Values of Q_S^* and Q_D^* can be determined experimentally for small targets (e.g. Arakawa *et al.*, 2002). Experiments are not possible in the gravity dominated regime, but Q_D^* can be estimated theoretically using numerical hydrodynamics (Benz & Asphaug, 1999; Leinhardt & Stewart, 2009) or (for rubble piles) rigid body

Table 5.1. *Parameters for the catastrophic disruption threshold fitting formula (Eq. 5.17), which describes how Q_D^* scales with the size of the target body. The quoted values were derived by Benz & Asphaug (1999) and Leinhardt & Stewart (2009) using numerical hydrodynamics simulations of collisions, which are supplemented in the strength dominated regime by experimental results.*

	v (km s^{-1})	q_s (erg g^{-1})	q_g (erg cm^3 g^{-2})	a	b
Ice (weak)	1.0	1.3×10^6	0.09	−0.40	1.30
Ice (strong)	0.5	7.0×10^7	2.1	−0.45	1.19
Ice (strong)	3.0	1.6×10^7	1.2	−0.39	1.26
Basalt (strong)	3.0	3.5×10^7	0.3	−0.38	1.36
Basalt (strong)	5.0	9.0×10^7	0.5	−0.36	1.36

dynamics simulations (Leinhardt & Richardson, 2002; Korycansky & Asphaug, 2006). The simplest parameterization of the numerical results is as a broken power-law that includes terms representing the strength and gravity regimes

$$Q_D^* = q_s \left(\frac{s}{1\,\mathrm{cm}} \right)^a + q_g \rho_m \left(\frac{s}{1\,\mathrm{cm}} \right)^b. \tag{5.17}$$

Often (but not always) Q_D^* is averaged over impact geometry, and q_s, q_g, a, and b are all constants whose values are derived by fitting to the results of numerical simulations.

Benz & Asphaug (1999) and Leinhardt & Stewart (2009) determined the values of the fitting parameters in Eq. (5.17) from the results of an ensemble of simulations of impacts into icy or rocky targets. Their results are given in Table 5.1 and plotted as a function of target size in Fig. 5.5. One observes immediately that the results for a particular target material vary with the impact velocity, and hence that Q_D^* is *not* the sole determinant of the outcome of collisions. There is, however, a clear transition between the strength and gravity dominated regimes, with the weakest bodies being those whose size is comparable to the cross-over point. The most vulnerable bodies are generally those with radii in the 100 m to 1 km range. Just how vulnerable such bodies are to catastrophic disruption depends sensitively on their make-up, and it would be unwise to place too much trust in precise numbers. As a rough guide, however, the weakest icy bodies have minimum $Q_D^* \sim 10^5$ erg g^{-1}, while the strongest conceivable planetesimals (unfractured rocky bodies) have minimum $Q_D^* > 10^6$ erg g^{-1}.

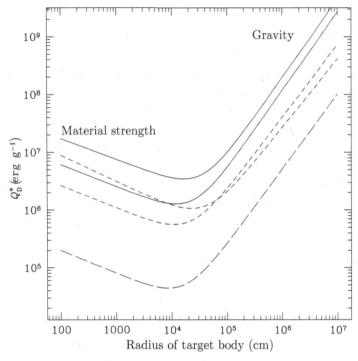

Fig. 5.5. The specific energy Q_D^* for catastrophic disruption of solid bodies is plotted as a function of the body's radius. The solid and short dashed curves show results obtained using fits to theoretical calculations for impacts into "strong" targets by Benz & Asphaug (1999). The long dashed curve shows the recommended curve for impacts into "weak" targets from Leinhardt & Stewart (2009), derived from a combination of impact experiments and numerical simulations. In detail the solid curves show results for basalt at impact velocities of $5\,\mathrm{km\,s^{-1}}$ (upper curve) and $3\,\mathrm{km\,s^{-1}}$ (lower curve). The short dashed curves show results for water ice at $3\,\mathrm{km\,s^{-1}}$ (the lower curve for small target sizes) and $0.5\,\mathrm{km\,s^{-1}}$ (upper curve for small target sizes). The long dashed curve shows results for normal impacts into weak water ice targets at $1\,\mathrm{km\,s^{-1}}$.

As a reality check, we may note that asteroids in the main belt with $e \simeq 0.1$ would be expected to collide today with typical velocities of the order of $2\,\mathrm{km\,s^{-1}}$. For a mass ratio $m/M = 0.1$ the specific energy of the collision is then around $Q = 2 \times 10^9\,\mathrm{erg\,g^{-1}}$, which from Fig. 5.5 is sufficient to destroy even quite large solid bodies with $s \simeq 100\,\mathrm{km}$. This is consistent with the observation of asteroid families, and the interpretation of such families as collisional debris. Evidently the random velocities that characterize collisions must have been *much* smaller during the epoch of planet formation if we are to build large planets successfully out of initially km-scale planetesimals.

5.2 Statistical models of planetary growth

The fundamental assumption underlying statistical models of planetary growth is that the number of bodies is large enough that the population can be described by one or more probability distributions that encode the probability that a body has a specified mass, inclination, and eccentricity. The details of actual orbits are assumed to be unimportant, and accordingly the distribution of the longitude of pericenter and the longitude of the ascending node is taken to be uniform. For bodies of a given mass the population is then described by the distribution of Keplerian orbital elements $f(e, i)$, which can be regarded as the analog of the Maxwellian distribution of velocities for particles in a gas. Numerical experiments show that the appropriate distribution that is set up as a result of gravitational interactions among a population of bodies is a Rayleigh distribution. Specifically, for a population of planetesimals of mass m and local surface density Σ_p, the probability distribution of eccentricity and inclination takes the form

$$f(e, i) = 4\frac{\Sigma_p}{m}\frac{ei}{\langle e^2\rangle\langle i^2\rangle}\exp\left[-\frac{e^2}{\langle e^2\rangle} - \frac{i^2}{\langle i^2\rangle}\right], \qquad (5.18)$$

where $\langle e^2\rangle$ and $\langle i^2\rangle$ are the mean square values of the eccentricity and inclination respectively (Lissauer, 1993). This distribution is equivalent to a Gaussian distribution of the *random* components (v_r, v_z, and $\delta v_\phi = v_\phi - v_K$) of planetesimal velocities

$$f(z, \mathbf{v}) = \frac{\Omega\Sigma_p}{2\pi^2 m\sigma_r^2\sigma_z^2}\exp\left[-\frac{(v_r^2 + 4\delta v_\phi^2)}{2\sigma_r^2} - \frac{(v_z^2 + \Omega^2 z^2)}{2\sigma_z^2}\right], \qquad (5.19)$$

where the velocity dispersions σ_r and σ_z in the radial and vertical direction are related to the mean square eccentricities and inclinations via

$$\sigma_r^2 = \frac{1}{2}\langle e^2\rangle v_K^2, \qquad (5.20)$$

$$\sigma_z^2 = \frac{1}{2}\langle i^2\rangle v_K^2. \qquad (5.21)$$

One should note that the exact conversion between orbital elements and the random components of the velocity depends upon the problem under study. The definitions given in Eq. (5.21) are appropriate for converting between mean square orbital elements and velocity dispersions in the planetesimal distribution function, but for other applications different formulae are needed. For a single planetesimal with eccentricity e and inclination i at least three slightly different definitions of the "random velocity" σ can be useful (Lissauer & Stewart, 1993):

(1) The planetesimal velocity relative to a circular orbit with $i = 0$ with the same semi-major axis

$$\sigma = (e^2 + i^2)^{1/2} v_K. \tag{5.22}$$

(2) The planetesimal velocity relative to the local circular orbit with $i = 0$

$$\sigma = \left(\frac{5}{8}e^2 + \frac{1}{2}i^2\right)^{1/2} v_K. \tag{5.23}$$

(3) The planetesimal velocity relative to other planetesimals

$$\sigma = \left(\frac{5}{4}e^2 + i^2\right)^{1/2} v_K. \tag{5.24}$$

Yet more variations on these expressions arise due to the need to average over the planetesimal distribution when calculating, for example, the collision rates of planetesimals with protoplanets.

Very often the physics of the excitation and damping of e and i is such that the velocity distribution closely approximates an isotropic form. In this limit

$$\langle e^2 \rangle^{1/2} = 2 \langle i^2 \rangle^{1/2}, \tag{5.25}$$

and the problem has one fewer degree of freedom. Whether or not we make this assumption, however, our task is to calculate, for a given distribution of e and i, the rate of collisions while accounting properly for the effects of both gravitational focusing and three-body dynamics. Having done this we then need to consider the mechanisms that determine appropriate values for $\langle e^2 \rangle$ and $\langle i^2 \rangle$.

5.2.1 Approximate treatment

Simple results for the growth rate of protoplanets can readily be derived for the dispersion dominated regime. Let us assume that a relatively massive body of mass M, physical radius R_s, and surface escape speed v_{esc} is embedded within a swarm of smaller planetesimals. The planetesimal swarm has a local surface density Σ_p and a velocity dispersion σ (assumed here to be isotropic so that no distinction between σ_r and σ_z is needed). The vertical scale-height of the swarm is then $h_p \simeq \sigma/\Omega$ and its volume density is

$$\rho_{sw} \simeq \frac{\Sigma_p}{2h_p}. \tag{5.26}$$

The statistical (or "particle in a box") estimate for the growth rate of the massive body is then simply the product of the density of the planetesimal swarm, the encounter velocity at infinity, and the collision cross-section Γ (Eq. 4.57)

$$\frac{dM}{dt} = \rho_{sw}\sigma \pi R_s^2 \left(1 + \frac{v_{esc}^2}{\sigma^2}\right). \tag{5.27}$$

Substituting for ρ_{sw} and h_p we find that

$$\frac{dM}{dt} = \frac{\sqrt{3}}{2}\Sigma_p\Omega\pi R_s^2 \left(1 + \frac{v_{esc}^2}{\sigma^2}\right), \tag{5.28}$$

where the numerical pre-factor, which we have not derived, is correct for an isotropic velocity dispersion (Lissauer, 1993). The inverse scaling of the density with the velocity dispersion results in a cancellation of the direct effect of the velocity dispersion, which enters only via the gravitational focusing term.

Two simple solutions to this equation give insight into the properties of more sophisticated models of planetary growth. First, we assume that the gravitational focusing factor

$$F_g = \left(1 + \frac{v_{esc}^2}{\sigma^2}\right), \tag{5.29}$$

is a constant, and that the surface density of the planetesimal swarm does not change with time. In this limit the equation for the growth of the protoplanet's radius,

$$\frac{dR_s}{dt} = \frac{\sqrt{3}}{8}\frac{\Sigma_p\Omega}{\rho_m}F_g, \tag{5.30}$$

where ρ_m is the density of the protoplanet, can be trivially integrated. We find that

$$R_s \propto t, \tag{5.31}$$

and the radius grows at a linear rate. For an icy body with $\rho_m = 1\,\mathrm{g\,cm^{-1}}$ at 5.2 AU (the current orbital radius of Jupiter),

$$\frac{dR_s}{dt} \approx 1 \left(\frac{\Sigma_p}{10\,\mathrm{g\,cm^{-2}}}\right)F_g\,\mathrm{cm\,yr^{-1}}. \tag{5.32}$$

Unless gravitational focusing is dominant this is a very slow growth rate. At this rate it would take 10 Myr – as long or longer than the lifetime of typical protoplanetary disks – to grow a solid body up to a size of 100 km. We conclude immediately that to build large bodies in the outer regions of the protoplanetary disk large gravitational focusing enhancements to the cross-section are unavoidable.

Given the necessity of gravitational focusing, the other obvious limit to consider is that in which $v_{esc} \gg \sigma$. In this regime

$$F_g \simeq \frac{v_{esc}^2}{\sigma^2} = \frac{2GM}{\sigma^2 R_s}, \tag{5.33}$$

and the rate of growth of the protoplanet mass becomes

$$\frac{dM}{dt} = \frac{\sqrt{3}\pi G \Sigma_p \Omega}{\sigma^2} M R_s = k M^{4/3}, \tag{5.34}$$

where k is a constant if the properties of the planetesimal swarm are fixed (i.e. if the growing protoplanet neither excites the velocity dispersion nor consumes a significant fraction of the planetesimals). If under these conditions we consider the growth of *two* bodies with masses M_1 and M_2 such that $M_1 > M_2$ we find that

$$\frac{dM_1/M_1}{dM_2/M_2} = \frac{R_1}{R_2} > 1. \tag{5.35}$$

The initially more massive body grows faster than its less massive cousin, both absolutely and as measured by the ratio of masses M_1/M_2. This phenomenon is called *runaway growth*, and it allows for much more rapid formation of large bodies. Indeed, if we formally integrate Eq. (5.34) for a fixed planetesimal velocity dispersion we obtain

$$M(t) = \frac{1}{(M_0^{-1/3} - k't)^3}, \tag{5.36}$$

where k' is a constant. The planet attains an infinite mass at a finite time. In reality this singularity is avoided because the feedback from the growing planet results in an increase of the planetesimal velocity dispersion, which slows growth. The physical lesson, however, is correct – gravitational focusing can drive runaway growth and rapid formation of large bodies.

5.2.2 Shear and dispersion dominated limits

In the dispersion dominated limit the approximate treatment leading to Eq. (5.28) suffices to expose most of the important physics. The shear dominated limit is trickier to analyze, and in general there is no alternative but to use the results of three-body numerical experiments or fitting formulae derived from such experiments. To gain some analytic insight into the important processes we content ourselves here with an approximate treatment that is a simplified version of that given by Greenberg *et al.* (1991). The calculation has two parts. First, we estimate the width of the annulus (described as the "feeding zone") surrounding the protoplanet within which planetesimals can be diverted on to trajectories that enter the Hill sphere. Second, we assess the fraction of incoming planetesimals that actually impacts the protoplanet.

To derive the width of the feeding zone we consider the limit in which the growing protoplanet, of mass M, has a circular orbit. We assume that the planetesimals likewise have very small eccentricities, so that we are firmly in the shear dominated regime. A planetesimal whose orbit differs from that of the protoplanet by Δa must then have its eccentricity excited to

$$e \approx \frac{\Delta a}{a}, \tag{5.37}$$

if it is to be diverted on to a trajectory that would approach the planet.

Let us now evaluate the impulse that the planetesimal receives as it passes by the protoplanet. The strongest perturbations occur while the two bodies are separated by an azimuthal distance of Δa or less, which persists for a time interval $\delta t = 4/\Omega$. The impulse imparted to the planetesimal is then

$$\delta v \approx \frac{4}{\Omega} \frac{GM}{(\Delta a)^2}, \tag{5.38}$$

which corresponds to an eccentricity of,

$$e \approx \frac{4}{\Omega^2 a} \frac{GM}{(\Delta a)^2}. \tag{5.39}$$

Equating this value of the eccentricity to that *required* in order to yield an approach trajectory (Eq. 5.37) we find that the outer edge of the feeding zone is delimited by a half-width[2]

$$\Delta a \simeq \left(\frac{4M}{M_*} \right)^{1/3} a = 2.3 \, r_{\mathrm{H}}, \tag{5.40}$$

where the Hill sphere radius r_{H} is defined by Eq. (5.15).

Planetesimals with orbits as far away from the protoplanet as $(a - \Delta a)$ and $(a + \Delta a)$ can be deflecting on to approach trajectories. However, *not all* orbits within the feeding zone permit close approaches. As illustrated in Fig. 5.3, those planetesimals whose orbits are too *close* to that of the protoplanet describe horseshoe orbits that never encounter the Hill sphere. Roughly speaking, those planetesimals with semi-major axes between $(a - \Delta a/2)$ and $(a + \Delta a/2)$ are protected from encounters. This reduces the effective width of the feeding zone from $2\Delta a$ to Δa. The typical planetesimal that can take part in the feeding flow is then at a radial distance of $0.75 \Delta a$ from the protoplanet. In the frame rotating with the protoplanet, the average relative velocity of approach for a planetesimal that will enter the Hill sphere is then

$$v_{\mathrm{shear}} = a \left| \frac{d\Omega}{da} \right| \frac{3}{4} \Delta a = \frac{9}{8} \Omega \Delta a. \tag{5.41}$$

---

[2] This is an approximate derivation that happens to give an answer that is quite close to that derived from numerical experiments. Equally valid approximate treatments will all yield slightly different numerical factors.

Combining the results for the effective width of the feeding zone and the approach velocity we obtain

$$\frac{dM_H}{dt} = \frac{9}{8} \Omega \Delta a \, \Sigma_p \Delta a, \tag{5.42}$$

as the rate at which mass flows towards the protoplanet from the planetesimal disk in the shear dominated regime. Although we have not proved it here, this mass flux that is diverted towards the planet is to a good approximation also the mass flux that enters the Hill sphere around the protoplanet.

The planetesimals that flow into the Hill sphere do so with typical encounter velocities that are (using Eq. 5.37) of the order of,

$$v_{enc} \sim ev_K \sim \Omega \Delta a. \tag{5.43}$$

Since Δa scales as the Hill sphere radius the mass ratio of the protoplanet to the star enters this expression only weakly, as the one third power. Once the planetesimals have entered the Hill sphere the collision cross-section with the protoplanet can be evaluated using two-body dynamics. Provided that the half-thickness of the incoming planetesimal flow exceeds the gravitational capture radius

$$ai > R_{capture}, \tag{5.44}$$

where

$$R_{capture} = R_s \left(1 + \frac{v_{esc}^2}{v_{enc}^2} \right)^{1/2}, \tag{5.45}$$

the collision cross-section takes the same form as in the dispersion dominated limit

$$\Gamma = \pi R_s^2 \left(1 + \frac{v_{esc}^2}{v_{enc}^2} \right). \tag{5.46}$$

Simple geometry, depicted in Fig. 5.6, then allows us to determine the *fraction* of planetesimals entering the Hill sphere that go on to impact the protoplanet. This fraction is

$$f \approx \frac{\Gamma}{(2r_H)(2ai)}. \tag{5.47}$$

Collecting together results, the rate at which the protoplanet grows via accretion of planetesimals in the shear dominated regime is the product of this fraction with the mass flow rate into the Hill sphere (Eq. 5.42).

$$\frac{dM}{dt} = \frac{9}{32} \frac{(\Delta a)^2}{air_H} \Sigma_p \Omega \pi R_s^2 \left(1 + \frac{v_{esc}^2}{v_{enc}^2} \right). \tag{5.48}$$

Although this expression is superficially similar to the result that we deduced in the dispersion dominated regime (Eq. 5.28) there are a number of important

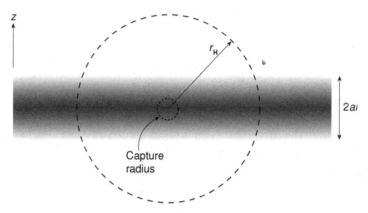

Fig. 5.6. An edge-on view showing the flow of planetesimals into the protoplanet's Hill sphere in the shear dominated regime. In the limit depicted here, the vertical thickness of the planetesimal disk exceeds the capture radius of the protoplanet and the three-dimensional formula for gravitational focusing describes the cross-section. If the disk is *very* thin a different regime of essentially two-dimensional accretion applies.

differences. First, the gravitational focusing term within the parentheses no longer depends upon the random velocities of the planetesimals, but rather on the ratio between the escape velocity and the speed at which planetesimals enter the Hill sphere. Since $v_{enc}^2 \propto r_H^2 \propto M^{2/3}$, while $v_{esc}^2 \propto M$, there is a partial cancelation of the mass dependence of the gravitational focusing term. Second, there is an explicit dependence on the vertical thickness of the planetesimal swarm, $dM/dt \propto 1/i$. The eccentricity is formally irrelevant, although it will often be the case that the eccentricities and inclinations of the planetesimals continue to obey the relation $e \sim 2i$.

The above analysis was predicated on the assumption that $ai > R_{capture}$. If the planetesimal disk is extremely cold and thin this assumption can be violated, and there is a further transition from the three-dimensional accretion geometry discussed above to an essentially two-dimensional planar flow. In the very thin disk limit the rate at which planetesimals enter the Hill sphere remains unaltered, but the fraction of those planetesimals that are accreted becomes

$$f \approx \frac{R_{capture}}{r_H}. \tag{5.49}$$

The protoplanet growth rate in this regime contains no dependence upon either e or i.

The parameters that control the growth rate of the protoplanet are different in the shear and dispersion dominated limits, and hence there is no universality to the transition between the different regimes. In particular, the switch from dispersion

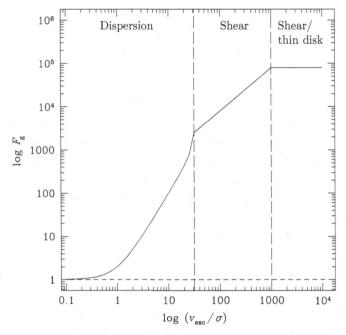

Fig. 5.7. How the gravitational enhancement factor (the ratio of the true colli-sion cross-section to the geometric cross-section) varies with v_{esc}/σ, the ratio of the escape speed from the protoplanet to the velocity dispersion of the planetes-imal swarm. In constructing this plot it is assumed that the protoplanet mass is fixed and that the velocity dispersion of the planetesimals is the quantity being varied.

to shear domination does not occur at a universal value of the gravitational focusing factor. Figure 5.7 shows the dependence of the collision cross-section (normalized to the geometric cross-section) as a function of the ratio v_{esc}/σ (after Greenberg *et al.*, 1991). This plot is constructed for specific protoplanet parameters assuming that v_{esc} remains fixed, while σ is varied maintaining $e = 2i$. In this case, and typically, four regimes can be identified:

- For $v_{esc}/\sigma < 1$ gravitational focusing is irrelevant, and growth occurs in the dispersion dominated regime according to the geometric cross-section.
- Once $v_{esc}/\sigma > 1$ gravitational focusing becomes significant and rapidly comes to dom-inate the cross-section, which increases quadratically with v_{esc}/σ whilst we remain dispersion dominated.
- Exiting the dispersion dominated regime the growth of the cross-section first steepens, before increasing more slowly (as $1/\sigma$) in the shear dominated regime.
- Finally, the thickness of the disk falls below the scale of the capture radius $R_{capture}$. In this thin disk regime the effective cross-section is constant.

Computations of the growth rate based on numerical integrations of test particles in the gravitational field of the protoplanet and the star show, not unexpectedly, that the transitions between these regimes occur quite smoothly.

5.2.3 Isolation mass

The growth of a protoplanet within the planetesimal disk is eventually limited by the feedback of the protoplanet on the properties of the planetesimals. The possible types of feedback include excitation of the random velocities of the planetesimals (which will slow growth by reducing the magnitude of gravitational focusing) and simple depletion of the local planetesimal surface density due to accretion. We can estimate the maximum mass that a protoplanet can attain during runaway growth by finding the mass of a single body that has consumed all of the planetesimals in its vicinity. This is known as the *isolation mass*. We work in the shear dominated regime, since for a massive body to have grown rapidly the planetesimal disk within which it is embedded must have had a low velocity dispersion.

To determine the isolation mass we make use of the fact that in the shear dominated regime a body of mass M can accrete only those planetesimals whose orbits lie within the feeding zone. The radial extent of the feeding zone Δa_{max} scales with the radius of the Hill sphere

$$\Delta a_{max} = C r_H. \tag{5.50}$$

For planetesimals on initially circular orbits the width of the feeding zone can be estimated by evaluating the maximum separation for which collisions are possible within the context of the restricted three-body problem. This yields $C = 2\sqrt{3}$ (Lissauer, 1993).[3] The mass of planetesimals within the feeding zone is

$$2\pi a \cdot 2\Delta a_{max} \cdot \Sigma_p \propto M^{1/3}. \tag{5.51}$$

Note the one third power of the planet mass, which arises from the mass dependence of the Hill radius. As a planet grows its feeding zone expands, but the mass of new planetesimals within the expanded feeding zone rises more slowly than linearly. We thus obtain the isolation mass by setting the protoplanet mass equal to the mass of the planetesimals in the feeding zone of the original disk

$$M_{iso} = 4\pi a \cdot C \left(\frac{M_{iso}}{3M_*}\right)^{1/3} a \cdot \Sigma_p, \tag{5.52}$$

---

[3] This is larger than the value ($C \simeq 2.3$) that we derived in Section 5.2.2 because here we consider all planetesimals that are not dynamically forbidden from eventually encountering the protoplanet, whereas previously we counted only those that would be deflecting into the Hill sphere on an orbital time scale.

which gives

$$M_{\text{iso}} = \frac{8}{\sqrt{3}} \pi^{3/2} C^{3/2} M_*^{-1/2} \Sigma_{\text{p}}^{3/2} a^3. \qquad (5.53)$$

Evaluating this expression in the terrestrial planet region, taking $a = 1$ AU, $\Sigma_{\text{p}} = 10$ g, cm^{-2}, $M_* = M_\odot$, and $C = 2\sqrt{3}$ we obtain

$$M_{\text{iso}} \simeq 0.07 \, M_\oplus. \qquad (5.54)$$

Isolation is therefore likely to occur late in the formation of the terrestrial planets. Repeating the estimate for the conditions appropriate to the formation of Jupiter's core, using $\Sigma_{\text{p}} = 10 \, \text{g cm}^{-2}$ as adopted by Pollack *et al.* (1996) gives

$$M_{\text{iso}} \simeq 9 \, M_\oplus. \qquad (5.55)$$

This estimate is comparable to or larger than the current best determinations for the mass of the Jovian core. Full isolation may or may not be relevant to the formation of Jupiter and the other giant planets, depending upon the adopted disk model.

5.3 Velocity dispersion

The random velocities of bodies during planet formation are determined by a competition between excitation and damping processes. Four main processes have been studied in detail:

- **Viscous stirring**. The random motions of a population of planetesimals that are initially in a cold disk will be excited by the cumulative effect of weak gravitational encounters. The increase in the random energy of the bodies comes ultimately from the orbital energy. This is the only excitation process that operates within a disk composed of equal mass bodies.
- **Dynamical friction**. Gravitational scattering behaves somewhat differently in the case where there is a spectrum of masses. The tendency for the system to seek equipartition of energy between the particles results in a transfer of energy from massive bodies to less massive ones, and the development of a mass-dependent velocity dispersion.
- **Gas drag**. Aerodynamic drag, although much weaker for planetesimals than for meter-scale bodies, continues to damp both e and i.
- **Inelastic collisions**. Physical collisions between bodies that result in energy dissipation also damp e and i.

The most general formulation of the problem of planetary growth requires writing time-dependent equations for de^2/dt and di^2/dt that take account of these processes. The eccentricities and inclinations (or equivalently the velocity dispersion in the radial and vertical directions) depend upon mass, and are coupled to

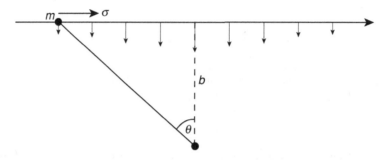

Fig. 5.8. A body of mass m passes by a second body with an impact parameter (i.e. a distance of closest approach) b and a relative velocity σ. In the regime of a weak encounter the deflection angle is small and the impulse can be calculated assuming that the trajectory is unperturbed.

the equations describing the growth of protoplanets. If the time scales for growth and damping are short enough, the problem can be simplified by solving for the equilibrium values of e and i.

5.3.1 Viscous stirring

We can estimate how quickly an initially cold disk of equal mass planetesimals will be heated by gravitational interactions by considering the amount of energy that is converted from ordered to disordered motion during a single encounter, and then summing over all encounters. We will work in the dispersion dominated regime in which two-body dynamics suffices to describe the encounters, and consider distant (or weak) encounters during which the perturbation to the initial trajectories is small. An individual flyby of two planetesimals with mass m, relative velocity σ, and impact parameter b then has the geometry shown in Fig. 5.8. The component of the gravitational force that is perpendicular to the trajectory is then

$$F_\perp = \frac{Gm^2}{d^2} \cos\theta, \tag{5.56}$$

where d is the instantaneous separation between the bodies. If we define $t = 0$ to coincide with the moment of closest approach, then the distance along the trajectory from the point of closest approach is just $x = \sigma t$ and we can write the time-dependent perpendicular component of the force as

$$F_\perp = \frac{Gm^2}{b^2} \left[1 + \left(\frac{\sigma t}{b} \right)^2 \right]^{-3/2}. \tag{5.57}$$

Integrating this force over the duration of the encounter yields the impulse felt by the planetesimals

$$|\delta v_\perp| = \int_{-\infty}^{\infty} \frac{F_\perp}{m} dt = \frac{2Gm}{b\sigma}, \tag{5.58}$$

and the change in kinetic energy

$$\delta E = \frac{1}{2} m |\delta v_\perp|^2 = \frac{2G^2 m^3}{b^2 \sigma^2}. \tag{5.59}$$

For consistency with the assumption that the trajectory remains almost unperturbed, we require that $|\delta v_\perp| \ll \sigma$, and this condition defines

$$b_{\min} = \frac{2Gm}{\sigma^2}, \tag{5.60}$$

as the value of the impact parameter that delineates the boundary between weak (small deflection angle) encounters and strong (large angle) encounters.

To sum up the effect that many individual weak encounters have on the random velocity of one planetesimal, we note that the rate of encounters with an impact parameter in the range between b and $(b + db)$ is

$$\Gamma = 2\pi b \, db \, n_{\text{sw}} \sigma, \tag{5.61}$$

where n_{sw} is the number density of the planetesimal swarm. The rate of change of the kinetic energy is then

$$\frac{dE}{dt} = \frac{4\pi G^2 m^3 n_{\text{sw}}}{\sigma} \int_{b_{\min}}^{b_{\max}} \frac{db}{b}, \tag{5.62}$$

which is conventionally written in the form

$$\frac{dE}{dt} = \frac{4\pi G^2 m^3 n_{\text{sw}}}{\sigma} \ln \Lambda, \tag{5.63}$$

where $\ln \Lambda$ is known as the "Coulomb logarithm." The value of the Coulomb logarithm depends upon the size of the system (which determines the most distant encounters that need to be summed over) and may change with time, though given the weak (logarithmic) dependence it can often be approximated as a constant. We observe that for a given surface density of planetesimals viscous stirring will be more efficient if the planetesimal mass is large, since large masses imply larger individual kicks whose effects do not cancel out to the same extent as smaller more numerous scattering events.

To apply Eq. (5.63) to the problem of viscous stirring of a planetesimal disk we note that

$$n_{\text{sw}} = \frac{\rho_{\text{sw}}}{m} \simeq \frac{\Sigma_{\text{sw}} \Omega}{2m\sigma}, \tag{5.64}$$

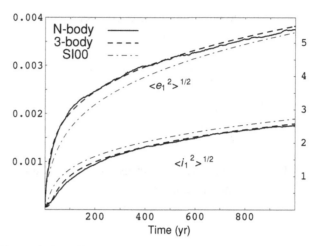

Fig. 5.9. The evolution of the root mean square eccentricity and inclination of a disk composed of 10^3 equal mass bodies orbiting at or near 1 AU. The bodies have $m = 10^{24}$ g and lie in a disk with surface density $\Sigma_p = 10\,\mathrm{g\,cm^{-2}}$. No damping processes are included. The solid lines show numerical N-body results, while the long-dashed lines labeled "3-body" and "S100" show different simplified results. Reproduced from Ohtsuki *et al.* (2002), with permission.

and identify the energy E as the kinetic energy associated with the random component of the planetesimals' velocities. Equation (5.63) can then be written in the form

$$\frac{d\sigma}{dt} = \frac{2\pi G^2 m \Sigma_{sw} \Omega \ln \Lambda}{\sigma^3},\qquad(5.65)$$

where all the terms in the numerator on the right-hand-side are either exactly or (in the case of $\ln \Lambda$) approximately independent of time. Integrating, we predict that the random velocity of the planetesimals (and, hence, the vertical thickness of the planetesimal disk) ought to increase with time as

$$\sigma(t) \propto t^{1/4}.\qquad(5.66)$$

Gravitational scattering is therefore an efficient mechanism for heating an initially cold thin disk, but the efficiency declines as the disk heats up and the encounter velocities increase. These properties are illustrated in Fig. 5.9, which shows the evolution of a single component disk under the action of viscous stirring calculated using N-body methods (Ohtsuki *et al.*, 2002). As noted previously, viscous stirring in this regime approximately maintains a ratio $e \simeq 2i$.

Our analysis is not sufficiently careful as to warrant any great faith in the accuracy of the pre-factor in Eq. (5.63). We can, however, estimate to an order of magnitude the time scale over which scattering will heat a planetesimal disk. To do so we adopt conditions that might be appropriate at 1 AU, where the surface density of 10 km radius planetesimals is $\Sigma_p = 10\,\mathrm{g\,cm^{-2}}$. We assume that the planetesimals have a mass $m = 10^{19}$ g and a random component of velocity $\sigma = 10^3$ cm s^{-1} (this is about an order of magnitude in excess of the Hill velocity). Taking b_{max} to be equal to the vertical thickness of the planetesimal disk yields an estimate for the Coulomb logarithm $\ln \Lambda \simeq 9$, and we predict that the disk ought to be heated via gravitational scattering on a time scale

$$t_{VS} = \frac{\sigma}{d\sigma/dt} \sim 6 \times 10^3 \text{ yr.} \tag{5.67}$$

This time scale is short enough that viscous stirring due to mutual gravitational perturbations will be an important source of heating for disks of planetesimals prior to the formation of any large bodies.

5.3.2 Dynamical friction

Identical arguments can be applied to a two-component disk of planetesimals made up of bodies with masses m and M. As before, gravitational scattering *among* bodies of the same mass results in a steady increase of the random component of the planetesimal velocities, and a corresponding thickening of the disk. There also is a new effect, however, which derives from the fact that an encounter between a low mass body of mass m and a larger one of mass M gives a greater impulse to the lower mass object. The result of many such scatterings is that the system tries to attain a state of energy equipartition in which

$$\frac{1}{2}m\sigma_m^2 = \frac{1}{2}M\sigma_M^2, \tag{5.68}$$

where σ_m and σ_M are respectively the random velocities of the low and high mass bodies. This process is known as dynamical friction, and it leads to a mass-dependence of the mean eccentricity and inclination of planetesimals and growing protoplanets. An example is shown in Fig. 5.10 from numerical results obtained by Ohtsuki *et al.* (2002). Although complete equipartition may not always be attained (Rafikov, 2003) this simple result is nonetheless of considerable import for studies of terrestrial planet formation. It implies that the relative velocity between a growing protoplanet and the surrounding planetesimals is *smaller* than that between two planetesimals. Lower velocities mean larger enhancements to the cross-section due to gravitational focusing, and hence dynamical friction tends to amplify the tendency toward runaway growth of the larger bodies.

5.3.3 Gas drag

The influence of aerodynamic drag on the orbital properties of planetesimals can be estimated by computing the friction time scale (Eq. 4.11) in the Stokes regime appropriate for large bodies. For a planetesimal of material density ρ_m the friction

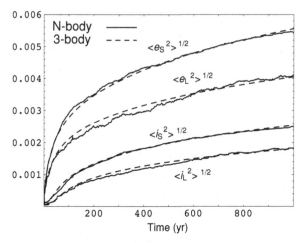

Fig. 5.10. The evolution of the root mean square eccentricity and inclination of bodies within a two-component disk made up of 800 bodies with $m = 10^{24}$ g and 200 bodies with $m = 4 \times 10^{24}$ g. The summed surface density of the two components is $\Sigma_p = 10 \, \mathrm{g \, cm^{-2}}$. The bodies are distributed randomly in a narrow annulus centered on 1 AU with initially small values of e and i. Viscous stirring heats both disks, but the eccentricity $\langle e_L^2 \rangle^{1/2}$ and inclination $\langle i_L^2 \rangle^{1/2}$ of the large bodies remains systematically lower than that of the smaller bodies due to dynamical friction. Reproduced from Ohtsuki *et al.* (2002), with permission.

time scale against a gas disk of density ρ is

$$t_{\mathrm{fric}} = \frac{8}{3C_D} \frac{\rho_m s}{\rho \sigma}, \tag{5.69}$$

where C_D is a dimensionless drag coefficient. It is obvious that this is (to an order of magnitude) the time scale on which gas drag will act to damp planetesimal inclination, and less obvious but still true that gas drag will also damp the eccentricity. Evaluating this expression under the same conditions that we used for the determination of the viscous stirring time scale – 10 km planetesimals of mass 10^{19} g orbiting with a random velocity of $10^3 \, \mathrm{cm \, s^{-1}}$ at 1 AU within a gas disk of density $\rho = 5 \times 10^{-10} \, \mathrm{g \, cm^{-3}}$ – we find that $t_{\mathrm{fric}} \sim 10^6$ yr. This is a long time scale, and all we appear to have shown is that aerodynamic drag can indeed be neglected once bodies have grown to the scale of planetesimals. The gas drag time scale, however, is a decreasing function of σ whereas the viscous stirring time scale increases very rapidly with σ. Equating these time scales we find that for our parameters the weak effect of aerodynamic drag is nonetheless able to offset the heating due to gravitational scattering for $\sigma \sim 3 \times 10^3 \, \mathrm{cm \, s^{-1}}$, which corresponds to an inclination that is only of the order of $i \sim 10^{-3}$. Gas drag is therefore sufficient to maintain the population of small bodies on almost circular orbits despite the heating due to gravitational scattering. Since larger bodies are, in turn, cooled by dynamical friction against the small bodies the overall result is to maintain relatively low random velocities and large gravitational focusing factors throughout the population of growing bodies.

5.4 Analytic formulae for planetary growth

The expressions derived above for the growth rate of protoplanets and for the velocity evolution of planetesimals are only suitable for order of magnitude estimates. More accurate semi-analytic formulae can be derived via two approaches: a kinetic formalism that is based on adding a collisional term to the collisionless Boltzmann equation describing the evolution of the phase-space density of bodies (Hornung *et al.*, 1985; Stewart & Wetherill, 1988), or a celestial mechanics treatment of three-body dynamics using Hill's equations (Ida, 1990). These approaches are equivalent (Stewart & Ida, 2000) and the approximate formulae that result can be further improved by reference to numerical integrations of the three-body problem. We quote here formulae appropriate for the most common situation in which a relatively massive body of mass M grows by accreting planetesimals whose velocity dispersion is controlled by a balance between viscous stirring by the large bodies and gas drag.

We consider a growing body of mass M and physical radius R_s orbiting in a circular orbit with semi-major axis a and angular velocity Ω (as noted above dynamical friction will tend to circularize the orbit of a large body embedded within a swarm of smaller ones, so the assumption that $e = 0$ is not a severe restriction). The body grows by accretion from a disk of planetesimals with surface density Σ_p, with the individual bodies having mass m and physical radius s. To describe the transition between the shear and dispersion dominated limits it is helpful to express the eccentricities and inclinations of the planetesimals in a normalized form

$$\tilde{e} \equiv \frac{a}{r_H} e, \tag{5.70}$$

$$\tilde{i} \equiv \frac{a}{r_H} i, \tag{5.71}$$

where the mutual Hill radius is defined as

$$r_H \equiv \left(\frac{M + m}{3M_*} \right)^{1/3} a. \tag{5.72}$$

A reduced eccentricity of unity ($\tilde{e} = 1$) then corresponds to planetesimal orbits whose radial excursions match the Hill radius of the growing protoplanet. Under the assumption that the planetesimal swarm is characterized by a Rayleigh distribution, Inaba *et al.* (2001), drawing on earlier results by Greenzweig & Lissauer (1992) and Ida & Nakazawa (1989), show that the growth rate of the large body can be written as

$$\frac{\mathrm{d}M}{\mathrm{d}t} = \Sigma_\mathrm{p}\Omega r_\mathrm{H}^2 P_{\mathrm{col}}, \tag{5.73}$$

where the collision probability P_{col} is given by a piecewise function of the root mean square reduced inclination and eccentricity,

$$P_{\mathrm{high}} = \frac{(R_\mathrm{s} + s)^2}{2\pi r_\mathrm{H}^2}\left[I_\mathrm{F}(\beta) + \frac{6r_\mathrm{H}I_\mathrm{G}(\beta)}{(R_\mathrm{s} + s)\langle \tilde{e}^2 \rangle}\right], \tag{5.74}$$

$$P_{\mathrm{med}} = \frac{(R_\mathrm{s} + s)^2}{4\pi r_\mathrm{H}^2 \langle \tilde{i}^2 \rangle^{1/2}}\left[17.3 + \frac{232 r_\mathrm{H}}{(R_\mathrm{s} + s)}\right], \tag{5.75}$$

$$P_{\mathrm{low}} = 11.3\left(\frac{R_\mathrm{s} + s}{r_\mathrm{H}}\right)^{1/2}. \tag{5.76}$$

The high velocity limit is appropriate for $\tilde{e}, \tilde{i} > 2$, the medium velocity case for $0.2 < \tilde{e}, \tilde{i} < 2$, and the low velocity limit for $\tilde{e}, \tilde{i} < 0.2$. The parameter $\beta = \langle \tilde{i}^2 \rangle^{1/2}/\langle \tilde{e}^2 \rangle^{1/2}$, while the functions I_F and I_G are defined as

$$I_\mathrm{F}(\beta) = 8\int_0^1 \frac{\beta^2 E(\theta)}{[\beta^2 + (1 - \beta^2)\lambda^2]^2}\mathrm{d}\lambda, \tag{5.77}$$

$$I_\mathrm{G}(\beta) = 8\int_0^1 \frac{K(\theta)}{\beta^2 + (1 - \beta^2)\lambda^2}\mathrm{d}\lambda, \tag{5.78}$$

$$\theta \equiv \frac{1}{2}\sqrt{3 - 3\lambda^2}. \tag{5.79}$$

In these expressions $K(\theta)$ and $E(\theta)$ are complete elliptic integrals of the first and second kind.

Although these expressions are rather complex the basic dependencies of the collision rate on the eccentricity and inclination of the planetesimals match those deduced earlier from simple arguments. In the high velocity regime the velocity dispersion (strictly all that matters is $\langle \tilde{e}^2 \rangle$) enters via the gravitational focusing term, in the intermediate regime the scaling goes as $1/\tilde{i}$ (cf. Eq. 5.48), while in the cold disk limit there is no dependence on either eccentricity or inclination. It is also worth noting that the cumbersome integration over elliptic integrals (Eq. 5.79) is only required to describe how the high velocity collision rate varies with the ratio of inclination to eccentricity β. In the common case where $\beta = 1/2$ the functions are just constants given by $I_\mathrm{F}(0.5) = 17.34$ and $I_\mathrm{G}(0.5) = 38.22$.

The rates of eccentricity and inclination evolution due to the combined effects of viscous stirring and dynamical friction have been calculated by Ohtsuki *et al.* (2002) using a combination of semi-analytic results and three-body numerical integrations. The results of Ohtsuki *et al.* (2002) can be applied to single-component systems (which evolve under the action of viscous stirring alone), two-component systems (where dynamical friction also occurs), and situations where there is a continuous distribution of sizes. For the sake of brevity we quote here what is often the most important result: the rate of viscous stirring of planetesimals of mass m by protoplanets of mass M. In this limit the rate of excitation of the planetesimals' eccentricity and inclination is given by

$$\frac{d\langle e^2 \rangle}{dt} = a^2 \Omega N_e \left(\frac{M+m}{3M_*} \right)^{4/3} \frac{M^2}{(M+m)^2} P_{VS}, \tag{5.80}$$

$$\frac{d\langle i^2 \rangle}{dt} = a^2 \Omega N_e \left(\frac{M+m}{3M_*} \right)^{4/3} \frac{M^2}{(M+m)^2} Q_{VS}, \tag{5.81}$$

where P_{VS} and Q_{VS} are described as the viscous stirring rates for eccentricity and inclination respectively, and $N_e = \Sigma_e/M$ is the surface number density of planetary embryos (the large bodies of mass M) that have a surface mass density Σ_e. For $m \ll M$ these expressions simplify considerably, so that the eccentricity evolution equation for example can be written as

$$\frac{d\langle e^2 \rangle}{dt} = \frac{a \Omega r_H \Sigma_e}{3M_*} P_{VS}, \tag{5.82}$$

where r_H is the Hill radius of the planetary embryos.

The viscous stirring rates for a Rayleigh distribution of planetesimal eccentricity and inclination are given as the sum of formulae that apply to the high and low velocity limits (Ohtsuki et al., 2002),

$$P_{VS} = \frac{73 \langle \tilde{e}^2 \rangle}{10 \Lambda^2} \ln(1 + 10\Lambda^2 / \langle \tilde{e}^2 \rangle)$$

$$+ \frac{72 I_{PVS}(\beta)}{\pi \langle \tilde{e}^2 \rangle^{1/2} \langle \tilde{i}^2 \rangle^{1/2}} \ln(1 + \Lambda^2), \tag{5.83}$$

$$Q_{VS} = \frac{4\langle \tilde{i}^2 \rangle + 0.2 \langle \tilde{e}^2 \rangle^{3/2} \langle \tilde{i}^2 \rangle^{1/2}}{10 \Lambda^2 \langle \tilde{e}^2 \rangle^{1/2}} \ln(1 + 10\Lambda^2 \langle \tilde{e}^2 \rangle^{1/2})$$

$$+ \frac{72 I_{QVS}(\beta)}{\pi \langle \tilde{e}^2 \rangle^{1/2} \langle \tilde{i}^2 \rangle^{1/2}} \ln(1 + \Lambda^2). \tag{5.84}$$

In these expressions

$$\Lambda = \frac{1}{12} \left(\langle \tilde{e}^2 \rangle + \langle \tilde{i}^2 \rangle \right) \langle \tilde{i}^2 \rangle^{1/2}, \tag{5.85}$$

the parameters β and θ (Eq. 5.79) are defined as before, and I_{PVS} and I_{QVS} are again integrals over elliptic integrals that are functions only of the ratio of mean

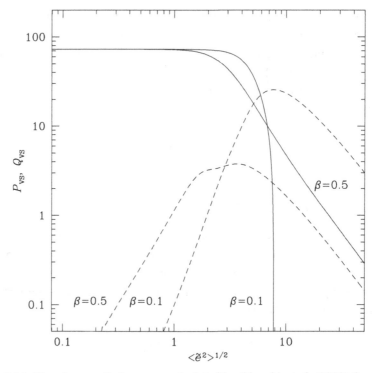

Fig. 5.11. The viscous stirring rates calculated by Ohtsuki *et al.* (2002) for planetesimals interacting with embedded planetary embryos are plotted as a function of the reduced eccentricity of the planetesimal population. The solid curves show the stirring rate of the eccentricity P_{VS}, while the dashed curves show the stirring rate of the inclination Q_{VS}. In each case two curves are plotted to demonstrate the effect of variations in the ratio of inclination to eccentricity β.

inclination to mean eccentricity,

$$I_{PVS}(\beta) = \int_0^1 \frac{5K(\theta) - 12(1 - \lambda^2)E(\theta)/(1 + 3\lambda^2)}{\beta + (\beta^{-1} - \beta)\lambda^2}\,d\lambda, \qquad (5.86)$$

$$I_{QVS}(\beta) = \int_0^1 \frac{K(\theta) - 12\lambda^2 E(\theta)/(1 + 3\lambda^2)}{\beta + (\beta^{-1} - \beta)\lambda^2}\,d\lambda. \qquad (5.87)$$

Such complex expressions are not very enlightening, but their evaluation for the purposes of calculating planetary growth rates is straightforward.[4] Figure 5.11 shows the form of these functions as the reduced eccentricity of the planetesimals is varied. As one could have anticipated based on the arguments in Section 5.2.2, viscous stirring of planetesimal eccentricity by embedded massive bodies is vastly more efficient than viscous stirring of inclination in the shear dominated regime.

---

[4] Polynomial fits to the integrals of sufficient accuracy for almost any purpose are given by Chambers (2006).

There are also order of magnitude variations in the stirring rates as β is reduced and the velocity dispersion of the planetesimals becomes increasingly anisotropic.

Damping of planetesimal eccentricity and inclination by gas drag is a great deal easier to model than excitation by gravitational scattering. Due to the fact that the gas is partially supported against gravity by a pressure gradient, planetesimals experience aerodynamic drag even if $e = i = 0$ (cf. Section 2.3 and Eq. 4.30), and this modifies the damping rates for eccentric and/or inclined bodies. Defining the fractional difference between the Keplerian and gas disk velocities as

$$\eta' = \frac{v_K - v_{\phi,\text{gas}}}{v_K},$$
(5.88)

Adachi *et al.* (1976) find that the damping rate of planetesimal eccentricity and inclination can be approximately described via

$$\frac{de}{dt} = -\frac{e}{t_{\text{drag}}} \left(\eta'^2 + \frac{5}{8}e^2 + \frac{1}{2}i^2 \right)^{1/2},$$
(5.89)

$$\frac{di}{dt} = -\frac{i}{2t_{\text{drag}}} \left(\eta'^2 + \frac{5}{8}e^2 + \frac{1}{2}i^2 \right)^{1/2}$$
(5.90)

where the characteristic time scale is defined as

$$t_{\text{drag}} = \frac{8}{3C_D} \frac{\rho_m s}{\rho v_K}.$$
(5.91)

Since η' is typically a few $\times 10^{-3}$ the term in the parentheses due to the background gas drag is dominant for very small values of e and i, and in this limit the damping time scale is independent of the actual values of the eccentricity and inclination. For low eccentricity planetesimals of mass 10^{19} g and radius 10 km orbiting within a gas disk of density $\rho = 5 \times 10^{-10}$ g cm^{-3} and $\eta' = 0.004$, for example, the damping time scale is of the order of 10^5 yr.

Within their domain of validity these semi-analytic formulae for the excitation and damping of planetesimal eccentricity (supplemented where necessary with the similar expressions for dynamical friction and stirring by planetesimal–planetesimal scattering) are quite accurate. Figures 5.9 and 5.10 show, for example, the comparison between the semi-analytic formulae and N-body results for the cases of one- and two-component disks of bodies evolving under the action of viscous stirring and dynamical friction, and it is clear that the agreement is impressively good. The semi-analytic expressions can be used to estimate the growth rate of planetary embryos or giant planet cores (e.g. Chambers, 2006), and also form the basis for computing the velocity evolution of bodies within more sophisticated treatments of growth based on the coagulation equation (Section 5.6). Calculations of terrestrial planet formation generally show that once a few massive bodies have formed via runaway growth, ongoing growth occurs in a high velocity regime where the planetesimal disk is heated by viscous stirring from the protoplanets and cooled by gas drag (Ida & Makino, 1993). This regime – which has been dubbed "oligarchic growth" – is partially self-limiting, since growth of any individual protoplanet increases the random velocities of planetesimals in its vicinity and decreases the gravitational focusing enhancement to the collision cross-section.

5.5 Collisional damping and turbulent excitation

The critical role of gravitational focusing in determining the growth rate of proto-planets means that any error in assessing the strength of the processes that excite or damp planetesimal random motions could qualitatively alter our picture of either terrestrial or giant planet formation. It is therefore worth considering whether there are mechanisms other than gravitational scattering and gas drag that might be significant. Two possibilities – damping via inelastic collisions and excitation by gravitational coupling to disk turbulence – have received recent theoretical attention, and although neither has yet been studied to the same level of detail as gravitational scattering and gas drag their potential importance merits a brief discussion.

Sufficiently inelastic collisions between planetesimals can damp eccentricity and inclination by (ultimately) converting some of the random motion into heat within the bodies. To estimate the potential importance of this effect we first observe that if the gravitational focusing factor is large most "encounters" between planetesimals result in gravitational scattering (which is a heating process) rather than physical collisions. Inelastic collisions are only potentially significant when $\sigma > v_{esc}$,[5] in which case the collision time scale is

$$t_{inelastic} = \frac{1}{n_{sw}\pi s^2 \sigma}.$$ (5.92)

Assuming an isotropic velocity dispersion the mid-plane number density of plan-etesimals n_{sw} is inversely proportional to σ, which cancels as usual to yield an estimate

$$t_{inelastic} \simeq \frac{8\rho_m s}{3\Sigma_p \Omega}.$$ (5.93)

---

[5] Since we have previously argued that gravitational focusing must be important if planetary growth is to occur on a reasonable time scale the reader may wonder whether this is not already sufficient cause to rule out inelastic collisions. To do so would be premature. Protoplanets could initially grow under conditions of strong gravitational focusing, and then subsequently excite the planetesimals so that *their* collisions lead to fragmentation. Such a scenario results in growth of protoplanets within a swarm of very small bodies whose collision rate is set by the physical rather than the gravitational focusing cross-section.

We can compare this time scale to the time scale for damping of planetesimal random motion by gas drag (Eq. 5.90), which in the limit of small e and i is

$$t_{gas} \simeq \frac{8\rho_m s}{3\eta' C_D \rho v_K}. \tag{5.94}$$

Equating $t_{inelastic}$ to t_{gas}, we find that up to factors of the order of unity the condition for inelastic collisions to dominate is

$$\Sigma_p \gtrsim \rho r \eta'. \tag{5.95}$$

This can be further simplified by noting that the deviation of the gas velocity from the Keplerian value $\eta' \sim (h/r)^2$. We find finally

$$\frac{\Sigma_p}{\Sigma} \gtrsim \frac{h}{r}, \tag{5.96}$$

as the condition for inelastic collisions to be more important than gas drag if, additionally, gravitational focusing is unimportant. Since Σ_p/Σ is assuredly smaller than the local dust to gas ratio – whose standard value of $f = 10^{-2}$ is in turn smaller than the typical $(h/r) \approx 0.05$ – this estimate justifies our neglect of inelastic collisions relative to aerodynamic drag in the preceding section. One can see, however, that the margin by which inelastic collisions are subdominant is not all that large, and as the gas disk evolves and dissipates it may be possible to encounter a regime in which inelastic collisions provide the main cooling mechanism. In this spirit Goldreich *et al.* (2004) have proposed a model for the formation of the ice giants that relies on the presence of a cold planetesimal disk made up of small bodies for which inelastic collisions furnish efficient damping. The collisional cooling increases the gravitational focusing enhancement to the collision cross-section of the larger bodies, thereby accelerating the formation of the ice giants.

Planetesimal excitation by gravitational coupling to turbulent fluctuations in the gas disk is a second poorly understood process that might be important enough to require modification of the standard theory of terrestrial planet formation. For planetesimal-scale bodies the dimensionless friction time $\tau \gg 1$ and *aerodynamic* drag on the bodies is essentially independent of the strength or nature of any intrinsic turbulence within the disk. Independent of the aerodynamic coupling, however, there could also be gravitational coupling. Many plausible sources of disk turbulence (including the magnetorotational instability) create spatial and temporal fluctuations in the gas surface density of the disk. Any solid body orbiting within the gas experiences stochastic forcing as a result of the gravitational forces from the nonaxisymmetric surface density fluctuations, and these randomly directed impulses act as an excitation mechanism for eccentricity and inclination. The amplitude and temporal coherence of the forcing can be calculated in principle from *ab initio* simulations of disk turbulence (Nelson & Papaloizou, 2003; Laughlin *et al.*, 2004). For terrestrial planet formation the main interest in this process derives from the possibility that excitation due to gravitational coupling to turbulence might dominate viscous stirring prior to the formation of large protoplanets, and thereby delay or (in some regions of the disk) even prevent collisional growth.

Since the mechanism or mechanisms that result in angular momentum transport within protoplanetary disks remain uncertain it will come as no surprise that precise calculations of the properties of stochastic forcing by this process remain lacking. Plausible estimates by Ida *et al.* (2008), however, show that even for minimum mass Solar Nebula values of the surface density, turbulence driven by the magnetoro-tational instability can be rather efficient at exciting planetesimal random motion. Their results suggest that in a turbulent disk the equilibrium value of planetesimal eccentricity is increased to the point that collisions of even 100 km bodies would result in fragmentation and disruption rather than accretion. Since it is clear that larger bodies do manage to form within the disk this argument can be considered to provide a limit on one or more of (a) the minimum size of planetesimals, (b) the strength of disk turbulence, or (c) the gas mass at the time when large bodies form.

It is worth noting that even if these processes prove to be unimportant for planet formation under Solar System conditions, the same need not be true elsewhere. Gas disks that were highly turbulent, for example due to the onset of self-gravity (Britsch *et al.*, 2008), would act to excite planetesimals more efficiently, while planets forming in systems where the gas disk was dispersed early on might grow more rapidly if the planetesimals in the disk were ground down to smaller sizes and cooled via inelastic collisions.

5.6 Coagulation equation

For much of the preceding discussion we have assumed that the distribution of masses of the growing bodies can be partitioned into two groups: large planetary embryos with mass M and negligible random velocities, and much less massive planetesimals with mass m and significant random velocities. If runaway growth occurs this simple approximation is actually quite a good way to think about the problem, since each annulus of the disk *will* contain one body that is much larger than all the rest. One may rightly worry, however, that attempting to demonstrate the existence of runaway growth using a two-groups approximation involves a suspiciously circular logic. The right framework for approaching the problem of whether runaway growth occurs requires dropping the two-groups model and treating the evolution of an arbitrary size distribution of bodies using the methods of coagulation theory (Smoluchowski, 1916). The application of this theory to planet formation was pioneered by Safronov (1969), and has subsequently been adopted by many authors. Clear descriptions of the basic method are given by Wetherill (1990) and Kenyon & Luu (1998), whose nomenclature is largely adopted here.

Coagulation theory is based on solutions to the coagulation equation, which can be written equivalently in either integral (Eq. 4.67) or discrete form. In the discrete representation, we assume that at some time t there are n_k bodies within some fixed volume that have a mass km_1 that is an integral multiple of some small mass m_1. We treat n_k as a *continuous function*.[6] Ignoring radial migration and fragmentation, we can then write the discrete coagulation equation for an annulus of the disk in the form

$$\frac{dn_k}{dt} = \frac{1}{2} \sum_{i+j=k} A_{ij} n_i n_j - n_k \sum_{i=1}^{\infty} A_{ik} n_i, \tag{5.97}$$

where the A_{ij}, known as the kernels, describe the probability of a collision that leads to accretion between bodies with masses im_1 and jm_1 per unit time. In general these will be nonlinear functions of the masses, random velocities, and physical properties (e.g. density) of the bodies involved. One observes that the number of bodies in the kth mass bin changes due to two processes. First, the number of bodies of mass km_1 *increases* whenever there is a collision between any pair of bodies whose total mass $(i + j)m_1$ sums to km_1 (the factor of $1/2$ is present to avoid double counting these collisions). Second, the number of bodies of mass km_1 *decreases* whenever there is any collision with a body of any other mass.

There are no known analytic solutions to the coagulation equation for kernels that encode the critical physics of planet formation – gravitational focusing and mass-dependent velocity dispersion. Numerical solutions provide the only way to handle these complexities. We can, however, gain some insight into the nature of general solutions to the coagulation equation by studying three known solutions for cases with particularly simple kernels. If

$$A_{ij} = \alpha, \tag{5.98}$$

where α is a constant, and there are initially n_0 bodies with mass m_1, then at time t

$$n_k = n_0 f^2 (1 - f)^{k-1}, \tag{5.99}$$

$$f = \frac{1}{1 + \frac{1}{2}\alpha n_0 t}. \tag{5.100}$$

Inspection of the units shows that f is a dimensionless measure of the time. The physical interpretation of f is that it is the fraction of bodies in the smallest mass bin that have yet to collide with any other body.

---

[6] This is reasonable provided that $n_k \gg 1$, but we must be cognizant that the statistical foundations of the method can break down if for some k of interest $n_k \sim 1$.

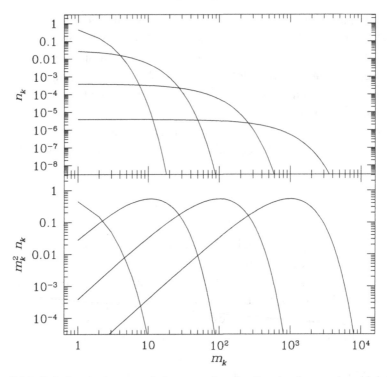

Fig. 5.12. Solution to the coagulation equation for the simple case in which the kernel $A_{ij} = \alpha$ is a constant. Initially all bodies have mass m_1. The solution is plotted for scaled times $t' \equiv \alpha n_0 t$ equal to 1, 10, 100, and 10^3. The upper panel shows the evolution (on an arbitrary vertical scale) of the number of bodies n_k as a function of mass, while the lower panel shows the evolution of the mass distribution.

The solution expressed in Eq. (5.100) is plotted in Fig. 5.12. This solution is an example of *orderly growth*. As time progresses the mean mass of the population increases, but the shape of the mass spectrum approaches an asymptotic form and its width (expressed in logarithmic units) does not increase. Qualitatively similar is the solution for the kernel

$$A_{ij} = \alpha(m_i + m_j), \tag{5.101}$$

which has the form

$$n_k = n_0 \frac{k^{k-1}}{k!} f(1-f)^{k-1} \exp[-k(1-f)], \tag{5.102}$$

$$f = \exp[-\alpha n_0 t]. \tag{5.103}$$

This solution also exhibits orderly growth.

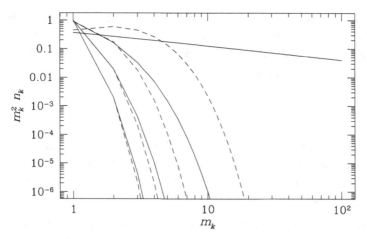

Fig. 5.13. Comparison of the analytic solution to the coagulation equation with the kernel $A_{ij} = \alpha$ (dashed lines) to that with $A_{ij} = \alpha m_i m_j$ (solid lines). The solutions are plotted for scaled times $t' \equiv \alpha n_0 t$ equal to 10^{-3}, 10^{-2}, 0.1, and 1. The collision rate for $i = j = 1$ is identical for the two solutions, so the very early time behavior is the same. Large differences develop just prior to the onset of runaway growth at a time $\alpha n_0 t = 1$.

The final known solution applies to the case of a kernel that is the product of the masses of the bodies

$$A_{ij} = \alpha m_i m_j. \tag{5.104}$$

In this case,

$$n_k = n_0 \frac{(2k)^{k-1}}{k!k} \left(\frac{1}{2} \alpha n_0 t \right)^{k-1} \exp[-\alpha n_0 k t]. \tag{5.105}$$

This solution is plotted in Fig. 5.13 together with a solution for the case of a constant kernel that has the same initial collision rate. Equation (5.105) represents a qualitatively different class of solution to the coagulation equation, in which *runaway growth* develops. The mass distribution develops a power-law tail toward high masses as a small number of bodies grow rapidly at the expense of all of the others. Solutions to the coagulation equation that display runaway growth generally apply only at early times, since once most of the mass accumulates into a single massive body the assumptions upon which the coagulation equation is based break down. Formally Eq. (5.105) is valid up until a time $\alpha n_0 t = 1$.

Realistic kernels for planet formation do not scale with mass in the same way as any of the analytically tractable forms. If gravitational focusing is unimportant, the collision cross-section scales with the geometric area and $A \propto R_s^2 \propto m^{2/3}$. This scaling is bracketed by the two analytic solutions that display orderly growth. Conversely, if gravitational focusing dominates then $A \propto m R_s \propto m^{4/3}$, which lies between analytic solutions exhibiting orderly and runaway growth. This observation does not allow us to draw any definitive conclusions, but numerical calculations suggest that runaway growth can occur for kernels that describe collisions during the early phases of planet formation.

5.7 Final assembly

Calculations of coagulation in the terrestrial planet zone based on the aforementioned physics have given us a standard scenario for the early growth of terrestrial planets. Starting from a large population of planetesimals (of uncertain size) the first two phases are:

- **Runaway growth**. Initially there are no large bodies, so the random velocity of planetesimals is set by a balance between viscous stirring among the planetesimals themselves and damping via gas drag. The combined influence of dynamical friction and gravitational focusing results in runaway growth of a small fraction of planetesimals.
- **Oligarchic growth**. Runaway growth ceases at the point when the rate of viscous stirring of the planetesimals by the largest bodies first exceeds the rate of self-stirring among the planetesimals. The resulting boost in the strength of viscous stirring increases the equilibrium values of planetesimal eccentricity and inclination, partially limiting the gravitationally enhanced cross-section of the protoplanets. In this regime the growth of protoplanets continues to outrun that of planetesimals, but the dominance is local rather than global. Across the disk many *oligarchs* grow at similar rates by consuming planetesimals within their own largely independent feeding zones.

These initial stages of terrestrial planet formation are rapid (of the order of 0.01– 1 Myr), and result in the formation of perhaps 10^2 to 10^3 large bodies across the terrestrial planet zone. These are massive objects (of the order of 10^{-2} $M_\oplus$ to 0.1 $M_\oplus$, so comparable to the mass of the Moon or Mercury) but they are not yet terrestrial planets.

The final assembly of terrestrial planets gets underway once the oligarchs have depleted the planetesimal disk to the point that dynamical friction can no longer maintain low eccentricities and inclinations of the oligarchs. Beyond this point the assumption that each oligarch grows in isolation breaks down, and the largest bodies start to interact strongly, collide, and scatter smaller bodies across a significant radial extent of the disk. Numerical N-body simulations (modern examples include Raymond *et al.*, 2006 and O'Brien *et al.*, 2006), an example of which is shown in Fig. 5.14, show that this final stage is by far the slowest, with large collisions continuing out to at least 10 Myr. The process is chaotic, and hence identical initial conditions can give rise to a range of outcomes that must be compared to the observed properties of the Solar System's terrestrial planets statistically. The level of agreement attained is by no means perfect, with the low eccentricity of

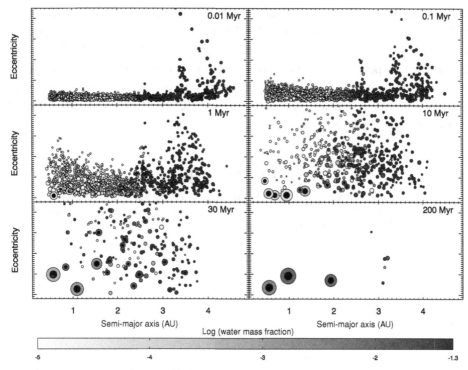

Fig. 5.14. Results from a high resolution simulation of the final assembly of terrestrial planets (Raymond *et al.*, 2006). The initial conditions for this simulation are representative of the late stages of oligarchic growth, with approximately 2000 planetary embryos (shaded according to their water content) distributed in a disk with a surface density profile of $\Sigma_p \propto r^{-3/2}$ and a surface density at 1 AU of 10 g cm^{-2}. Jupiter is assumed to have a circular orbit at 5.5 AU during the duration of the final assembly phase, which takes around 100 Myr. Reproduced in modified form from Raymond *et al.* (2006), with permission.

the Earth and the small mass of Mars (compared to that of the Earth and Venus) being two oft-quoted discrepancies. Viewed on a coarser level, however, reasonable values of the disk surface density in the inner Solar System do appear roughly to reproduce properties of the terrestrial planets, and this gives confidence that the basic collisional scenario for the growth of these objects is valid.

Extending calculations of terrestrial planet formation to study extrasolar planets requires a substantial expansion in the range of parameters investigated. In addition to the obvious additional variable due to different stellar masses, the disk properties and giant planet environment during the formation of extrasolar terrestrial planets may also have been utterly unlike nominal models for the Solar System. Existing calculations suggest that the typical outcome of terrestrial planet formation varies depending upon the surface density of the planetesimal disk – higher Σ_p typically yields a smaller number of more massive planets (Wetherill, 1996; Kokubo *et al.*, 2006). Both the presence and the orbital properties of giant planets also impact the outcome of terrestrial planet formation (Levison & Agnor, 2003; Raymond, 2006). These dependencies – which have potentially dramatic implications for the predicted abundance and habitability of terrestrial planets in extrasolar planetary systems – will be tested with the advent of samples of lowmass planets detected via transits over the next few years.

5.8 Further reading

A self-contained summary of the processes that control terrestrial and giant planet formation can be found in the review article "Planet formation by coagulation: A focus on Uranus and Neptune," P. Goldreich, Y. Lithwick, & R. Sari (2004), *ARA&A*, **42**, 549.

6

Giant planet formation

Understanding the formation of giant planets with substantial gaseous envelopes forces us to confront once again the physics of the gas within the protoplanetary disk. Unlike the case of terrestrial planet formation, two qualitatively different theories have been proposed to account for the formation of massive planets. In the core accretion theory of giant planet formation, the acquisition of a massive envelope of gas is the final act of a story that begins with the formation of a core of rock and ice via the identical processes that we discussed in the context of terrestrial planet formation. The time scale for giant planet formation in this model – and to a large extent its viability – hinges on how quickly the core can be assembled and on how rapidly the gas in the envelope can cool and accrete on to the core. In the competing disk instability theory, giant planets form promptly via the gravitational fragmentation of an unstable protoplanetary disk – a purely gaseous analog of the Goldreich–Ward mechanism for planetesimal formation that we discussed in Chapter 4. Fragmentation turns out to require that the disk be able to cool on a relatively short time scale that is comparable to the orbital time scale, and whether these conditions are realized within disks is the main theoretical issue that remains unresolved. Drawing on our prior results on gravitational instabilities in disks and on terrestrial planet formation, the goal in this chapter is to describe the physical principles behind both models and to provide a summary of some of the relevant observational constraints. It is worth emphasizing at the outset that although partisans on both sides of the debate sometimes view the two theories as mutually exclusive rivals, there is no physical reason why this should be the case. It is conceivable, for example, that the Solar System's gas and ice giants formed via core accretion while gravitational instability yields a population of massive planets at large orbital radii in some extrasolar planetary systems.

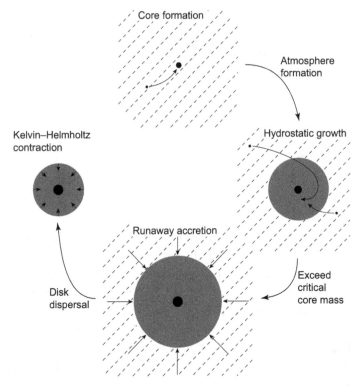

Fig. 6.1. Stages in the formation of giant planets via core accretion.

6.1 Core accretion

The core accretion model for gas giant formation rests on one assumption: that a seed planet or *core* grows via two-body collisions rapidly enough that it can exceed a certain critical mass prior to the dissipation of the gas disk. If this condition is satisfied, it can be shown (Perri & Cameron, 1974; Mizuno, 1980) that the core triggers a hydrodynamic instability that results in the onset of rapid gas accretion on to the core. Since the critical core mass is typically of the order of 10 $M_\oplus$, the end result is a largely gaseous but heavy element enriched planet that at least qualitatively resembles Jupiter or Saturn.

Figure 6.1 illustrates the four main phases in the formation of giant planets via core accretion:

- **Core formation**. A solid protoplanet (which henceforth we will rename a "core") grows via a succession of two-body collisions until it becomes massive enough to retain a significant gaseous atmosphere or envelope. The physics during this initial phase is identical to that discussed in the context of terrestrial planet formation in Chapter 5, with the rate of growth being controlled by the initial surface density of rocky and icy bodies and by the extent of gravitational focusing.

- **Hydrostatic growth**. Initially the envelope surrounding the solid core is in hydrostatic equilibrium. Energy liberated by planetesimals impacting the core, together with gravitational potential energy released as the envelope itself contracts, must be transported through the envelope by radiative diffusion or convection before it is lost to the large gas reservoir of the protoplanetary disk. Over time both the core and the envelope grow until eventually the core exceeds a critical mass. The critical mass is not a constant but rather a calculable function of (primarily) the planetesimal accretion rate and opacity in the envelope.
- **Runaway growth**. Once the critical mass is exceeded a runaway phase of gas accretion ensues. The rate of growth is no longer demand limited (defined by the cooling properties of the envelope) but instead supply limited and defined by the hydrodynamic interaction between the growing planet and the disk. For massive planets the bulk of the planetary envelope is accreted during this phase, which is typically rather brief – of the order of 10^5 yr.
- **Termination of accretion**. Eventually the supply of gas is exhausted, either as a consequence of the dissipation of the entire protoplanetary disk or, more probably, as a consequence of the planet opening up a local *gap* in the disk (the physics of gap-opening will be discussed more fully in Chapter 7). Accretion tails off and the planet commences a long phase of cooling and quasi-hydrostatic contraction.

We can readily estimate some of the masses that characterize the transitions between these phases. The weakest condition that must be satisfied if a planet embedded within a gas disk is to hold on to a bound atmosphere is that the escape speed v_{esc} at the surface of the planet exceeds the sound speed c_s within the gas.[1] A solid body of mass M_p and material density ρ_m has a radius

$$R_s = \left(\frac{3}{4\pi} \frac{M_p}{\rho_m} \right)^{1/3},$$

(6.1)

and a surface escape speed

$$v_{esc} = \sqrt{\frac{2GM_p}{R_s}}.$$

(6.2)

Noting that the sound speed in the protoplanetary disk can be written in terms of the disk thickness (h/r) and Keplerian velocity v_K via

$$c_s = \left(\frac{h}{r} \right) v_K,$$

(6.3)

---

[1] A planet that marginally satisfies this condition would lose its atmosphere on approximately the dynamical time scale in the absence of the gas disk, so this is *not* the same condition as for an isolated planet to be able to retain an atmosphere. To avoid atmospheric erosion via the process known as Jeans escape, an isolated planet must have a much higher surface escape speed that suffices to retain molecules far out in the Maxwellian tail of the particle velocity distribution.

the condition that $v_{esc} > c_s$ can be expressed as

$$M_p > \left(\frac{3}{32\pi}\right)^{1/2}\left(\frac{h}{r}\right)^3\frac{M_*^{3/2}}{\rho_m^{1/2}a^{3/2}},\tag{6.4}$$

where a is the orbital radius. This mass is very small. Substituting numbers appropriate for an icy body at 5 AU in a disk around a Solar mass star with $(h/r) = 0.05$ we find that *some* atmosphere will be present provided that $M_p \gtrsim 5 \times 10^{-4}\, M_\oplus$.

The existence of a tenuous wisp of an atmosphere will not have any dynamical importance. A more pertinent question is what is the minimum core mass able to maintain an envelope with a mass that is a nonnegligible fraction of the mass of the core? To estimate this mass we assume that the envelope makes up a small fraction ϵ of the total planet mass, in which case the equation expressing hydrostatic equilibrium for the envelope can be written as

$$\frac{dP}{dr} = -\frac{GM_p}{r^2}\rho,\tag{6.5}$$

where r is the distance from the center of the planet and M_p can be considered to be a constant. To keep the problem simple (this is after all only an estimate) we also assume that the envelope is isothermal, so that the pressure $P = \rho c_s^2$. Equation (6.5) can then be integrated immediately to yield an expression for the radial density profile of the gas in the envelope

$$\ln\rho = \frac{GM_p}{c_s^2}\frac{1}{r} + \text{const.}\tag{6.6}$$

The constant of integration on the right-hand-side can be evaluated by matching the envelope density ρ to the density ρ_0 in the unperturbed disk at the radius r_{out} where the escape velocity from the planet matches the disk sound speed

$$r_{out} = \frac{2GM_p}{c_s^2}.\tag{6.7}$$

With the constant thereby determined, the envelope density profile is

$$\rho(r) = \rho_0\exp\left[\frac{GM_p}{c_s^2}\frac{1}{r} - \frac{1}{2}\right].\tag{6.8}$$

The density at the disk mid-plane ρ_0 can itself be written in terms of the surface density Σ and vertical scale-height h as $\rho_0 = (1/\sqrt{2\pi})(\Sigma/h)$ (Eq. 2.9).

Most of the mass of an envelope with the above density profile lies in a shell close to the surface of the solid core. We can therefore approximate the envelope mass as

$$M_{env} \approx \frac{4}{3}\pi R_s^3\rho(R_s),\tag{6.9}$$

where $\rho(R_s)$ is the envelope density evaluated at the surface of the core. Substituting for both the density profile and R_s, the condition that the envelope makes up a non-negligible fraction of the total mass

$$M_{env} > \epsilon M_p, \tag{6.10}$$

reduces to

$$M_p \gtrsim \left(\frac{3}{4\pi\rho_m}\right)^{1/2} \left(\frac{c_s^2}{G}\right)^{3/2} \left[\ln\left(\frac{\epsilon\rho_m}{\rho_0}\right)\right]^{3/2}. \tag{6.11}$$

The dominant dependence is on the disk sound speed, which is a decreasing function of radius. A growing protoplanet will therefore start to acquire a substantial gaseous envelope at a lower mass if it is located in the cool outer regions of the disk. To give a concrete example, we consider an icy body ($\rho_m = 1\,\mathrm{g\,cm^{-3}}$) growing at 5 AU in a disk with $\rho_0 = 2 \times 10^{-11}\,\mathrm{g\,cm^{-3}}$ and sound speed $c_s = 7 \times 10^4\,\mathrm{cm\,s^{-1}}$. Taking $\epsilon = 0.1$ to represent the threshold above which the envelope could be said to be significant, we predict that this will occur for planet masses

$$M_p \gtrsim 0.2\,M_\oplus. \tag{6.12}$$

At 1 AU, on the other hand, where the disk parameters might be a density $\rho_0 = 6 \times 10^{-10}\,\mathrm{g\,cm^{-3}}$ and a sound speed $c_s = 1.5 \times 10^5\,\mathrm{cm\,s^{-1}}$, a rocky protoplanet ($\rho_m = 3\,\mathrm{g\,cm^{-3}}$) must grow to $M_p \sim M_\oplus$ before we would expect to find it enshrouded in a dense massive envelope.

In light of the rather crude analysis that we have employed, Eq. (6.11) should only be trusted to an order of magnitude. Taking the estimate at face value, however, we can give a plausible argument for why gas giant planets ought to form only in the cool outer regions of the protoplanetary disk. As we have seen, a protoplanet must grow to something like the mass specified in Eq. (6.11) before it starts to acquire a massive gaseous envelope, and this must occur *prior* to the dispersal of the protoplanetary disk. If we assume that the maximum protoplanet mass that can be attained prior to disk dispersal is comparable to the isolation mass (Eq. 5.53), we can determine for any particular disk model where in the disk planets will grow fast enough to capture envelopes. As an example, let us suppose that the radial profiles of the gas surface density and solid surface density are those specified by the minimum mass Solar Nebula model given in Section 1.1.1, and that the sound speed is that appropriate to a disk with a constant geometric thickness $(h/r) = 0.05$.[2] Equations (5.53) and (6.11) then yield estimates for the isolation

---

[2] This value is adopted for consistency with most of the other estimates given in this section. It is not strictly consistent with the temperature profile usually assumed for the minimum mass Solar Nebula.

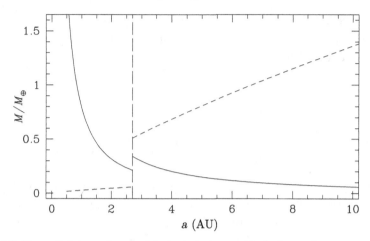

Fig. 6.2. The radial dependence of the isolation mass (dashed curve) as compared to the minimum planet mass needed to sustain a massive envelope (solid curve). Both curves are computed for a Solar mass star surrounded by a disk with $(h/r) = 0.05$. The surface density of the gas and solid component is taken to follow Hayashi's minimum mass Solar Nebula model, with a snowline at 2.7 AU. In computing the minimum mass necessary for envelope capture it has been assumed that interior to the snowline $\rho_m = 3\,\mathrm{g\,cm}^{-3}$, while beyond the snowline $\rho_m = 1\,\mathrm{g\,cm}^{-3}$.

mass and the minimum mass necessary for envelope capture, and these estimates are plotted in Fig. 6.2. The result is suggestive. Interior to the snowline the isolation mass is *smaller* than the minimum mass required for envelope capture. In this region it is unlikely that protoplanets will grow fast enough to capture envelopes prior to disk dispersal, and the ultimate outcome of planet formation will instead be terrestrial planets. At orbital radii beyond the snowline, conversely, the isolation mass *exceeds* the minimum envelope capture mass. Giant planet formation is much more probable in this case, though as yet we cannot say whether the outcome is planets with a modest but still significant envelope (akin to Uranus and Neptune) or true gas giants whose mass is dominated by the contribution from the envelope.

It is worth emphasizing that although the isolation mass generically jumps upward at the snowline, the critical radius beyond which this argument predicts that giant planet formation may occur need *not* always coincide with the radius of the snowline. Using a different disk model (or indeed a more accurate model for when envelope capture commences) one might instead predict a critical radius for giant planet formation that lies either inside or outside the radius of the snowline. The only safe generalization is that envelope capture becomes more likely at larger orbital radii, and detailed calculations (such as those presented by Bodenheimer *et al.*, 2000) are needed before one can decide whether a particular giant planet is or is not likely to have formed *in situ* at its observed orbital radius.

6.1.1 Core/envelope structure

The physical underpinning of the core accretion model for gas giant planet formation is the existence of a critical core mass – a core mass beyond which it is not possible to find a hydrostatic solution for the structure of a surrounding gaseous envelope. To understand the origin of the critical core mass, we modify some standard results from the theory of stellar structure to describe the structure of a giant planet envelope. We assume that the planet of total mass M_p has a well-defined solid core of mass M_{core} and an envelope mass M_{env}. Rotation is neglected and the envelope is taken to be in hydrostatic and thermal equilibrium. Conservation of mass and momentum then yield two differential equations for the structure of the envelope

$$\frac{dM}{dr} = 4\pi r^2 \rho,$$

$$\frac{dP}{dr} = -\frac{GM}{r^2}\rho. \tag{6.13}$$

Here P is the pressure and $M = M(r)$ is the total mass enclosed within radius r. We have now dropped the simplifying assumption that $M \simeq M_{core}$ so the above equations are valid regardless of the mass of the envelope.

Although these equations only involve P and ρ, the temperature invariably enters as a third variable because the pressure $P = P(\rho, T)$ depends upon it. The temperature is determined by the requirement that the radial temperature *gradient* must be sufficient to transport the luminosity of the planet – generally produced at or near the surface of the core by the accretion of planetesimals – from the deep interior to the surface. This transport can occur via radiative diffusion or convection. In the radiative case the resulting temperature gradient is

$$\frac{dT}{dr} = -\frac{3\kappa_R \rho}{16\sigma T^3}\frac{L}{4\pi r^2}, \tag{6.14}$$

where κ_R is the Rosseland mean opacity (defined in terms of the frequency dependent opacity via Eq. 2.58), σ is the Stefan–Boltzmann constant and L could in principle be a function of radius. The temperature gradient is directly proportional to the local flux $L/(4\pi r^2)$, so if the luminosity originates from planetesimal bombardment of the core we expect that a large flux and a correspondingly steep temperature gradient will be present for $r \sim R_s$. This immediately suggests that we need to be alert to the possibility that the radiative temperature gradient, which we conventionally define by combining Eq. (6.14) with the equation of hydrostatic equilibrium to yield

$$\nabla_{rad} \equiv \left(\frac{d\ln T}{d\ln P}\right)_{rad} = \frac{3\kappa_R L P}{64\pi\sigma GMT^4}, \tag{6.15}$$

may be unstable to the onset of convection.

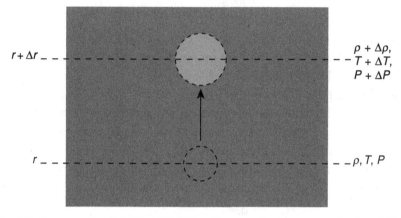

Fig. 6.3. To assess the stability of the planetary envelope to convective instability, we imagine displacing a notional blob of fluid from r to $r + \Delta r$ under adiabatic conditions. Stability requires that the displaced fluid be denser than its new surroundings.

We can establish the stability of the envelope to the onset of convection with the aid of a well-known thought experiment (e.g. Kippenhahn & Weigert, 1990, whose treatment we follow here) illustrated in Fig. 6.3. Convection is a buoyancy instability that occurs when the radial structure of an initially stationary atmosphere or envelope is prone to break up into upward moving underdense blobs and downward moving overdense streams. To assess whether this will happen, we imagine that the gas in the envelope is initially at rest with some specified gradient of density $(\mathrm{d}\rho/\mathrm{d}r)_{\mathrm{env}}$, temperature $(\mathrm{d}T/\mathrm{d}r)_{\mathrm{env}}$ and pressure $(\mathrm{d}P/\mathrm{d}r)_{\mathrm{env}}$. We now displace a notional blob of fluid upwards from r to $r + \Delta r$ *slowly* (so that the fluid within the blob remains in pressure equilibrium with the surrounding envelope) and *adiabatically* (so that no exchange of energy between the blob and the envelope occurs). It is immediately obvious that if the displaced blob finds itself denser than the envelope at $r + \Delta r$ it will tend to sink back down, whereas if it is less dense then buoyancy will cause it to rise and we conclude that the initial equilibrium is unstable. Mathematically the equilibrium is *stable* if

$$\left(\frac{\mathrm{d}\rho}{\mathrm{d}r}\right)_{\mathrm{ad}} > \left(\frac{\mathrm{d}\rho}{\mathrm{d}r}\right)_{\mathrm{env}}, \qquad (6.16)$$

where the left-hand-side represents the density change within the blob under adiabatic conditions and the right-hand-side is the density gradient in the surrounding envelope. The fact that both gradients are negative quantities occasions a good deal of confusion, but the physical argument in terms of buoyancy is clear.

The equations for the structure of the envelope do not directly specify $(\mathrm{d}\rho/\mathrm{d}r)$ and hence it is useful to express the stability condition in terms of the temperature gradient. This requires only routine mathematical manipulation. For any equation of state of the form $\rho = \rho(P, T, \mu)$, where μ is the molecular weight, we can write

$$\frac{d\rho}{\rho} = \alpha \frac{dP}{P} - \delta \frac{dT}{T} + \varphi \frac{d\mu}{\mu}, \tag{6.17}$$

where the α, δ, and φ are defined via,

$$\alpha \equiv \left(\frac{\partial \ln \rho}{\partial \ln P} \right)_{T,\mu}, \tag{6.18}$$

$$\delta \equiv - \left(\frac{\partial \ln \rho}{\partial \ln T} \right)_{P,\mu}, \tag{6.19}$$

$$\varphi \equiv \left(\frac{\partial \ln \rho}{\partial \ln \mu} \right)_{P,T}. \tag{6.20}$$

These parameters are specified by the equation of state. For an ideal gas one has that $\rho \propto P\mu/T$ and hence $\alpha = \delta = \varphi = 1$.

We can now convert the convective stability condition given by Eq. (6.16) into an equivalent condition based on temperature with the aid of Eq. (6.17). Noting that the molecular weight of the gas within the blob does not change as it is displaced, the condition for stability becomes

$$\left(\frac{\alpha}{P} \frac{dP}{dr} \right)_{ad} - \left(\frac{\delta}{T} \frac{dT}{dr} \right)_{ad} > \left(\frac{\alpha}{P} \frac{dP}{dr} \right)_{env} - \left(\frac{\delta}{T} \frac{dT}{dr} \right)_{env} + \left(\frac{\varphi}{\mu} \frac{d\mu}{dr} \right)_{env}. \tag{6.21}$$

This expression can be further simplified by noting that the two terms involving the pressure gradient vanish on account of the assumed pressure balance between the blob and its surroundings. Canceling these terms and multiplying through by $-P(dr/dP)$ yields the Ledoux criterion for the stability of the envelope against the onset of convection

$$\left(\frac{d \ln T}{d \ln P} \right)_{env} < \left(\frac{d \ln T}{d \ln P} \right)_{ad} + \frac{\varphi}{\delta} \left(\frac{d \ln \mu}{d \ln P} \right)_{env}. \tag{6.22}$$

If the composition is uniform, the simpler Schwarzschild criterion is adequate

$$\left(\frac{d \ln T}{d \ln P} \right)_{env} < \left(\frac{d \ln T}{d \ln P} \right)_{ad}. \tag{6.23}$$

We are now in a position to answer the question of when the envelope of our growing giant planet will develop convection. All we need to do is to compute the radiative gradient ∇_{rad} (Eq. 6.15) that would exist *in the absence* of convection and compare it to the adiabatic gradient defined by

$$\nabla_{ad} \equiv \left(\frac{d \ln T}{d \ln P} \right)_{ad}. \tag{6.24}$$

If (assuming for simplicity uniform composition)

$$\nabla_{\text{rad}} < \nabla_{\text{ad}}, \tag{6.25}$$

then the envelope is stable to convection and the radiative temperature gradient defined by Eq. (6.15) is self-consistent. If, conversely

$$\nabla_{\text{rad}} > \nabla_{\text{ad}}, \tag{6.26}$$

then convection and bulk fluid motion will set in and at least some of the luminosity will be transported by convection. An approximate theory known as "mixing-length" theory exists which can be used to calculate *how much* of the luminosity is carried convectively along with the value of the actual temperature gradient in convectively unstable regions. The details of this theory can be found in Kippenhahn & Weigert (1990) or in any other stellar structure textbook. In many circumstances, however, it turns out that convection, once it is established, is extremely efficient at transporting energy – so efficient in fact that it almost succeeds in erasing the unstable temperature gradient that set up convection in the first place! Simply replacing the radiative gradient ∇_{rad} with the adiabatic gradient ∇_{ad} in convectively unstable zones is thus often a good approximation that suffices for constructing simple envelope models.

Equations (6.13), along with the appropriate expression for the temperature gradient (either ∇_{rad} or ∇_{ad}) form a coupled set of differential equations that describe the envelope. They must be supplemented by expressions for the equation of state (i.e. the functional relationship $P = P(\rho, T, \mu)$), Rosseland mean opacity and luminosity, and solved subject to appropriate boundary conditions. A simple case to consider is that in which the luminosity is entirely due to the collision of planetesimals with the core. In that case

$$L \simeq \frac{GM_{\text{core}}\dot{M}_{\text{core}}}{R_{\text{s}}}, \tag{6.27}$$

is a constant throughout the envelope. One inner boundary condition is obvious: at $r = R_{\text{s}}$ we require that $M = M_{\text{core}}$. The outer boundary conditions require slightly more thought. Physically we require that the envelope ought to match on smoothly to the disk at the "outer" radius of the planet, but how exactly should this outer radius be defined? There is no precise answer to this question (and indeed different authors use slightly different definitions) but there are two basic possibilities, the accretion radius

$$r_{\text{acc}} = \frac{GM_{\text{p}}}{c_{\text{s}}^2}, \tag{6.28}$$

which is a measure of the maximum distance at which gas in the disk with sound speed c_s will be bound to the planet, and the Hill sphere radius

$$r_{\rm H} = \left(\frac{M_{\rm p}}{3M_*}\right)^{1/3} a, \tag{6.29}$$

which is a measure of the distance beyond which shear in the Keplerian disk will unbind gas from the planetary envelope. For gas in the envelope to be bound to the planet, it must satisfy both $r < r_{\rm acc}$ and $r < r_{\rm H}$, so a logical choice for the location to specify the outer boundary conditions is at

$$r_{\rm out} = \min\left(r_{\rm acc}, r_{\rm H}\right). \tag{6.30}$$

At $r_{\rm out}$ we have that $M = M_{\rm p}$, $P = P_{\rm disk}$, and $T \simeq T_{\rm disk}$.[3] Given these boundary conditions the envelope structure can be computed numerically using standard methods developed for stellar structure calculations.

6.1.2 Critical core mass

An intuitive argument for the existence of a critical core mass follows from consideration of the simplest situation: a hydrostatic envelope surrounding a core that is *not* being bombarded by planetesimals. In this limit, elementary thermodynamics tells us that the envelope – given enough time – must eventually lose any excess heat left over from its accretion and cool down to match the temperature of the gas in the neighboring protoplanetary disk. The resulting density profile for a low mass envelope is exponential (Eq. 6.8) with the pressure that supports the outer envelope being furnished by the high density close to the surface of the core. Now imagine slowly increasing the core mass. As we have already demonstrated, this causes the *fraction* of the total mass that is contained in the envelope to increase. At least initially this more massive envelope can be supported against gravity by simply increasing the density (and hence pressure) near the core. Once $M_{\rm env} \sim M_{\rm core}$, however, this equilibrium ceases to exist, since for still more massive envelopes increasing the base density also significantly increases the mass and hence the gravitational force. For higher core masses no hydrostatic solution exists, since any possible boost to the pressure by adding yet more gas to the envelope fails to compensate for the additional mass.

Real envelopes of giant planets are not isothermal, but the order of magnitude conclusion that hydrostatic equilibrium is possible only for $M_{\rm env} \lesssim M_{\rm core}$ carries over to physical models. In general, solutions to the equations for envelope structure

---

[3] The approximate equality reflects the fact that the envelope temperature T will slightly exceed $T_{\rm disk}$ on account of the luminosity flowing through the planet (see e.g. Papaloizou & Terquem, 1999).

given in Section 6.1.1 fall into two classes depending upon the energy transport mechanism that dominates near the outer boundary of the planet (for a detailed discussion see Rafikov (2006) and references therein). Both classes admit the existence of a critical core mass. One class of solution is fully convective, with an envelope entropy that is set by the external boundary conditions imposed by the protoplanetary disk. The second class has a radiative layer separating the inner envelope (which is usually convective) from the disk. The radiative layer decouples the structure of the planet from the disk, and as a consequence the critical core mass of planets described by these solutions is almost independent of the pressure and temperature in the surrounding disk. The critical core mass depends instead upon the luminosity of the planet and upon the opacity within the radiative layer. It is this second class of solution that probably describes planets forming in the outer regions of the protoplanetary disk, and accordingly we will focus exclusively on models of this class when describing the predicted sequence of giant planet formation in Section 6.1.3.

Accurate models of the structure of giant protoplanets are by necessity numerical, and it is not possible to calculate accurate values of the critical core mass analytically. The rather simple explanation that we gave above for the *existence* of a critical core mass suggests, however, that this ought to be a generic feature of any gaseous envelope in hydrostatic equilibrium around a solid core. This intuition is correct, and in fact we can gain considerable additional insight into the critical core mass by considering a simple model of a massive protoplanet with a purely radiative envelope. This model, unlike the fully realistic models that have both convective and radiative zones, is amenable to an approximate analytic treatment (Stevenson, 1982).

The goal of Stevenson's (1982) analysis is to obtain an approximate analytic form for the density profile within a radiative envelope, which can then be integrated to give an expression for the envelope mass. We first seek a relation between the temperature and the pressure. Starting from the equation of hydrostatic equilibrium (Eq. 6.13) and the expression for the radiative temperature gradient (Eq. 6.14), we eliminate the density by dividing one equation by the other

$$\frac{\mathrm{d}T}{\mathrm{d}P} = \frac{3\kappa_R L}{64\pi\sigma GMT^3}.$$

(6.31)

We now integrate this equation inward from the outer boundary, making the approximation that $M(r) \approx M_p$ and taking L and κ_R to be constants

$$\int_{T_{\mathrm{disk}}}^{T} T^3\,\mathrm{d}T = \frac{3\kappa_R L}{64\pi\sigma GM_p} \int_{P_{\mathrm{disk}}}^{P} \mathrm{d}P.$$

(6.32)

Once we are well inside the planet we expect that $T^4 \gg T_{\text{disk}}^4$ and that $P \gg P_{\text{disk}}$, so the integral yields, approximately

$$T^4 \simeq \frac{3}{16\pi} \frac{\kappa_R L}{\sigma G M_p} P.$$

(6.33)

Substituting P in this equation with an ideal gas equation of state,

$$P = \frac{k_B}{\mu m_p} \rho T,$$

(6.34)

we eliminate T^3 in favor of the expression involving dT/dr from Eq. (6.14) and integrate once more with respect to radius to obtain

$$T \simeq \left(\frac{\mu m_p}{k_B}\right) \frac{G M_p}{4r},$$

(6.35)

$$\rho \simeq \frac{64\pi\sigma}{3\kappa_R L} \left(\frac{\mu m_p G M_p}{4 k_B}\right)^4 \frac{1}{r^3}.$$

(6.36)

Having derived the density profile, the mass of the envelope follows immediately

$$M_{\text{env}} = \int_{R_s}^{r_{\text{out}}} 4\pi r^2 \rho(r) dr$$

$$= \frac{256\pi^2\sigma}{3\kappa_R L} \left(\frac{\mu m_p G M_p}{4 k_B}\right)^4 \ln\left(\frac{r_{\text{out}}}{R_s}\right).$$

(6.37)

The right-hand-side of this equation has a strong dependence on the total planet mass M_p and a weaker dependence on the core mass M_{core} via the expression for the luminosity

$$L = \frac{G M_{\text{core}} \dot{M}_{\text{core}}}{R_s} \propto M_{\text{core}}^{2/3} \dot{M}_{\text{core}}.$$

(6.38)

In principle there are further dependencies to consider since r_{out} is a function of M_p and R_s is a function of M_{core}, but these enter only logarithmically and can be safely ignored. Noting that

$$M_{\text{core}} = M_p - M_{\text{env}},$$

(6.39)

we find that,

$$M_{\text{core}} = M_p - \left(\frac{C}{\kappa_R \dot{M}_{\text{core}}}\right) \frac{M_p^4}{M_{\text{core}}^{2/3}},$$

(6.40)

where we have shown explicitly the dependence on the envelope opacity and planetesimal accretion rate but have swept all the remaining constants (and near-constants) into a single constant C.

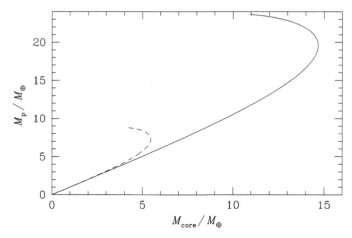

Fig. 6.4. The total planet mass M_p is plotted as a function of the core mass M_{core} using Stevenson's (1982) approximate analytic solution for a fully radiative envelope. For any given choice of envelope opacity and planetesimal accretion rate there is a maximum core mass beyond which no hydrostatic solution exists. The solid curve is plotted for typical values of the opacity and accretion rate, while the dashed curve shows the lower critical core mass that results from reducing $\kappa_R \dot{M}_{core}$ by a factor of ten.

Solutions to Eq. (6.40) are plotted in Fig. 6.4. For fixed values of the core accretion rate $\dot{M}_{core}$ and envelope opacity κ_R there is a maximum or critical core mass M_{crit} beyond which no solution is possible. The exact values of the critical core mass derived from this toy model should not be taken too seriously, but characteristic masses of the order of 10 $M_\oplus$ are obtained using plausible estimates of κ_R and $\dot{M}_{core}$ (Stevenson, 1982). One also observes that M_{crit} is a function (though not a very rapidly varying one) of the product $\kappa_R \dot{M}_{core}$ – reducing either the opacity or the accretion rate results in a lower value of the critical core mass. This property of the analytic model is also a feature of more complete numerical models of giant protoplanet structure.

6.1.3 Growth of giant planets

The equations describing hydrostatic envelope solutions can be readily modified to model the *time-dependent* growth of a giant planet up to the point where rapid accretion of the envelope begins. The basic assumption of time-dependent models is that the planet mass (and other time-variable quantities such as the core accretion rate) varies slowly enough that the envelope is always in hydrostatic equilibrium. Since hydrostatic equilibrium is established on a time scale that is of the order of the sound crossing time scale this is a very good approximation which permits us to treat the growth of the planet as a slowly changing sequence of hydrostatic models. The fact that the envelope mass changes with time means that we must modify Eq. (6.38) to include the luminosity that derives from contraction of the envelope, but apart from this the mathematical description of time-dependent models is identical to that of hydrostatic ones. Taking the enclosed mass M as the dependent variable (as is often convenient) Ikoma *et al.* (2000), for example, employ the following set of equations

$$\frac{\partial P}{\partial M} = -\frac{GM}{4\pi r^4}, \tag{6.41}$$

$$\frac{\partial r}{\partial M} = \frac{1}{4\pi r^2 \rho}, \tag{6.42}$$

$$\frac{\partial T}{\partial M} = \begin{cases} -\frac{3\kappa_R}{64\pi\sigma r^2 T^3}\frac{L}{4\pi r^2} & \text{if radiative,} \\ \left(\frac{\partial T}{\partial P}\right)_S \left(\frac{\partial P}{\partial M}\right) & \text{if convective,} \end{cases} \tag{6.43}$$

$$\frac{\partial L}{\partial M} = \epsilon_{\text{acc}} - T\frac{dS}{dT}, \tag{6.44}$$

where S is the specific entropy of the gas in the envelope and the energy released by infalling planetesimals is assumed to be liberated at the core/envelope interface

$$\epsilon_{\text{acc}} = \frac{\delta(r - R_s)}{4\pi r^2 \rho}\dot{M}_{\text{core}}\int_{R_s}^{r_{\text{out}}}\frac{GM}{r^2}dr. \tag{6.45}$$

As with the hydrostatic solutions, different authors adopt slight variations of these equations that differ, for example, in the assumed location of the outer boundary and in where within the envelope the energy of accreted planetesimals is released. These differences are rarely consequential. Of greater importance is the fact that the equations for the planetary structure must be supplemented with a model for how the planetesimal accretion rate $\dot{M}_{\text{core}}$ varies with time. The core accretion rate can be calculated using methods similar to those employed in the study of terrestrial planet growth (Section 5.2) provided that we allow for the fact that the planet mass must now include both the core and the envelope contributions, and that the cross-section for accretion may be modified by aerodynamic processes within the bound envelope. Any uncertainties in the magnitude and evolution with time of $\dot{M}_{\text{core}}$ (for example due to different assumptions as to the degree of gravitational focusing) propagate directly into the calculation of giant planet growth, and influence both the time scale and feasibility of planet formation at a particular location within the disk.

The first fully consistent time-dependent models of giant planet growth were published by Pollack *et al.* (1996), and their calculation – reproduced in slightly simplified form in Fig. 6.5 – remains a benchmark for all subsequent studies. For their baseline model Pollack *et al.* (1996) considered the formation of Jupiter

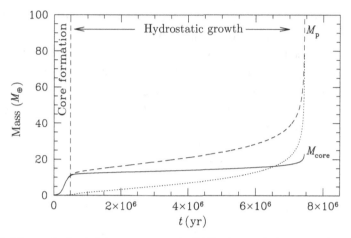

Fig. 6.5. The evolution of the core mass (solid line), envelope mass (dotted line) and total mass (dashed line) from a time-dependent calculation of giant planet formation via core accretion (Rice & Armitage, 2003). In this illustrative calculation, which is based on a slightly simplified version of the physics described in Pollack *et al.* (1996), a core grows at a fixed radius of 5.2 AU in a disk with a solid surface density $\Sigma_p = 10\,\mathrm{g\,cm^{-2}}$ and a gaseous surface density $\Sigma = 7 \times 10^2\,\mathrm{g\,cm^{-2}}$. With this choice of parameters and input physics one obtains a relatively short-lived phase of core formation that is followed by an extended period of slow coupled growth of the core and envelope. The critical core mass is exceeded and runaway growth starts after about 7–8 Myr, at which time the core mass is approximately $20\,M_{\oplus}$.

from a core at 5.2 AU in a disk with a solid surface density $\Sigma_p = 10\,\mathrm{g\,cm^{-2}}$ and a gaseous surface density $\Sigma = 7 \times 10^2\,\mathrm{g\,cm^{-2}}$. They adopted an opacity in the outer envelope that would be appropriate for a solar mixture of small grains that follow an interstellar size distribution. With these assumptions the growth of Jupiter proceeds through three well-separated phases:

- **Core formation**. Initially the core grows very rapidly due to runaway accretion of planetesimals within its feeding zone. The envelope mass remains small, and the growth of the total mass is dominated by the growth of the core. This phase is relatively brief ($\approx 0.5\,\mathrm{Myr}$) and comes to an end once the core approaches its isolation mass.
- **Hydrostatic growth**. This phase is characterized by the slow accretion of gas from the protoplanetary disk. The increasing mass of the planet drives a slow expansion of the feeding zone which suffices to maintain continued accretion of planetesimals, albeit at a lower rate than in the first phase. The hydrostatic phase continues for $\approx 7\,\mathrm{Myr}$ and ends only when the mass of the envelope starts to approach the mass of the core.
- **Runaway growth**. Once $M_{\mathrm{env}} \gtrsim M_{\mathrm{core}}$ the rate of accretion accelerates dramatically and runaway growth of the envelope sets in. The rate of accretion becomes limited by the rate at which the disk can supply gas, and is eventually shut off either by the onset of local tidal effects (to be discussed in more detail in Chapter 7) or by global disk dispersal.

As shown in Figure 6.5 this model yields a formation time scale for Jupiter of 7–8 Myr and a primordial core mass of $M_{core} \simeq 20\,M_{\oplus}$.

The Pollack *et al.* (1996) calculation is justly famous, but it is important to recognize that it represents only one illustrative realization of a very broad class of "core accretion" models, some of which yield substantially different formation time scales and core masses. We can gain considerable insight into the range of possibilities by considering the analytic fits to numerical models of hydrostatic planet envelopes calculated by Ikoma *et al.* (2000). For envelope opacities $\kappa_R \geq 10^{-2}\,\mathrm{cm^2\,g^{-1}}$ the critical core mass can be approximated by a power law in the planetesimal accretion rate $\dot{M}_{core}$ and envelope opacity κ_R

$$M_{crit} \sim 7 \left(\frac{\dot{M}_{core}}{10^{-7}\,M_{\oplus}\,\mathrm{yr^{-1}}} \right)^q \left(\frac{\kappa_R}{1\,\mathrm{cm^2\,g^{-1}}} \right)^s M_{\oplus}, \tag{6.46}$$

where the power-law indices q and s are both estimated to lie in the range 0.2–0.3. Ikoma *et al.* (2000) also calculated the time scale on which the envelope would grow in the complete absence of planetesimal accretion. Without ongoing core accretion the growth of the envelope is still limited by its ability to radiate away thermal energy as it contracts, which becomes more difficult for a lower mass core. Defining a growth time scale

$$\tau_{grow} \equiv \left(\frac{1}{M_{env}} \frac{dM_{env}}{dt} \right)^{-1}, \tag{6.47}$$

Ikoma *et al.* (2000) estimate that

$$\tau_{grow} \sim 10^8 \left(\frac{M_{core}}{M_{\oplus}} \right)^{-2.5} \left(\frac{\kappa_R}{1\,\mathrm{cm^2\,g^{-1}}} \right)\,\mathrm{yr}. \tag{6.48}$$

Several important conclusions follow from these results. First, as we could have anticipated from the analytic analysis in Section 6.1.2, the primordial core mass that results from core accretion can vary substantially between otherwise identical models that vary in their assumed planetesimal accretion rate or envelope opacity. Core masses substantially larger than the fiducial 20 $M_{\oplus}$ could be formed within planetesimal disks whose higher surface density allowed for larger values of $\dot{M}_{core}$. It is somewhat harder to tweak the model to yield lower core masses, because although reducing $\dot{M}_{core}$ has the effect of lowering M_{crit} it has the side effect of increasing the time ($\sim M_{crit}/\dot{M}_{core}$) required to attain the critical mass. A better way to produce a model with a lower core mass is to reduce the envelope opacity, for example by postulating either agglomeration or settling of dust so that the envelope opacity is much lower than the value calculated assuming that the grains have the same properties as interstellar dust. Second, the core mass is limited from below by the fact that envelopes surrounding low mass cores cool and contract rather slowly, and this means that we cannot form giant planets with arbitrarily low mass cores simply by postulating sharp termination of planetesimal accretion. Adopting $\kappa_R = 1\,\text{cm}^2\,\text{g}^{-1}$, for example, an envelope around a 4 $M_\oplus$ core is estimated to contract on the rather leisurely time scale of 3 Myr – too slow to plausibly result in the formation of a fully-fledged gas giant before the gas disk dissipates. As before, lower envelope opacities ameliorate this time scale problem, but do not change the basic conclusion that *extremely* low core masses (significantly below around 5 $M_\oplus$) are unlikely to result from core accretion.

Although we have discussed the range of outcomes of core accretion from the starting point of approximate analytic fits to hydrostatic models, the conclusions are supported by fully time-dependent calculations (representative examples include Hubickyj *et al.*, 2005 and Alibert *et al.*, 2005). These authors allow themselves the freedom to independently vary the two critical parameters that determine the outcome of core accretion, first by reducing the envelope opacity by up to two orders of magnitude from the values used by Pollack *et al.* (1996), and second by modifying the average rate and time dependence of the planetesimal accretion rate (either by assuming a sharp cut-off that might be caused physically by the sudden overlap of two growing planets' feeding zones, or by allowing the core to drift slowly through the disk). It is debatable to what extent these modifications are physically justified. There is a plausible theoretical basis (based on calculations of grain growth by Podolak, 2003) for substantially reducing the envelope opacity from its interstellar value, but the problem is complex and it is currently not feasible to attempt an accurate *ab initio* calculation of κ_R. The approach adopted by recent workers of treating this quantity as a free parameter is thus justified. The situation is rather different when it comes to including migration of the core through the disk into the calculation of the planetesimal accretion rate, since there is a well-developed theory (to be discussed in Chapter 7) that predicts the rate of such migration. Unfortunately, attempts to date to model giant planet formation with the theoretically predicted rate of migration invariably fail – the predicted drift is too fast to permit successful planet growth – and hence calculations that include migration are forced to treat its rate as yet another free parameter. There is little justification for such an approach, and the apparent inconsistency between theoretical calculations of planet migration and core accretion models is clearly a serious issue. Nonetheless, if one accepts the validity of lower envelope opacities and variable planetesimal accretion histories, it is possible to combine these two effects to yield a wide variety of outcomes, including cases in which planets form either much more rapidly than the fiducial 7–9 Myr (time scales as short as 1 Myr) or with substantially smaller initial core masses (as low as about 5 $M_\oplus$). As we will discuss in Section 6.3, changes in these directions may be essential if the core accretion model is to reproduce the observed properties of Jupiter.

6.2 Disk instability

The disk instability model for giant planet formation is predicated on the assumption that the gaseous protoplanetary disk was massive enough to be unstable to instabilities arising from its own self-gravity, and that the outcome of these instabilities is fragmentation into massive planets. The fundamental difference between this scenario and the core accretion model arises from the fact that in the disk instability model the solid component of the disk is a bystander which plays only an indirect role (via its contribution to the opacity) in the process of planet formation. Historically the earliest discussions of disk instability as a mechanism for planet formation predate any serious work on core accretion (Kuiper, 1951; Cameron, 1978). Despite this long history, computational methods have only relatively recently advanced to the point of being able to reliably assess the viability of the disk instability theory, and much of the recent work in the field is an offshoot of an influential numerical simulation by Boss (1997).

We have already derived the conditions needed for a protoplanetary gas disk to become unstable to its own self-gravity in Sections 3.4.4 and 4.6.1. Globally the disk mass must satisfy

$$\frac{M_{\rm disk}}{M_*} \gtrsim \frac{h}{r}, \tag{6.49}$$

while locally

$$Q \equiv \frac{c_{\rm s}\Omega}{\pi G \Sigma} < Q_{\rm crit}, \tag{6.50}$$

where $Q_{\rm crit}$ is a dimensionless measure of the threshold below which instability sets in. It lies in the range $1 < Q_{\rm crit} < 2$. If we compare these requirements to observational determinations of protoplanetary disk properties the global condition suggests that widespread gravitational instability (i.e. instability that extends across a large range of disk radii) must be limited to disks at the upper end of the observed range, with masses of around a tenth of the stellar mass. Such massive disks may be commonly present early in the evolution of pre-main-sequence stars (e.g. Eisner *et al.*, 2005). For a disk around a Solar mass star the local condition can be written in the form

$$\Sigma \gtrsim 3.8 \times 10^3 \left(\frac{Q_{\rm crit}}{1.5}\right)^{-1} \left(\frac{h/r}{0.05}\right) \left(\frac{r}{5\,{\rm AU}}\right)^{-2} {\rm g\,cm}^{-2}. \tag{6.51}$$

One immediately observes that the surface densities required for gravitational instability are large – more than an order of magnitude in excess of the minimum mass Solar Nebula value at 5 AU – but such high surface densities are neither observationally excluded nor unreasonable on theoretical grounds. High surface densities are likely at early epochs when the disk accretion rate is large, especially if angular momentum transport within the disk is rather inefficient (for example in the "layered disk" model discussed in Section 3.5.1), and these conditions provide the most fertile ground for the development of gravitational disk instabilities.

Assuming for the time being that gravitational instability results in fragmentation, we can estimate the masses of the objects that would be formed using an identical argument to that given for planetesimals forming out of unstable *particle* layers in Section 4.6.2. Noting that the most unstable scale in a gravitationally unstable disk is $\lambda \sim 2c_s^2/(G\Sigma)$, we expect that fragmentation will result in objects whose characteristic mass is of the order of $M_p \sim \pi \lambda^2 \Sigma$. For a disk around a Solar mass star with $(h/r) = 0.05$ and $Q_{crit} = 1.5$ this characteristic mass is independent of orbital radius and equal to $M_p \approx 8\ M_J$, where M_J is the mass of Jupiter. This estimate is evidently on the high side for Jupiter and for the majority of known extrasolar planets. That said, it does suffice to establish that disk instability could result in the formation of substellar objects (massive planets or brown dwarfs) and it is easy to imagine – given the crude nature of the estimate – that a more sophisticated calculation might yield objects that populate a large fraction of the mass spectrum of gas giant planets.

6.2.1 Fragmentation conditions

A disk will become gravitationally unstable if $Q < Q_{crit}$, but satisfying this condition is *not* sufficient to guarantee that the result of the instability will be fragmentation. The first linearly unstable modes in a gravitationally unstable disk are generally nonaxisymmetric ones, which develop into a pattern of spiral structure which is able to transport angular momentum outward via gravitational torques. The fact that gravitational instabilities within a disk transport angular momentum is critically important, because it implies that self-gravity is able to transport matter inward and thereby tap into the reservoir of free energy available to the system. Dissipation of the accretion energy, in turn, can then act to heat the disk and (by raising Q) mitigate the strength of the instability. The fact that such a stabilizing feedback loop can exist means that within a disk geometry gravitational instability has two qualitatively distinct outcomes:

- **Stable angular momentum transport**. Heating of the flow due to the dissipation of gravitational potential energy as gas flows inward balances radiative cooling in such a way that the disk attains a quasi-steady state. Gravitational instability never becomes violent enough to result in fragmentation.
- **Fragmentation**.

In most circumstances the additional parameter (beyond Q) that determines which of these outcomes actually occurs is related to the ability of the disk to radiate thermal energy. One can make a plausible argument for this choice by thinking about the two routes by which an initially stable ($Q > Q_{\text{crit}}$) disk might become unstable: an increase in the local surface density or alternatively a decrease in the local mid-plane temperature and sound speed. Assuming no peculiar circumstances (such as rapid mass infall), the characteristic time scale for changes in the surface density of a disk is just the viscous time scale (Eq. 3.12), which can be written with the aid of the Shakura–Sunyaev α-prescription (Eq. 3.46) as

$$t_\nu \simeq \frac{1}{\alpha\Omega}\left(\frac{h}{r}\right)^{-2}. \tag{6.52}$$

If we compare this to the local thermal time scale of the disk (the time scale on which the disk, if suddenly deprived of its heat source, would radiate away its thermal energy)

$$t_{\text{th}} \simeq \frac{1}{\alpha\Omega}, \tag{6.53}$$

(e.g. Pringle, 1981), we find that for protoplanetary disk conditions $t_{\text{th}} \ll t_\nu$. Cooling rather than radial redistribution of mass is then likely to determine the strength of gravitational instability, and we should seek a criterion for fragmentation in terms of the disk cooling time.

We can express the cooling criterion for fragmentation in two equivalent ways, which, taken together, expose most of the important physics. The essential argument is based upon the fact that if cooling is very rapid gravitational instabilities must be rather violent to generate heating that is sufficient to mitigate these losses. Quantitatively we can define a local cooling time scale via

$$t_{\text{cool}} \equiv \frac{U}{|dU/dt|}, \tag{6.54}$$

where U is the thermal energy content of the disk per unit area, and ask what is the equivalent value of α that would be required to generate enough offsetting heating. The answer can be shown analytically[4] (Gammie, 2001) to be

$$\alpha = \frac{4}{9\gamma(\gamma - 1)\Omega t_{\text{cool}}}, \tag{6.55}$$

where γ is the two-dimensional adiabatic index. The equivalent α (a measure of the strength of the "self-gravitating turbulence" within the disk) is thus inversely proportional to the cooling time. Noting that when $\alpha \sim 1$ the velocity perturbations within the disk become supersonic, we might guess that this would be

---

[4] There are various technical caveats to this rather remarkable result, of which the most important is that the derivation is local. Since gravitational torques can act over long ranges one might legitimately worry about this, but in practice subsequent work shows that the formally local result is approximately valid for most conditions of interest for protoplanetary disks.

when transient clumps would become dense enough to collapse and trigger fragmentation. This intuition is roughly correct. Numerical simulations by Gammie (2001) and by Rice *et al.* (2005) show that the critical value of the equivalent α that divides the regimes of stable angular momentum transport from fragmentation is

$$\alpha_{\text{crit}} \simeq 0.1. \tag{6.56}$$

A locally self-gravitating disk cannot sustain a larger α without fragmenting. An equivalent statement of the result is to note that for a particular equation of state there is a critical cooling time scale below which fragmentation occurs. For $\gamma = 2$,

$$t_{\text{cool, crit}} \simeq 3\Omega^{-1}. \tag{6.57}$$

The existence of the threshold in this form is also intuitively obvious – if the cooling time scale is shorter than the orbital period, clumps cool and contract on a time scale that is shorter than that on which collisions between clumps can occur and generate compensating heating (Shlosman & Begelman, 1989).

The disk mass above which a disk becomes gravitationally unstable (Eq. 6.49) is equal (up to numerical factors) to the mass above which self-gravity provides the dominant contribution to the vertical acceleration (Eq. 2.14). It follows immediately that *if* fragmentation occurs, the collapse of the disk into dense clumps takes place on the orbital time scale. This result is supported by numerical simulations. Figure 6.6 depicts the onset of fragmentation in a disk that has been set up (rather artificially) with a cooling time that is everywhere slightly shorter than the critical cooling time needed to allow fragmentation. Fragmentation occurs promptly as soon as the disk has cooled to the point where $Q \lesssim Q_{\text{crit}}$, forming a number of clumps which, if they can survive and contract further, could be progenitors of massive planets or brown dwarfs. Simulations of this kind are best regarded as idealized numerical experiments, since in a real disk it is more likely that only a limited range of radii would become unstable at any one time. Nonetheless the basic physical conclusion is correct – if fragmentation can occur at all, the time scale on which it forms giant planets is very short compared to any evolutionary time scale of the protoplanetary disk.

6.2.2 Disk cooling time scale

The fruit of our work up to this point is a restatement of the question "can giant planets form via disk instability?," which we have parlayed into the twin conditions $Q < Q_{\text{crit}}$ and $t_{\text{cool}} < t_{\text{cool, crit}}$. It remains to determine whether the second condition, in particular, is often or ever satisfied within protoplanetary disks, and if so at

Fig. 6.6. An image of the surface density of a simulated protoplanetary disk that is subject to gravitational instability (adapted from Rice *et al.* 2003a). The non-linear outcome of gravitational instability in disks is generally a nonaxisymmetric pattern of transient spiral arms. If the cooling time of the gas within the disk is short enough – as in this example – the disk fragments into dense clumps. Provided that these clumps are able to survive and contract further they may form substellar objects (massive planets or brown dwarfs).

what radii. This requires knowledge of the thermodynamics and energy transport processes at work within the disk, since physically it is these properties that specify the cooling time scale via U and dU/dt.

Given a particular disk model, the computation of the cooling time scale is straightforward. Suppose, for example, that the run of surface density is described by the marginally unstable profile of Eq. (6.51) with $Q_{crit} = 1.5$ and $(h/r) = 0.05$. Knowledge of (h/r) immediately specifies the radial profile of the sound speed c_s and central temperature T_c, and gives sufficient information to estimate (in a vertically averaged approximation) the thermal energy content per unit area $U = c_s^2 \Sigma / (\gamma - 1)$. If the disk is optically thick and energy is transported via radiative diffusion, the central temperature is related to the effective temperature at the disk surface T_{disk} by Eq. (3.38)

$$\frac{T_c^4}{T_{disk}^4} \sim \tau, \tag{6.58}$$

where τ is the optical depth to the disk mid-plane. With the effective temperature in hand we can then proceed to calculate the cooling rate ($2\sigma T_{disk}^4$) and cooling time scale as functions of radius, and compare the latter to the critical time scale needed for fragmentation. Following through with these steps one finds for the disk model specified by Eq. (6.51) that

$$t_{cool}\Omega \sim 10\tau \left(\frac{r}{5\,AU}\right)^{-1/2}. \tag{6.59}$$

The pre-factor in this equation appears to suggest that the prospects for fragmentation occurring are not too bad, *until* one realizes that in the inner disk the optical depth is typically very large. At 5 AU, for example, the temperature is around 100 K and the opacity is of the order of $\kappa_R \sim 1\,cm^2\,g^{-1}$. The optical depth to the mid-plane is then $\tau \sim 10^3$ or more, and the cooling time scale is many orders of magnitude longer than that needed for fragmentation.

The above analysis seems to suggest that fragmentation is inconceivable, but there are two loopholes to our argument. We first note that low temperature opacities fall dramatically as the temperature decreases, and this effect works together with the declining surface density to yield much more promising conditions at larger radii. If the disk continued out to 50 or 100 AU, the outer regions would be close enough to the threshold value of the cooling time as to demand a more careful calculation. Second, our failure to obtain conditions conducive to fragmentation at smaller radii occurs largely because of the factor of τ that arose because we assumed vertical energy transport via radiative diffusion. If alternatively energy was either dissipated close to the photosphere, or transported there by a process (such as convection) that was more efficient than radiation, we could narrow the disparity between T_c and T_{disk} and bring the cooling time within range of the critical value.

Our conclusions up to this point – that fragmentation cannot take place at moderate disk radii without very efficient vertical energy transport but may be feasible much further out – are basically correct. They were derived, however, by calculating the cooling time of a single disk model whose properties were specified in an ad hoc manner. Fortunately it is possible to eliminate this defect and deduce a robust limit on the disk conditions that could permit fragmentation *without* specifying (h/r) or Σ in advance. The argument, due to Rafikov (2005), relies on rewriting the two conditions needed for fragmentation in the form of upper and lower limits on the disk sound speed, and then solving for the minimum value of the surface density such that both limits can be satisfied simultaneously.[5] We begin by noting that for a disk to be gravitationally unstable at all we require that

$$c_s \lesssim \pi Q_{crit} \frac{G\Sigma}{\Omega}. \tag{6.60}$$

This gives an upper limit to the sound speed. Next, we make use of the fact that – irrespective of the energy transport mechanism or form of the opacity within the disk – the effective temperature of the radiation from the surface cannot be larger than the central temperature. This implies that the local cooling time cannot be less than

$$t_{cool} = \frac{U}{2\sigma T_c^4} = \frac{\Sigma c_s^2}{(\gamma - 1)} \frac{1}{2\sigma T_c^4}, \tag{6.61}$$

which we equate to the expression for the critical cooling time scale $t_{cool,\,crit} = \xi \Omega^{-1}$. This gives a lower limit on the sound speed

$$c_s \gtrsim \left[\left(\frac{k_B}{\mu m_p} \right)^4 \frac{\Sigma \Omega}{2\sigma \xi (\gamma - 1)} \right]^{1/6}. \tag{6.62}$$

This lower limit on the sound speed is almost independent of the surface density, while the upper limit derived previously increases linearly with Σ. A little thought (or a quick sketch) is enough to show that the two conditions cannot be simultaneously satisfied unless $\Sigma \gtrsim \Sigma_{crit}$, with the critical value of the surface density being given by

$$\Sigma_{crit} = (\pi G Q_{crit})^{-6/5} \left[\left(\frac{k_B}{\mu m_p} \right)^4 \frac{1}{2\sigma \xi (\gamma - 1)} \right]^{1/5} \Omega^{7/5}. \tag{6.63}$$

This expression is very simple – almost all the terms are constants with the stellar mass and orbital radius entering only via the combination Ω. Numerically, this result is not tremendously more constraining of the conditions needed for fragmentation than our earlier estimate. Adopting plausible values of the parameters ($\mu = 2.3$, $\gamma = 2, \xi = 3$, $Q_{crit} = 1.5$) for a disk around a Solar mass star the minimum surface density needed for fragmentation varies with radius as

$$\Sigma_{crit} \simeq 5 \times 10^3 \left(\frac{r}{5\,\text{AU}} \right)^{-21/10} \text{g\,cm}^{-2}. \tag{6.64}$$

A disk with a surface density that followed this profile across a wide range of radii (and which would therefore be vulnerable to fragmentation everywhere) would

---

[5] Although couched here in the context of protoplanetary disks, an essentially identical analysis applies to any self-gravitating disk system. Levin (2007), for example, has studied the conditions necessary for (and outcome of) fragmentation in disks around supermassive black holes.

undeniably be massive – perhaps unrealistically so. Integrating the above surface density between 0.1 AU and 30 AU, for example, gives a disk mass of $\approx 0.5 \ M_\odot$, which is large enough to be globally as well as locally unstable. A narrow range of disk radii could, however, attain Σ_{crit} without violating any obvious observational bounds on disk properties. What is more interesting is that this derivation yields an absolute lower bound on the surface density of a disk that might fragment, independent of any significant assumptions as to the disk structure or energy transport mechanism. As previously, much stronger limits – strong enough essentially to rule out fragmentation as the origin of giant planets except at very large radii – can be deduced if one assumes that the disk is radiative (Rafikov, 2005).

Let us summarize the conclusions to be drawn from these analytic considerations. Simultaneously satisfying the conditions that the disk be both gravitationally unstable (low Q) *and* vulnerable to fragmentation (short t_{cool}) is difficult. Much the most likely site for successful formation of substellar objects via disk instability is the outer disk – perhaps at radii of 50 to 100 AU – where fragmentation is a plausible outcome of gravitational instability provided that the disk is massive enough at such large distances from the star. Any planets formed via this channel would almost certainly populate the upper end of the planetary mass function. At smaller radii of 10 AU and less, fragmentation is not possible if the disk cools via radiative diffusion of energy from the mid-plane to the photosphere, but the disk could approach the fragmentation boundary if the efficiency of cooling is almost as great as that allowed by the thermodynamic requirement that the mid-plane should be hotter than the photosphere. The answer to the original question of whether a planet such as Jupiter can form via disk instability then rests, ultimately, on the highly technical question of how efficiently energy is transported within the protoplanetary disk, and here one might hope for an answer from numerical simulations. Current simulations, unfortunately, yield only a confused picture, with different groups finding variously rapid cooling and consequent fragmentation or somewhat slower cooling and an absence of fragmentation. Further work is needed to elucidate the origin of these disparate results and determine firm bounds on the regions of the disk that are able to fragment.

6.3 Comparison with observations

Purely theoretical considerations do not give an unambiguous answer as to whether giant planets can form from disk instability, and they also fail to specify which of the many possible variants of core accretion is most commonly realized in real systems. What guidance can observations provide? It is worth noting at the outset that although the core accretion and disk instability models share the same relationship to each other as the two competing models for planetesimal formation

(Chapter 4), prospects for an observational determination of the giant planet formation mechanism are markedly better. Whereas both planetesimal formation theories end up producing very similar objects, core accretion and disk instability make distinct predictions for the initial structure and formation epoch of giant planets that are in principle observable. Indeed, over the years, many different observations have been interpreted as giving support to one or the other model. Here, we classify (somewhat subjectively) the various possible constraints in descending order of robustness as either "direct," "strong," or "model-dependent."

Direct constraints are observations that are sensitive to the most fundamental distinctions between disk instability and core accretion, and which can be interpreted with the minimum of reliance on theoretical models. Probably the most direct observational constraint of all would be a measurement of when and where giant planets form. This observation is not possible now but is in principle feasible via indirect detection of massive planets around young disk-bearing stars or direct imaging of planets orbiting within protoplanetary disks in nearby star forming regions. Disk instability forms giant planets very rapidly within disks that are massive enough to be gravitationally unstable and which can cool efficiently. The requirement that the disk be massive fixes the predicted epoch of planet formation to be early – possibly even prior to the optically visible Class II phase of Young Stellar Object evolution – while the need for rapid cooling is almost certainly easiest to meet at large radii. The generic predictions of core accretion models are exactly the opposite. A significant envelope cannot be accreted until we have assembled a core of several Earth masses. The time scale for core assembly tends to increase with orbital radius, and is not negligible (compared to the disk lifetime) even in the inner disk. Giant planet formation via core accretion is thus a mid- or late-stage event in the lifetime of protoplanetary disks (though whether it typically occurs after 1 Myr or several Myr is less certain). It is likely to occur in the inner disk beyond the snowline and be much more difficult beyond some maximum radius that may be of the order of 10–20 AU.

Strong constraints are derived from observations that probe fundamental distinctions between disk instability and core accretion, but which cannot be interpreted without the aid of additional theory. Much the most important of these constraints are the bounds placed on the core masses of Jupiter and Saturn by comparing theoretical models of the planets' internal structures to measurements of their gravitational fields. The principle of the method relies upon the fact that rotating giant planets[6] are axially rather than spherically symmetric bodies. The gravitational field *outside* the planet then depends not just on the planet mass and distance (as is the case for spherical bodies), but also on the distribution of mass *within* the planet.

---

[6] Jupiter has a rotation period of approximately 10 hours.

This connection between the external potential and the internal structure can be formalized by expanding the potential of a rotating planet in Legendre polynomials (e.g. Guillot *et al.*, 2004)

$$\Phi(r, \theta) = \frac{GM_{\mathrm{p}}}{r} \left[1 - \sum_{i=1}^{\infty} \left(\frac{R_{\mathrm{eq}}}{r} \right)^{2i} J_{2i} P_{2i}(\cos \theta) \right]. \qquad (6.65)$$

Here R_{eq} is the planet's equatorial radius, θ is the angular distance measured from the pole, and the P_{2i} are Legendre polynomials. The J_{2i} are called the gravitational moments, and these can be measured by modeling the trajectories of spacecraft that make flybys or enter orbit about the planet. In practice the series converges rapidly and only the first few moments are required to yield an accurate description of the gravitational field. For Jupiter the zeroth moment (the total mass) is measured to a precision of 10^{-6}, J_2 to a precision of 10^{-4}, and J_4 to about one percent (Guillot *et al.*, 2004). Note that J_6 cannot be measured with high significance.

To make use of the measured moments to constrain the internal structure, we note that they can be written as a volume integral over the planet's density distribution (e.g. Zharkov & Trubitsyn, 1974)

$$J_{2i} = -\frac{1}{M_{\mathrm{p}} R_{\mathrm{eq}}^{2i}} \int \rho r^{2i} P_{2i}(\cos \theta) \mathrm{d}V. \qquad (6.66)$$

A set of measured moments then yields multiple integral constraints on the density distribution that must be satisfied by any physical planet model. The weighting function inside the integral means that the higher order moments are sensitive primarily to structure near the surface of the planet, and accordingly the observational difficulty in measuring J_6 and above does not handicap efforts to deduce the core mass as much as one might naively suspect.

The limited number of observational constraints – primarily the total mass, radius, and surface rotation rate, together with two higher order gravitational moments and knowledge of the atmospheric composition – force us to exercise careful judgement in determining what type of theoretical model to compare the data against. There is no hope of measuring the core mass within highly flexible models that have many free parameters that must be empirically estimated, but equally a model with too *few* degrees of freedom may be overconstrained and lead to unrealistically precise measures.

Figures 6.7 and 6.8 show joint constraints on the core mass M_{core} and on the total mass of heavy elements within the envelope M_Z for Jupiter and Saturn respectively (Saumon & Guillot, 2004; Guillot, 2005). The particular model whose interior structure is constrained in these calculations assumes that these planets have three layers: an inner solid core, an inner envelope of hydrogen and helium with a protosolar helium fraction, and an outer envelope whose composition matches that

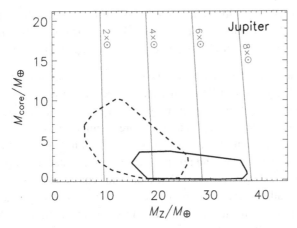

Fig. 6.7. Joint constraints on Jupiter's core mass M_{core} and on the total mass of heavy elements within the envelope M_Z (Saumon & Guillot, 2004). The assumed interior model has three layers: an outer hydrogen–helium envelope, an inner hydrogen–helium envelope (within which there is a different helium abundance), and a central core. Elements heavier than helium are assumed to be uniformly mixed throughout the envelope. The solid and dashed contours are computed for different equations of state, and encircle the range of parameters that match all available observational constraints. The straight lines indicate the values of M_Z that correspond to heavy element enrichment in the envelope by specified factors of the Solar value. Reproduced from Guillot (2005), with permission.

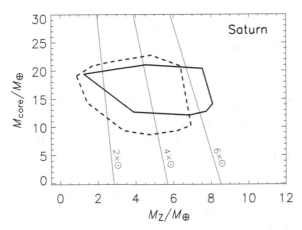

Fig. 6.8. As Fig. 6.7, but for Saturn. The lower pressures in the interior of Saturn greatly reduce the systematic uncertainties that arise in the case of Jupiter from different treatments of the high pressure equation of state. Reproduced from Guillot (2005), with permission.

of the deep atmosphere. Although one can certainly imagine other models for the internal structure the results obtained by Guillot (2005) provide a fair estimate of the current range of uncertainty and reveal several important features:

- Accurate knowledge of the equation of state in the deep interior of Jupiter is of paramount importance. A core mass of 10 $M_\oplus$ is just a few percent of the total mass of Jupiter, so it is essential to be able to relate pressure (in the equation of hydrostatic equilibrium) to density with percent-level or better precision. There are formidable challenges to determining the high pressure equation of state of hydrogen–helium mixtures either experimentally or from *ab initio* calculations, and this is reflected in the systematic differences between constraints computed with different (but plausible) equations of state. This source of uncertainty is smaller for Saturn, whose interior conditions are less extreme.
- The present core mass of Jupiter is probably less than about 10 $M_\oplus$. Models of Jupiter that have no core at all cannot be excluded.
- Saturn almost certainly does have a core with a mass of around 15 $M_\oplus$.
- The total heavy element abundances of Jupiter and Saturn are poorly constrained by the data.

The best near-term prospect for improvement to these results lies in the possibility of substantially reducing the uncertainties in the equation of state via new *ab initio* theoretical calculations, which have recently been completed and applied to planetary structure by two groups (Nettelmann *et al.*, 2008; Militzer *et al.*, 2008). The initial applications of these equations of state to Jovian models yield incompatible results (Nettelmann *et al.*, 2008 constrain the core mass to be in the range 0–7 $M_\oplus$, whereas Militzer *et al.*, 2008 find a core of 14–18 $M_\oplus$), most probably due to the different assumptions made by the modelers as to the class of model to constrain. New data (perhaps from a planned Jupiter orbiter) may therefore be required as well in order definitively to reduce the range of allowed parameter space in Fig. 6.7.

Even with current uncertainties, the constraints on the core masses of Jupiter and Saturn yield important information. Before drawing inferences about what the measurements imply for the formation mechanism, however, we must first digress to ask whether the core mass of a giant planet is necessarily constant over time. This depends, first, on whether the material of the core is soluble in the planetary envelope at the densities and temperatures experienced in the interior. If the core material is insoluble then the mass of the core today (after several billion years of evolution) is at least as large as the mass of the primordial core, and the constraints derived for Jupiter represent firm upper limits to the core mass at the time of formation. Matters are more complicated if the materials making up the core and envelope are mutually soluble (as is probably the case, Stevenson, 1982; Lissauer & Stevenson, 2007). In this case, as illustrated in Fig. 6.9, the conceivable evolutionary sequences include full or partial erosion of a primordial core, or a more complicated

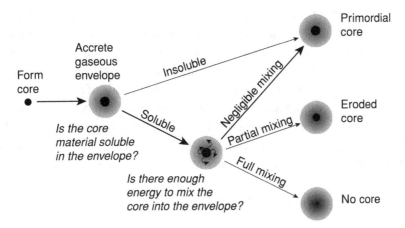

Fig. 6.9. Illustration of the different evolutionary paths that can be conceived for the primordial solid core present in giant planets that form via core accretion (after Stevenson, see e.g. Lissauer & Stevenson, 2007, for the physical arguments). Under conditions where the core material is soluble in the envelope the mass of the core at late times is only a reliable guide to the primordial mass *if* processes that would mix the core into the envelope can be demonstrated to be inefficient.

internal structure in which the boundary between core and envelope is not well-defined. Simple energetic arguments (e.g. Lissauer & Stevenson, 2007) suggest that convection in the interior of Jupiter over the planet's lifetime has probably not been powerful enough to efficiently mix much of the core into the envelope, so it is a reasonable working hypothesis to assume that constraints on the current core mass do have relevance to the problem of Jupiter's formation. Adopting this viewpoint, the unambiguous presence of a core within Saturn – at roughly the mass predicted by first-generation core accretion models (Pollack *et al.*, 1996) – yields considerable credence to the core accretion theory. Any conclusions that can be drawn for Jupiter are less definitive. If the Jovian core mass lies toward the upper end of the allowed range (about 10 $M_\oplus$ as shown in Fig. 6.7) that would be consistent with models that lie within the generalized core accretion family, though it could not be counted as a successful prediction of the first-generation models that suggested higher core masses.[7] A very low core mass, on the other hand – below about 5 $M_\oplus$ – would be consistent with the simplest predictions of disk instability and would pose a significant challenge for core accretion.

Finally for this section, we note that there are also a number of *model-dependent* constraints on the giant planet formation mechanism whose usefulness hinges on the uncertain question of whether the models used to derive them are unique. One

---

[7] An exception would be if the core mass of 14–18 $M_\oplus$ derived by Militzer *et al.* (2008) is confirmed by subsequent studies.

of these concerns the formation time scale of the Solar System's ice giants, Uranus and Neptune. If we assume that these planets formed at their current distances from the Sun, and that their (rather small) envelopes are gas that was captured from the protoplanetary disk, then a necessary condition for their formation is that it was possible to grow a core of around 10 $M_\oplus$ within a time scale of a few million years. This is difficult. The simple formula for the growth rate of a body of radius R_s within a planetesimal disk of surface density Σ_p (Eq. 5.28), when scaled to conditions appropriate to Neptune's location ($\Omega = 1.2 \times 10^{-9}\,\mathrm{s}^{-1}$, $\Sigma_p = 1\,\mathrm{g\,cm}^{-2}$), allows us to estimate the growth time scale for an icy body with a mass of 1 $M_\oplus$ as

$$t_{\mathrm{grow}} = \frac{M_p}{\mathrm{d}M_p/\mathrm{d}t} \approx 5 \times 10^{10} F_g^{-1}\ \mathrm{yr}, \tag{6.67}$$

where F_g is the gravitational focusing factor. We require $F_g \gtrsim 10^4$ in order to form Neptune in a reasonable time, and although this is not physically impossible, more typical estimates of the formation time scale are of the order of 100 Myr – substantially too long. On the face of it this time scale problem represents a strong constraint on core accretion (in this case favoring the subset of models that feature efficient damping of planetesimal velocities and consequently large values of F_g), but the constraint vanishes if we drop the assumption that Uranus and Neptune formed *in situ*. In fact, as we will discuss further in Chapter 7, there is abundant evidence that Neptune at least migrated outward during the early history of the Solar System, and it is quite possible that both ice giants formed at substantially smaller radii. If this is the case then not only is there no time scale problem, but (unless there is some way of divining the *actual* formation sites) there is also little hope of using the observed properties of Uranus and Neptune to constrain core accretion models.

Model-dependent constraints may fare better for extrasolar planetary systems, whose large numbers allow for a statistical comparison between theory and observations. It is clear that the observed correlation between planet frequency and host star metallicity (Fischer & Valenti, 2005) is *qualitatively* consistent with core accretion, since a disk enriched with heavy elements should yield a more massive planetesimal disk within which cores will grow faster, and simplified statistical models of giant planet formation via core accretion within an evolving disk provide a reasonable quantitative match (Ida & Lin, 2004). If this is interpreted as evidence that the bulk of presently known extrasolar planets are the product of core accretion, the question persists as to whether any of these planets instead formed via disk instability. Since disk instability is likely to yield very massive planets this question may also be answerable statistically given a large enough sample, by searching for a distinct metallicity distribution among those stars bearing the most massive planets (Rice *et al.*, 2003b).

6.4 **Further reading**

"Formation of giant planets," by Jack Lissauer & David Stevenson, 2007, in *Protostars and Planets V*, B. Reipurth, D. Jewitt, & K. Keil (eds.), Tucson: University of Arizona Press p. 591.

"Gravitational instabilities in gaseous protoplanetary disks and implications for giant planet formation," by Richard Durisen *et al.*, 2007, in *Protostars and Planets V*, B. Reipurth, D. Jewitt, & K. Keil (eds.), Tucson: University of Arizona Press p. 607.

"The interiors of giant planets: Models and outstanding questions," by Tristan Guillot, 2005, *Annual Review of Earth and Planetary Sciences*, **33**, 493.

7

Early evolution of planetary systems

The classical theory of giant planet formation described in the preceding chapter predicts that massive planets ought to form on approximately circular orbits, with a strong preference for formation in the outer disk at a few AU or beyond. Most currently known extrasolar planets have orbits that are grossly inconsistent with these predictions and, irrespective of the still open question of what the *typical* planetary system looks like, their existence demands an explanation. Even within the Solar System the existence of a large resonant population of Kuiper Belt Objects, and the time scale problem for the formation of Uranus and Neptune, suggest that the classical theory is at best incomplete.

In this chapter we describe a set of physical mechanisms – gas disk migration, planetesimal scattering, and planet–planet scattering – that promise to reconcile the observed properties of extrasolar planetary systems with theory. The common feature of all of these mechanisms is that they result in energy and angular momentum exchange either among newly formed planets, or between planets and leftover solid or gaseous debris in the system. The exchange of energy and angular momentum drives evolution of the planetary semi-major axis and eccentricity, which can be substantial enough to make the final architecture of the system unrecognizable from its state immediately after planet formation.

When discussing planetary system evolution two cautions are mandatory. First, although the foundations for the ideas we will study are old (very old in the case of planetary system stability, which is one of the classic problems of mathematical physics) much of their detailed development has occurred in response to observational discoveries that are not much more than a decade old. The references in this chapter are primarily to recent work, and in some cases that work is too new for a solid theoretical consensus to have developed. When it comes to the *application* of the theory to observations, these theoretical uncertainties are compounded by what is often sparse (although rapidly improving) data. Thus, although individually these processes can provide a reasonably satisfactory explanation for several otherwise mysterious properties of planetary systems, there is currently no unambiguous evidence as to which evolutionary processes are *always* important and which play a role in only a fraction of systems. Additional observations and modeling are needed to determine their relative importance.

7.1 Migration in gaseous disks

Gas disk migration refers to a change in the semi-major axis of a planet that is caused by exchange of angular momentum between the planet and the surrounding gaseous protoplanetary disk. The exchange of angular momentum is mediated by gravitational torques between the planet and the disk. No torque is exerted on a planet by an axisymmetric disk, so gas disk migration can only take place if the planet (or some other process, such as turbulence) excites nonaxisymmetric structure. In addition to angular momentum, energy is also exchanged between the planet and the disk, and depending upon the details of these exchanges the net result may be changes not just in a but also in e and i.

Migration is potentially important whenever a fully-formed planet (very roughly of the mass of Mars or above – smaller bodies do not interact strongly with the disk) co-exists with a gaseous disk. Such co-existence is inevitable for newly formed gas giants and for the cores of giant planets forming via core accretion, and the bulk of studies investigating gas disk migration focus on these stages of planet formation. In the conventional scenario for terrestrial planet formation in the Solar System, on the other hand, the gas disk is assumed to have dissipated *prior* to the final assembly of the terrestrial planets, and if this is the case then fully-formed planets never co-exist with a significant gas disk. Accordingly migration is often ignored in studies of terrestrial planet formation. One cannot easily exclude, however, the possibility that in some systems terrestrial planet formation occurs more rapidly, in which case gas disk migration may occur.

The detailed physics of the gravitational interaction between a planet and a surrounding gas disk is subtle, and several important aspects remain poorly understood. To gain some insight into the interaction it is useful to begin by considering a gaseous analog of the two-body dynamics discussed in the context of terrestrial planet formation in Section 5.3.1. We assume that a particle in the gas disk is initially on an unperturbed circular orbit, and calculate the angular momentum change that occurs as the particle is impulsively deflected during a close approach to the planet. The deflection is calculated as if the particle in the gas disk were a freely moving test particle, and hydrodynamics is ignored entirely except for the implicit assumption that the disk is able to "smooth out" the trajectories so that particles resume unperturbed orbits prior to making their next encounter with the planet. Despite these simplifications, calculation of the planet–disk interaction via the so-called *impulse approximation* (Lin & Papaloizou, 1979) yields correct scalings and better than order of magnitude estimates of the rate of angular momentum transport.

Working in a frame of reference moving with the planet, we consider the gravitational interaction between the planet and gas flowing past with relative velocity Δv and impact parameter b. We have already derived (in the two-body, free particle limit) the change to the perpendicular velocity that occurs during the encounter. It is (Eq. 5.58)

$$|\delta v_\perp| = \frac{2GM_p}{b\Delta v},$$ (7.1)

where M_p is the planet mass. This velocity is directed radially, and hence does not correspond to any angular momentum change. The interaction in the two-body problem is however conservative, so the increase in the perpendicular velocity implies a reduction (in this frame) of the parallel component. Equating the kinetic energy of the gas particle well before and well after the interaction has taken place we have that

$$\Delta v^2 = |\delta v_\perp|^2 + (\Delta v - \delta v_\parallel)^2,$$ (7.2)

which implies (for small deflection angles)

$$\delta v_\parallel \simeq \frac{1}{2\Delta v}\left(\frac{2GM_p}{b\Delta v}\right)^2.$$ (7.3)

If the planet has a semi-major axis a the implied angular momentum change per unit mass of the gas is

$$\Delta j = \frac{2G^2 M_p^2 a}{b^2 \Delta v^3}.$$ (7.4)

It is worth pausing at this juncture to fix the *sign* of the angular momentum change experienced by the gas and by the planet firmly in our minds. Gas exterior to the planet's orbit orbits the star more slowly than the planet, and is therefore "overtaken" by the planet. The decrease in the parallel component of the relative velocity of the gas therefore corresponds to an *increase* in the angular momentum of the gas exterior to the planet. Since the gravitational torque must be equal and opposite for the planet the sign is such that:

- Interaction between the planet and gas exterior to the orbit increases the angular momentum of the gas, and decreases the angular momentum of the planet. The planet will tend to migrate inward, and the gas will be repelled from the planet.
- Interaction with gas interior to the orbit decreases the angular momentum of the gas and increases that of the planet. The interior gas is also repelled, but the planet tends to migrate outward.

In the common circumstance where there is gas both interior and exterior to the orbit of the planet, the net torque (and sense of migration) will evidently depend upon which of the above effects dominates.

The total torque on the planet due to its gravitational interaction with the disk can be estimated by integrating the single particle torque over all the gas in the disk. For an annulus close to but exterior to the planet, the mass in the disk between b and $b + db$ is

$$dm \approx 2\pi a \Sigma db, \tag{7.5}$$

where Σ (assumed to be a constant) is some characteristic value of the gas surface density. If the gas in the annulus has angular velocity Ω and the planet has angular velocity Ω_p, all of the gas within the annulus will encounter the planet in a time interval

$$\Delta t = \frac{2\pi}{|\Omega - \Omega_p|}. \tag{7.6}$$

Approximating $|\Omega - \Omega_p|$ as

$$|\Omega - \Omega_p| \simeq \frac{3\Omega_p}{2a} b, \tag{7.7}$$

which is valid provided that $b \ll a$, we obtain the total torque on the planet due to its interaction with gas outside the orbit by multiplying Δj by dm, dividing by Δt, and integrating over impact parameters. Formally we would have that

$$\frac{dJ}{dt} = -\int_0^\infty \frac{8G^2 M_p^2 a \Sigma}{9\Omega_p^2} \frac{db}{b^4}, \tag{7.8}$$

but this integral is clearly divergent – there is what must be an unphysically infinite contribution from gas passing very close to the planet. Sidestepping this problem for now, we replace the lower limit with a minimum impact parameter b_{min} and integrate. The result is

$$\frac{dJ}{dt} = -\frac{8}{27} \frac{G^2 M_p^2 a \Sigma}{\Omega_p^2 b_{min}^3}. \tag{7.9}$$

It is possible to tidy up this calculation somewhat, for example by taking proper account of the rotation of the planet frame around the star, and if this is done the result is that the expression derived above must be multiplied by a correction factor (see e.g. Papaloizou & Terquem, 2006, and references therein).

Aside from the sign of the torque, the most important result that we can glean from this calculation is that the torque on the planet due to its interaction with the gas scales as the *square* of the planet mass. This scaling can be compared to the orbital angular momentum of the planet, which is of course linear in the planet mass. We conclude that if all other factors are equal – in particular if neither Σ in the vicinity of the planet nor b_{min} varies with M_p – the time scale for the planet to change its orbital angular momentum significantly will scale inversely with the planet mass. The migration velocity in this limit will then be proportional to M_p – more massive planets will migrate faster.

Finally, we can estimate the magnitude of the torque for parameters appropriate to different stages of giant planet formation. First, let us consider a rather low mass core ($M_p = M_\oplus$) growing at 5 AU in a gas disk of surface density $\Sigma = 10^2$ g cm^{-2} around a Solar mass star. Our treatment of the interaction has assumed that the disk can be treated as a two-dimensional sheet, so it is arguably natural to take as a minimum impact parameter $b_{min} = h \approx 0.05a$. Using these numbers we find that the exterior torque would drive inward migration on a time scale

$$\tau = \frac{J}{|dJ/dt|} \sim 1 \text{ Myr.} \qquad (7.10)$$

Of course this will be partly offset by the interior torque – which has the opposite sign – but unless there is some physical reason why these two torques should cancel to high precision, we would predict changes to the semi-major axis on a time scale of the order of a Myr. This is already rapid enough to be a potentially important effect during giant planet formation via core accretion. Second, we can evaluate the torque for a fully-formed gas giant. A Jupiter mass planet has a Hill sphere $r_H > h$, so it seems reasonable to argue that the value of b_{min} that we adopted for an Earth mass core is too small in this case. Picking a modestly larger value $b_{min} = 0.2a$ results in a time scale[1]

$$\tau \sim 2 \times 10^5 \text{ yr,} \qquad (7.11)$$

that is an order of magnitude shorter than typical protoplanetary disk lifetimes. If these estimates can be trusted to even an order of magnitude the conclusion is inescapable – gas disk migration ought to be an important process for the early evolution of the orbits of giant planets.

7.1.1 Resonant torques

The intuition that the impulse approximation gives for the planet–disk interaction is correct and valuable, but the details of the calculation will undoubtedly have left any mathematically inclined readers with a sour taste. Rather than try to patch up

---

[1] Physically the need to pick a larger value of b_{min} for a massive planet comes about precisely because the interaction close to the planet is strong – strong enough in fact to repel all the nearby gas so that there is nothing left for the planet to interact with. We will quantify this effect later, but for now the precise value of b_{min} can either be considered as an arbitrary but plausible guess, or as having been chosen with the bogus wisdom that comes from knowing the "right" answer so as to yield sensible estimates.

the impulse approximation, most recent analyses of the interaction take off from a different starting point: the equations for the evolution of linear perturbations within a fluid disk (akin to the calculation of disk stability developed in Section 4.6.1, though there, ironically, we ultimately applied the gas disk calculation to the stability of a *particle* disk). This approach, which traces a lineage back to theoretical studies of the stability of galactic disks, was first applied to the closely related problems of planet–disk and satellite–planetary ring interactions by Goldreich & Tremaine (1979, 1980). In outline, the calculation has two steps. First, the perturbation to the gravitational potential caused by the planet is decomposed into Fourier modes that vary azimuthally as $\exp[im(\phi - \Omega_p t)]$, where m is an azimuthal wavenumber. Second, the response of the gas disk to the perturbations is calculated (normally using the linearized hydrodynamic equations), from which can be derived the torque.

Actually carrying out this calculation is a substantial endeavor, which can be found in the original papers (Goldreich & Tremaine, 1979, 1980) and in subsequent work that has refined and improved the method (an incomplete list of important references includes Artymowicz, 1993a, 1993b; Ward, 1986, 1997; Korycansky & Pollack, 1993; Tanaka *et al.*, 2002). Here we will limit ourselves to a statement of the key results, of which the most fundamental is the demonstration that the angular momentum exchange can be expressed as the sum of torques that are exerted at discrete resonant locations within the gas disk. These resonances correspond to the locations within the disk at which the perturbation due to the planet excites waves, but the torque is largely independent of the details of how those waves propagate and dissipate within the fluid.

To go further, we need first to distinguish between the different types of important resonance and calculate where they fall within the disk. In general a resonance occurs when a characteristic frequency of the planet matches a frequency within the disk. The simplest case to consider is that where the planet has a circular orbit with angular frequency Ω_p and the gas disk has an orbital frequency $\Omega(r)$. The condition for a **co-rotation** resonance is simply that these frequencies match,

$$\Omega(r) = \Omega_p. \tag{7.12}$$

If we ignore (for the moment at least) the small changes in $\Omega(r)$ from its Keplerian value $\Omega(r) = \sqrt{GM_*/r^3}$ caused by radial pressure gradients or disk self-gravity, the co-rotation resonance is co-orbital with the location of the planet. The condition for a **Lindblad** resonance[2] is similar, except that this resonance occurs when gas

---

[2] The terminology used to classify these resonances is borrowed from the theory of galactic structure. Lindblad resonances are named after the Swedish astronomer Bertil Lindblad (1895–1965), who (simultaneously with Jan Oort) did much to advance our understanding of the rotation of the Milky Way.

in the disk is excited at its natural frequency for radial (or epicyclic) oscillations $\kappa(r)$,

$$m[\Omega(r) - \Omega_p] = \pm \kappa(r), \tag{7.13}$$

where m is an integer. For a Keplerian disk the epicyclic and orbital frequencies are identical and the nominal locations of Lindblad resonances are given by

$$r_L = \left(1 \pm \frac{1}{m}\right)^{2/3} a, \tag{7.14}$$

where a is the semi-major axis of the planetary orbit. In summary, for a planet on a circular orbit embedded within a Keplerian gas disk there is just one co-orbital co-rotation resonance together with a "comb" of Lindblad resonances that become closely spaced near to the location of the planet.

This census of resonances becomes more involved if the planet has an eccentric orbit, since in this case we cannot define a single rotating frame within which the perturbing potential remains fixed. Instead, we must decompose the potential due to the planet into a set of rigidly rotating components that have different pattern speeds. Defining Ω_p now to be the *mean* angular velocity of the planet, these components have frequencies

$$\Omega_p^{l,m} = \Omega_p + \frac{(l-m)}{m}\Omega_p, \tag{7.15}$$

where l is a second integer. The condition for resonance is then that *any* of the $\Omega_p^{l,m}$ match one of the natural frequencies of the gas disk (i.e. Ω_p in the above expressions is replaced by $\Omega_p^{l,m}$). Although in principle this leads to an enormous proliferation of resonances, it is often the case that only a modest number of the additional resonances prove to be important. In particular, if the planet eccentricity is small the amplitude of the perturbation scales as the $|l-m|$th power of the eccentricity. In this limit – which would be appropriate for example if we wish to study whether the eccentricity of an initially circular planet can be excited by resonant disk interactions – only the principal component (Ω_p, the mean motion) and the first-order components with $|l-m| = 1$ are likely to prove important.

Once the resonant locations are known, the total torque on the planet can be calculated by summing up the individual resonant contributions. At each resonance the torque depends upon two factors, the intrinsic strength of the resonance and the amount of disk gas present at that location. Of these factors, the intrinsic strength is the basic quantity that can be derived from a linear calculation of the disk response. It is largely independent of details of the disk physics such as the viscosity. The amount of gas present, on the other hand, will depend not just upon the unperturbed disk structure but also on how the disk evolves as angular momentum is gained or lost to the planet. There are two limiting cases:

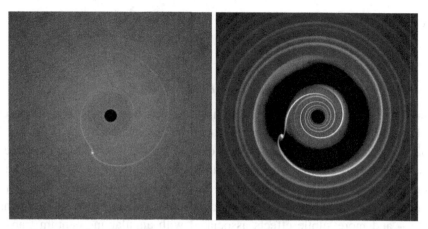

Fig. 7.1. Two-dimensional hydrodynamic simulations depicting the interaction between a planet and a viscous protoplanetary disk in (left panel) the Type 1 regime appropriate to low-mass planets and (right panel) the Type 2 regime relevant to giant planets. In both cases angular momentum exchange is the result of gravitational interaction with spiral waves set up within the disk as a consequence of the planetary perturbation. In the Type 1 regime the interaction is weak enough that the local surface density is approximately unperturbed, while in the Type 2 regime a strong interaction repels gas from the vicinity of the planet producing an annular gap.

- **Type 1** migration occurs for low-mass planets whose interaction with the disk is weak enough as to leave the disk structure almost unperturbed. This will certainly be true if the local exchange of angular momentum between the planet and the disk is negligible compared to the redistribution of angular momentum due to disk viscosity. The planet remains fully embedded within the gas disk and material is present at all resonant locations.
- **Type 2** migration occurs for higher mass planets whose gravitational torques locally dominate angular momentum transport within the disk. As we have already noted, gravitational torques from the planet act to repel disk gas away from the orbit of the planet, so in this regime the planet opens an annular gap within which the disk surface density is reduced from its unperturbed value. Resonances close to the planet are severely depleted of material and contribute little or nothing to the total torque.

A visual impression of the two regimes is shown in Fig. 7.1, whose two panels show the simulated surface density structure of a disk interacting with a planet in the Type 1 and Type 2 regimes. That the interaction is concentrated at resonances cannot easily be discerned. Rather, one sees that the interaction results in the formation of a wake of enhanced surface density that trails the planet outside its orbit and leads inside. It is the gravitational back-reaction of this wake on the planet that produces the positive torque[3] from material at $r < a$ and the negative torque from material at $r > a$.

7.1.2 Type 1 migration

In the Type 1 regime of planetary migration the perturbation the planet causes is small enough that it does not alter the background structure of the gas disk. The problem is then to calculate, for a given disk structure (specified for example via the surface density profile $\Sigma(r)$ and central temperature profile $T_c(r)$), the total torque on the planet. Viscosity plays no explicit role – though we often implicitly assume that the scaling of ν with radius sets the steady-state surface density profile – and more subtle effects associated with angular momentum transport such as turbulence and magnetic fields are customarily ignored. In a linear calculation the perturbation to the disk is proportional to the planet mass, and hence the torque (which arises from the gravitational force between the induced disk asymmetries and the planet) is bound to follow the M_p^2 scaling that we derived earlier.

If we choose to calculate the torque as a sum over discrete resonances, the total torque on a planet in a circular orbit can be written as

$$\Gamma = \sum_{m=1}^{\infty} \Gamma_{OLR}(m) + \sum_{m=2}^{\infty} \Gamma_{ILR}(m) + \Gamma_{CR}, \qquad (7.16)$$

where $\Gamma_{OLR}(m)$, $\Gamma_{ILR}(m)$, and Γ_{CR} are respectively the partial torques exerted at outer and inner Lindblad resonances and at the co-rotation resonance. Since gas is present at all radii in the disk, the sums over the Lindblad resonances formally extend to ∞, raising again the issues of (non) convergence that plagued our calculation of the torque in the impulse approximation. In fact this is not a problem. The sums converge because although the nominal resonant locations for test particles in a Keplerian disk (Eq. 7.14) crowd ever-closer to the planet at high m, the *effective* location of these resonances is shifted by pressure effects in a gas disk. The actual resonant locations are excluded from an annulus around the planet

$$r = a \pm \left(\frac{2}{3}\right) h, \qquad (7.17)$$

---

[3] Simulations yield what is in principle a full solution to the hydrodynamic equations, including nonlinear effects. However, one can also compute the linear torque either by summing over resonances, or by computing the back-reaction of the excited waves as they propagate through a model of the disk. These approaches are equivalent.

whose width is roughly the disk scale-height h. The most important Lindblad resonances prove to be those with

$$m \sim \left(\frac{h}{r}\right)^{-1}, \tag{7.18}$$

while the infinity of higher m Lindblad resonances are strongly suppressed and do not yield a divergent contribution to the torque (Ward, 1988; Artymowicz, 1993a).

General physical considerations imply that the planet gains angular momentum from the disk at the inner Lindblad resonances (summed over m, Γ_{ILR} is positive) while losing angular momentum to the disk at the outer Lindblad resonances. Ignoring the contribution from the co-rotation resonance for the moment, migration of the planet will occur if there is a mismatch in the magnitude of these two opposing torques. The degree of any mismatch can be expressed via a parameter

$$f = \frac{\Gamma_{\mathrm{ILR}} + \Gamma_{\mathrm{OLR}}}{|\Gamma_{\mathrm{ILR}}| + |\Gamma_{\mathrm{OLR}}|}. \tag{7.19}$$

If $f < 0$ loss of angular momentum to the outer disk dominates over that gained from the inner disk, and the planet migrates inward, while if $f > 0$ exactly the inverse holds true. The dividing line when $f = 0$ represents an interesting special case – in this case angular momentum is transported from the inner disk to the outer disk *via* the agency of the planet but the planet itself neither migrates inward nor outward.[4]

Since the torque at any given resonance depends upon the surface density of gas at that location, it is natural to suppose that the magnitude and sign of f ought to depend upon the surface density gradient. This intuition, however, proves to be false. Detailed calculations show that the net Lindblad torque experienced by a planet in the Type 1 regime is negative (implying inward migration) for almost all plausible combinations of disk surface density and temperature profiles (Ward, 1997). Several effects play a role in this surprising result, of which the simplest to grasp is the trade-off that occurs between the amount of gas at the inner and outer Lindblad resonances and the position of those resonances. Suppose that, in an attempt to reverse the generic trend of inward migration, we consider what happens in disks with steeper and steeper surface density profiles. A steeper surface density profile reduces the amount of gas at a fixed location exterior to the planet's orbit relative to a location interior to the orbit, and if the resonant locations were independent

---

[4] This raises the amusing possibility that a population of planets for which $f \approx 0$ (when the contribution from the co-rotation torque was also folded in) could act not just as a sort of macroscopic source of "viscosity" in the disk, but perhaps furnish the *only* source of viscosity in cool disks lacking intrinsic turbulence. The undeniable elegance of this idea has a considerable allure to theorists (Goodman & Rafikov, 2001), although it should be noted that current calculations of f do *not* suggest that it is near zero.

of the gradient in the surface density this would indeed weaken the outer Lindblad resonances as compared to the inner Lindblad resonances. Offsetting this effect, however, is the fact that a steeper surface density profile also implies a larger radial pressure gradient. This results in a more sub-Keplerian rotation profile for the gas (Section 2.3) and shifts the outer resonances closer to the planet, thereby increasing their strength. Although these compensating effects do not identically cancel each other out, the result is that the net Lindblad torque is almost always negative (except perhaps in localized regions where the disk has a strong temperature gradient) and is a considerable fraction of the total (or one-sided) Lindblad torque (Ward, 1997; Tanaka *et al.*, 2002).

Tanaka *et al.* (2002) have computed the linear Type 1 torque on a planet in a circular orbit due to both Lindblad and co-rotation torques. Their calculation is fully three-dimensional, but assumes that the disk is isothermal not only vertically but also in the radial direction. The only structural parameter that characterizes the disk is then the surface density gradient, which is taken to be the index of a power-law profile

$$\Sigma(r) \propto r^{-\alpha}. \tag{7.20}$$

For a planet of mass M_p, orbiting at radius a and angular frequency Ω_K in a disk with local surface density Σ and thickness (h/r), around a star of mass M_*, Tanaka *et al.* (2002) find the following expressions for the net Lindblad, co-rotation, and total torque on the planet

$$\Gamma_{LR} = -(2.34 - 0.10\alpha) \left(\frac{M_p}{M_*}\right)^2 \left(\frac{h}{r}\right)^{-2} \Sigma a^4 \Omega_K^2, \tag{7.21}$$

$$\Gamma_{CR} = (0.98 - 0.64\alpha) \left(\frac{M_p}{M_*}\right)^2 \left(\frac{h}{r}\right)^{-2} \Sigma a^4 \Omega_K^2, \tag{7.22}$$

$$\Gamma_{total} = -(1.36 + 0.54\alpha) \left(\frac{M_p}{M_*}\right)^2 \left(\frac{h}{r}\right)^{-2} \Sigma a^4 \Omega_K^2. \tag{7.23}$$

Comparing the scalings here to those derived via the impulse approximation, we find that the more sophisticated calculation of the linear theory torque yields a dependence on the disk thickness that scales as $(h/r)^{-2}$ rather than $(h/r)^{-3}$. The reason for this difference is that the intrinsic asymmetry between the inner and outer Lindblad torques is itself an increasing function of the disk thickness. The remaining scalings – with planet mass, disk surface density, and orbital radius – are identical to those deduced from elementary arguments.

The torque values quoted above are frequently employed as the fiducial estimates of the Type 1 torque experienced by a planet orbiting within a gas disk. We will use them ourselves in Section 7.1.4 to estimate the migration rate of growing giant planet cores. One should note, however, that these are current best estimates rather than definitive numbers. In addition to the obvious caveats – real disks are not isothermal and probably are turbulent – it is also the case that the calculation of the co-rotation torque remains subject to considerable uncertainty. There may be regimes where thermal or nonlinear effects that are not included in the Tanaka *et al.* (2002) analysis are important. Although the nominal co-rotation torque is considerably smaller than the net Lindblad torque it is by no means negligible, and a substantial change to its magnitude could qualitatively affect predictions that rely on accurate knowledge of the Type 1 migration rate.

7.1.3 Type 2 migration

As planets become more massive the Type 1 torque that they exert on the disk increases as M_p^2. Inevitably this torque eventually ceases to be a mere perturbation to the internal viscous stresses within the disk, and starts to modify the disk structure in the neighborhood of the planet. Since the interaction adds angular momentum to the disk exterior to the planet, and removes it from the interior gas, the overall effect is that a strong torque repels gas from the vicinity of the planet's orbit creating a gap (Fig. 7.1). Once a gap starts to form the assumption that the net torque on the planet can be calculated using an unperturbed disk model breaks down, and the torque must be calculated self-consistently together with a model for the viscous evolution of the gas disk.

The mass above which a planet is able to open a gap within the gas can be estimated by considering the competing effects that try to open and close annular gaps within the disk. The planetary torque *always* acts in such a way as to try and open a gap, but this tendency is opposed by viscosity (internal angular momentum transport processes that would be present in the absence of a planet) which acts diffusively to smooth out sharp radial gradients in the disk surface density. This competition is illustrated in Fig. 7.2.

The threshold mass above which a planet succeeds in opening a gap can be estimated via several more or less equivalent arguments, none of which should be trusted to yield precise values. A simple approach that draws on our analysis of the torque in the impulse approximation is based on calculating the time scale for opening a gap between $a \pm h$ in the disk. It is plausible that a gap of half-width h is roughly the smallest gap that we can conceive of opening, first because Lindblad resonances are at their most effective at about this distance from the planet, and second because gaps whose radial width was much smaller than the vertical thickness of the disk might well be unstable.

To estimate the gap opening time scale t_{open} we note that the amount of angular momentum that must be added to evacuate all of the gas between a and $(a + h)$

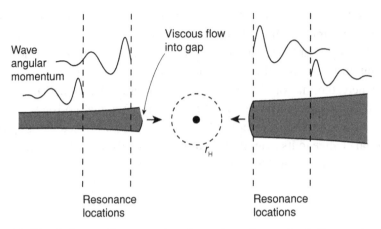

Fig. 7.2. The balance of torques that determines (in part) whether a planet is able to open a gap within the disk. Waves excited at resonant locations act to remove angular momentum from the disk interior to the planet, and add angular momentum to the disk outside, thereby opening a gap. Viscous flow counteracts this tendency.

out of the annulus is

$$\Delta J = 2\pi a h \Sigma \cdot \left.\frac{\mathrm{d}l}{\mathrm{d}r}\right|_a \cdot h, \tag{7.24}$$

where $l = \sqrt{GM_* r}$ is the specific angular momentum of gas in a Keplerian orbit. The gap opening time scale can then be estimated as

$$t_{\mathrm{open}} = \frac{\Delta J}{|\mathrm{d}J/\mathrm{d}t|}, \tag{7.25}$$

with $\mathrm{d}J/\mathrm{d}t$ given by the impulse equation formula (Eq. 7.9) with $b_{\mathrm{min}} = h$. The expression that this yields is not terribly enlightening. To make progress, we compare the gap opening time scale to the time scale on which viscosity would act to close a gap of characteristic scale h within the gas disk. The gap closing time scale is (Eq. 3.11)

$$t_{\mathrm{close}} = \frac{h^2}{\nu}, \tag{7.26}$$

where ν is the kinematic viscosity. Making use of the Shakura–Sunyaev α-prescription, $\nu = \alpha c_s h$ (Equation 3.46), we then equate t_{open} and t_{close}. The result is an estimate for the critical mass ratio $q \equiv M_{\mathrm{p}}/M_*$ between the planet and the star above which a gap can be opened

$$q_{\mathrm{crit}} \simeq \left(\frac{27\pi}{8}\right)^{1/2} \left(\frac{h}{r}\right)^{5/2} \alpha^{1/2}. \tag{7.27}$$

In this expression (h/r) is to be interpreted as the geometric thickness that the disk would have in the absence of the planet. For typical disk parameters ($\alpha = 10^{-2}$, $h/r = 0.05$) we obtain

$$q_{\text{crit}} \simeq 2 \times 10^{-4}, \tag{7.28}$$

which corresponds to a planet around a Solar mass star with a mass comparable to that of Saturn. The gap opening mass has no explicit dependence on the disk surface density, though there is an implicit dependence which arises from the fact that in disk models that include nonnegligible viscous heating h/r will vary with surface density.

In deducing this *viscous* criterion for gap-opening we have used formulae derived under the assumption that the planet–disk interaction can be treated two-dimensionally. This is valid provided that the Hill sphere (Eq. 5.15) of the planet is larger than the scale-height of the disk

$$r_{\text{H}} \gtrsim h, \tag{7.29}$$

and this can be viewed as a second criterion for gap opening. In terms of the mass ratio it yields

$$q_{\text{crit}} \gtrsim 3 \left(\frac{h}{r}\right)^3, \tag{7.30}$$

which evaluates to a rather similar number – $q_{\text{crit}} \approx 4 \times 10^{-4}$ – for fiducial disk parameters. This similarity is nothing more than a numerical coincidence, but it reinforces the conclusion that the critical planet mass for gap opening is likely to be a fraction of a Jupiter mass. Numerical simulations are broadly consistent with this analytic expectation.

A planet whose mass lies close to the critical value for gap opening will typically succeed only in depressing, rather than cleanly evacuating, the surface density at radii around its orbit. The rate (and possibly even the direction) of migration of planets around this Type 1/Type 2 boundary is hard to calculate, because in this regime no aspect of the planet–disk interaction is negligible. The gas disk cannot be assumed to have its unperturbed structure (as is done in calculations of Type 1 migration), the co-orbital region still contains gas and may yield a large contribution to the torque, and partial torques from all Lindblad resonances must be evaluated. Accounting reliably for all of these effects remains an open research problem.

The Type 2 regime of migration is much easier to analyze if we restrict our attention to massive planets for which $q \gg q_{\text{crit}}$ (in practice this means Jupiter mass planets and above). The interaction of such massive planets with the disk is strong enough to clear a clean gap in the gas around the orbit (Fig. 7.1), so that the only significant torques (for a planet on a circular orbit) are those exerted at a small number of low-m Lindblad resonances. The edges of the gap are defined by the location of the lowest m resonances that are just able to hold back the viscous inflow of the disk into the gap region. A useful conceptual model to understand how the disk–planet system behaves in this limit is to imagine that the planet is surrounded by "brick walls" that define the inner and outer edges of the gap. At the edge of the gap interior to the planet, the wall removes exactly enough angular momentum from gas to prevent the disk overflowing the wall, whereas at the outer edge of the gap the wall (or, physically, the planetary torque) adds the appropriate amount of angular momentum. If we now assume that in the region near the planet most of the angular momentum resides in the *gas rather than in the planet* it is fairly clear how the coupled system of the planet plus the disk will evolve. As gas from the outer disk flows inward due to ordinary viscous evolution it runs up against the barrier imposed by the planetary torque at the outer edge of the gap. To prevent the gas encroaching into the gap the planet must give angular momentum to the outer disk, which must be balanced by a loss of angular momentum from the inner disk. If the inner disk loses angular momentum, however, its gas must move inward and away from the inner wall. To maintain the edges of the gap at the resonant locations where torque is transmitted the planet itself must move inward.

This thought experiment demonstrates that the role of a relatively low-mass planet migrating in the Type 2 regime is simply that of a catalyst whose torques bridge the gap in the disk surface density without otherwise altering the angular momentum flux that would be present in the disk in the absence of the planet. The gas then flows inward at the same rate as in a planet-free disk, and the planet migrates at the same rate as the local disk gas in order to remain at the center of the gap. We therefore define the nominal Type 2 migration rate as being equal to the velocity of gas inflow in a steady disk

$$v_{\text{nominal}} = -\frac{3v}{2r}. \tag{7.31}$$

Using the α-prescription once more we obtain

$$v_{\text{nominal}} = -\frac{3}{2}\alpha \left(\frac{h}{r}\right)^2 v_{\text{K}}, \tag{7.32}$$

where v_{K} is the Keplerian orbital velocity. The speed of Type 2 migration depends upon poorly known disk parameters (in particular, it depends linearly on the efficiency of angular momentum transport as parameterized via α) but it is generically rather rapid, with nominal migration time scales from 5 AU of the order of 10^5 yr being typical. One should note that although migration is always inward in a steady disk of large radial extent, our argument actually implies only that the planet ought to track the local motion of the surrounding gas. A planet that forms near the outer edge of a disk that is *expanding* viscously would therefore migrate with the gas to larger orbital radii (Veras & Armitage, 2004).

Although one will sometimes see Eq. (7.32) used as *the* expression for the Type 2 migration rate, this is incorrect. Massive planets are predicted to migrate at the nominal Type 2 rate only if the condition that we emphasized above – that the disk near the planet contains most of the angular momentum – is satisfied. For a rough estimate we can express this condition as requiring that

$$f \equiv \frac{M_p}{\pi a^2 \Sigma} \lesssim 1, \tag{7.33}$$

where Σ is some characteristic surface density that the disk would have at that radius in the absence of the planet. In many circumstances of interest the planet will be too massive to satisfy this condition. For example, if we adopt a surface density profile

$$\Sigma(r) = 1.5 \times 10^3 \left(\frac{r}{1\ \text{AU}}\right)^{-1} \text{g cm}^{-2}, \tag{7.34}$$

migration will occur at the nominal Type 2 rate only for masses

$$M_p \lesssim 0.6 \left(\frac{a}{1\ \text{AU}}\right) M_J. \tag{7.35}$$

Massive extrasolar planets migrating in toward the hot Jupiter region at *much* less than 1 AU will thus be too massive to be considered as passive tracer particles within the gas disk.

The actual Type 2 migration rate for $f > 1$ can be estimated, although such estimates depend upon aspects of the viscosity in the disk that are even less certain than the viscous inflow time scale required to evaluate the nominal rate. A simple guess follows from noting that a planet that has a mass (and angular momentum) f times that of the local unperturbed disk will take $\sim f$ viscous times to export that angular momentum to the disk and migrate inward. This estimate probably overstates the degree to which migration is suppressed, since gas flowing in toward a very massive planet will tend to accumulate near the tidal barrier at the outer gap edge. The enhanced surface density (above the unperturbed disk value) at the gap edge increases the torque, so that the suppression of migration is by less than the linear factor of f implied by our simple argument. Better estimates of the actual Type 2 migration rate in the $f > 1$ limit can be derived for specific disk models (Syer & Clarke, 1995; Ivanov *et al.*, 1999), and these predict a lesser but still significant reduction in the Type 2 migration rate for massive planets that migrate to small orbital radii.

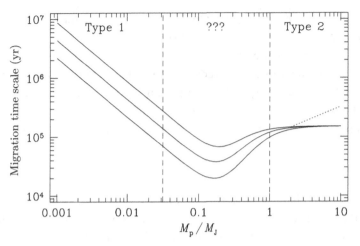

Fig. 7.3. The nominal migration time scale from 5 AU as a function of planet mass, based on linear theory calculations by Tanaka *et al.* (2002) and numerical simulations by Bate *et al.* (2003). Results are plotted for disks around a Solar mass star with $(h/r) = 0.05$ and surface densities at 5 AU of (from top down at left) 50 g cm^{-2}, 100 g cm^{-2}, and 200 g cm^{-2}. In the Type 1 regime the migration time scale is inversely proportional to both the planet mass and the surface density. In the Type 2 regime the migration time scale is of the order of the viscous time scale of the disk, which is formally independent of planet mass and surface density if (h/r) is taken to be fixed. More careful analyses predict a sub-linear increase in the migration time scale with planet mass (indicated schematically as the dotted line) once the planet mass becomes comparable to a local estimate of the disk mass.

7.1.4 Applications

By combining our results for the rate of Type 1 and Type 2 migration it is possible to sketch out how the predicted migration time scale

$$t_{\text{migrate}} \equiv \frac{a}{|v_{\text{migrate}}|}, \tag{7.36}$$

varies across the entire range of planet masses of interest. Such a sketch is shown in Fig. 7.3. No observations furnish any direct constraints on migration time scales, so for the time being one can only note three predictions of migration theory:

- In the Type 1 regime the net torque scales as M_{p}^2, while the orbital angular momentum of the planet is directly proportional to the mass. The migration time scale is thus inversely proportional to mass – more massive planets migrate *faster*.
- In the Type 2 regime the nominal migration time scale is *independent* of mass. It is determined instead by the angular momentum transport properties of the disk.
- The most rapid migration is predicted to occur at the boundary between the Type 1 and Type 2 regimes. For typical disk models this corresponds to planet masses of the order of $0.1 M_{\text{J}}$.

This last prediction is particularly worrisome – we have concluded that the most dramatic effects occur precisely in the regime where acknowledged inadequacies in our theory are the worst! In fact, although we have drawn a smooth curve to interpolate between the better-determined Type 1 and Type 2 time scales almost *any* behavior is possible in this middle region of the plot.

The primary interest in Type 2 migration is as a candidate mechanism for explaining the population of massive extrasolar planets in short-period orbits. Both the analytic arguments summarized in Section 6.1 and detailed calculations (Bodenheimer *et al.*, 2000) show that it is difficult to form gas giants at orbital radii of ∼1 AU and below. Finding a mechanism for the *in situ* formation of true hot Jupiters interior to 0.1 AU is even harder to fathom. The existence of *some* type of migration process that allows massive planets to form at larger radii before being transported inward is thus a robust (though still indirect) observational inference. Since gas giants inescapably form in an environment that is conducive to Type 2 migration, gas disk processes could well be the dominant mechanism.

The coupled evolution of a massive planet embedded within an evolving disk can be approximately modeled with straightforward generalizations of the one-dimensional disk evolution equation discussed in Section 3.2 (Lin & Papaloizou, 1986). As an illustration of such a calculation, Fig. 7.4 shows the predicted evolutionary track followed by a Jupiter mass planet that forms at 5 AU within a disk that is still quite massive. The planet opens a gap and migrates inward while remaining locked to the viscous evolution of the disk, in the manner described above for Type 2 migration. The time scale for the planet to reach quite small orbital radii is similar to that estimated above, and for a variety of disk parameters is typically in the range between 10^5 yr and a few times that value. Such rapid migration is clearly consistent with the idea that short-period planets may have arrived at their current locations following a phase of Type 2 migration. Indeed the predicted time scales are perhaps *shorter* than one might like. Migration time scales that are substantially less than a Myr imply that some planets may form "too early" and suffer the fate of being swept all the way into the star. If this is the case the observed massive planets in short-period orbits are the subset made up of survivors whose migration was interrupted by the dispersal of the protoplanetary disk.

Many short-period extrasolar planets are measured to have significantly eccentric orbits. An important question is whether these eccentricities were acquired as a consequence of planet–disk interactions during Type 2 migration, or whether they arise from entirely unrelated processes. This issue can be investigated within the same resonant framework used to compute the net torque, by evaluating the net damping or excitation of planetary eccentricity that results from the interaction of a slightly eccentric planet with the disk at all significant resonant locations

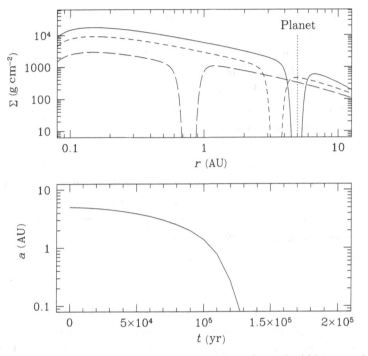

Fig. 7.4. Predicted Type 2 migration of a 1 M_J planet formed within an evolving protoplanetary disk at 5 AU, based on an approximate one-dimensional treatment of the interaction (Armitage *et al.*, 2002). The upper panel shows the disk surface density as a function of radius at three different epochs, while the lower panel shows the evolution of the planetary semi-major axis. The model assumes that the planet formed at an early epoch while the gas disk was still massive. The planet migrates inward within a gap on a short time scale, reaching small radii after approximately 10^5 yr.

(Goldreich & Tremaine, 1980). Table 7.1 lists the nine most important resonances,[5] along with the sign of the eccentricity change they induce on an embedded planet (Goldreich & Sari, 2003; Ogilvie & Lubow, 2003; Masset & Ogilvie, 2004). The interaction at some resonances excites eccentricity, while that at others damps it. For a massive planet that clears a gap (so that the three co-orbital resonances listed in the table play no role) the balance between excitation and damping at the remaining resonances is delicate, with damping winning by only a narrow margin. This conclusion could be reversed in a number of ways, most of which involve suppressing the damping effect of the first-order co-rotation resonances by at least a small amount. This suppression could occur if the planet clears a *very* wide gap (as happens for binary stars, Artymowicz *et al.*, 1991), or if the co-rotation resonances

---

[5] The nomenclature of "fast" and "slow" denotes resonances that occur where the pattern speed is either faster or slower than the planet's orbital frequency.

Table 7.1. *The principal and first-order Lindblad and co-rotation resonances, together with their effect on the eccentricity of an embedded giant planet (after Masset & Ogilvie, 2004).*

Resonance	$(r/a)^{-2/3}$	Effect on e
Principal OLR	$m/(m+1)$	excite
Principal ILR	$m/(m-1)$	damp
Co-orbital CR	1	unclear
Fast first-order OLR	1	damp
Fast first-order ILR	$(m+1)/(m-1)$	excite
Fast first-order CR	$(m+1)/m$	damp
Slow first-order OLR	$(m-1)/(m+1)$	excite
Slow first-order ILR	1	damp
Slow first-order CR	$(m-1)/m$	damp

are intrinsically slightly weaker due to saturation at less than the nominal strength (Goldreich & Sari, 2003; Ogilvie & Lubow, 2003). Testing the different analytic expectations via numerical simulation is challenging, but it appears that sufficiently massive planets of a Jupiter mass and above do suffer at least moderate eccentricity growth during Type 2 migration (Papaloizou *et al.*, 2001; D'Angelo *et al.*, 2006). Excitation of eccentricity has not been reported for sub-Jupiter mass planets, nor have simulations (of any mass planet) demonstrated that planet–disk interactions can produce the large eccentricities characteristic of at least some extrasolar planets.

As Fig. 7.3 makes clear, lower mass planets whose masses fall into the 1–10 $M_\oplus$ range are predicted to suffer Type 1 rather than Type 2 migration due to their interactions with surrounding disks. Unlike the case of Type 2 migration, however, there are no observed phenomena that can be more easily explained by appealing to significant Type 1 migration. In fact, the rapid rate of orbital decay predicted to occur toward the high mass limit of the Type 1 regime is in an apparent and unresolved conflict with the relatively slow time scale of giant planet growth via core accretion. For a disk whose surface density scales with radius as $\Sigma \propto r^{-\alpha}$, the nominal Type 1 migration time scale implied by the torque in Eq. (7.23) is (Tanaka *et al.*, 2002)

$$t_{\text{migrate}} = (2.7 + 1.1\alpha)^{-1} \frac{M_*^2}{M_p \Sigma a^2} \left(\frac{h}{r}\right)^2 \Omega_K^{-1}. \tag{7.37}$$

The exact rate of Type 1 migration for a growing giant planet core depends upon its mass and on the surface density of the disk, but it is typically rapid. To derive a fairly extreme limit we assume that cores must grow to 5 $M_\oplus$ before rapid gas accretion sets in, and that the amount of gas in the disk is close to the absolute minimum needed to form the envelope. With these parameters, the above formula becomes

$$t_{\text{migrate}} \simeq 1.4 \times 10^6 \left(\frac{M_{\text{p}}}{5\ M_{\oplus}}\right)^{-1} \left(\frac{a}{5\ \text{AU}}\right)^{-1/2}$$

$$\times \left(\frac{\Sigma}{20\ \text{g cm}^{-2}}\right)^{-1} \left(\frac{h/r}{0.05}\right)^{2} \ \text{yr.} \qquad (7.38)$$

The surface density value used here is that appropriate for a disk with $\alpha = 1$ that has a mass of only $2\ M_{\text{J}}$ between 0.1 AU and 30 AU.

It might just be possible to form the core of a giant planet given a migration time scale comparable to this $\sim$Myr estimate. Indeed, a *small* amount of migration can hasten the growth of planetary cores by allowing the core to accrete planetesimals across a wider region of the disk and thereby avoid isolation. The problem arises from the fact that we would *typically* expect the surface density to be much higher than the bare minimum value adopted above (even for a minimum mass Solar Nebula, the surface density at 5 AU is almost an order of magnitude larger than the 20 g cm^{-2} value used above). Core formation cannot proceed to successful envelope accretion in the presence of undiluted Type 1 migration within such disks. From detailed calculations Alibert *et al.* (2005) find that a joint model of core accretion in the presence of Type 1 migration is only viable if the Type 1 migration rate is approximately an order of magnitude (or more) lower than the nominal value. The logical conclusion is that this conflict points to some flaw in our understanding of core accretion, of Type 1 migration, or of both, with most suspicion falling upon the computation of the co-orbital torque. This aspect of the calculation certainly appears to be the weakest theoretical link, though even if the co-orbital torque is in error it would seem to require a coincidence for the correct co-orbital torque to cancel almost exactly the other better understood torques. Together with determining the mechanism of planetesimal formation, resolving the question of how giant planet cores form in the presence of Type 1 migration is one of the most important unsolved problems of planet formation theory.

7.2 Resonant evolution

In our survey of the Solar System (Section 1.1) we noted that a resonance occurs when there is a near-exact commensurability among the characteristic frequencies of one or more bodies. The use of the term "characteristic frequency" is deliberately vague, since it must encompass a wide variety of possibilities that include the orbital frequency, the spin frequency of a planet's rotation, the precession frequency of the orbit, and more. We can start, however, by considering the simplest case where the commensurability is between the orbital frequencies (or periods) of two planets on circular orbits. We have already noted (Eq. 1.3, here expressed in an equivalent form) that the condition for resonance can be written as

$$\frac{P_{\text{in}}}{P_{\text{out}}} \simeq \frac{p}{p+q}, \tag{7.39}$$

where P_{in} and P_{out} are the orbital periods of the two planets and p and q are integers. The definition, of course, raises the immediate question of what is meant by the approximate equality sign – how close do the two planets have to be to the exact commensurability for them to be as a practical matter "in resonance." To address this we first assume that the planets are in exact resonance and rewrite the above expression in terms of the mean motions $n \equiv 2\pi/P$ of the planets

$$\frac{n_{\text{out}}}{n_{\text{in}}} = \frac{p}{p+q}. \tag{7.40}$$

If we can ignore any perturbations between the planets, the angle λ between the radius vector to one of the planets and a reference direction advances linearly with time. Defining $t = 0$ and $\lambda = 0$ to coincide with a moment when the two planets are in conjunction, we have that

$$\lambda_{\text{in}} = n_{\text{in}}t, \tag{7.41}$$
$$\lambda_{\text{out}} = n_{\text{out}}t, \tag{7.42}$$

and the resonance condition becomes

$$(p+q)\lambda_{\text{out}} = p\lambda_{\text{in}}. \tag{7.43}$$

Finally, we can define a *resonant argument*

$$\theta = (p+q)\lambda_{\text{out}} - p\lambda_{\text{in}}, \tag{7.44}$$

which will evidently remain zero for all time if the planets are in exact resonance. For planets on general circular orbits, on the other hand, λ_{in} and λ_{out} still advance linearly with time, but no small p and q can be found so that θ remains a constant. Sampled at random intervals, in fact, θ takes on all values in the range $[0, 2\pi]$. It is this basic distinction that furnishes a definition of what it means for two planets to be in resonance. A resonance occurs when one or more resonant arguments is bounded (though it need not be exactly constant), while there is no resonance if the argument takes on all possible values.

The above discussion is devoid of dynamical content, and useful only as an introduction to the concept of a resonant angle. A clear discussion of the dynamics of resonance can be found in Murray & Dermott (1999), who analyze the conditions for resonance within the context of the circular restricted three-body problem.

Following their treatment, we first define some nomenclature for eccentric orbits. The *mean anomaly* is given by

$$M = n(t - t_{\text{peri}}), \tag{7.45}$$

where t_{peri} is the time of pericenter passage. The *mean longitude* is

$$\lambda = M + \varpi, \tag{7.46}$$

where ϖ is the longitude of pericenter. We now consider the condition for exact resonance between a planet on a circular orbit ($e_{\text{in}} = 0$) and a low mass body on an eccentric orbit ($e_{\text{out}} \neq 0$). The potential experienced by the low mass body is not quite Keplerian, but to a good approximation we can assume that the orbit is an instantaneous Keplerian orbit whose longitude of pericenter varies with time at some rate $\dot{\varpi}_{\text{out}} \neq 0$. Working in a frame that rotates along with the drift in the longitude of pericenter of the outer body, the condition for exact resonance is

$$\frac{n_{\text{out}} - \dot{\varpi}_{\text{out}}}{n_{\text{in}} - \dot{\varpi}_{\text{out}}} = \frac{p}{p+q}, \tag{7.47}$$

and the resonant argument is,

$$\theta = (p+q)\lambda_{\text{out}} - p\lambda_{\text{in}} - q\varpi_{\text{out}}. \tag{7.48}$$

There is *still* no dynamical content to this statement. However when the dynamics is studied in detail (Murray & Dermott, 1999) it is found that for a system that is close to the exact resonance, perturbations between the bodies act such as to keep θ bounded even if the instantaneous periods are not exactly commensurable. We say that the system is:

- In resonance if θ *librates*, by which we mean that the resonant argument may be time-dependent but varies only across some limited range of angles.
- Out of resonance if θ *circulates*, taking on all values between 0 and 2π.

In fact for this model problem it can be shown that the behavior of θ over time is remarkably simple – the angle oscillates in a manner that is mathematically identical to that of a pendulum

$$\frac{d^2\theta}{dt^2} + \omega_0^2 \sin\theta = 0. \tag{7.49}$$

The oscillation frequency ω_0 is known in this context as the libration frequency, and the corresponding time scale

$$t_{\text{lib}} \equiv \frac{2\pi}{\omega_0}, \tag{7.50}$$

as the libration time scale. The final basic quantity of interest is the range of semi-major axis values for which θ librates rather than circulates. This quantity is called the width of the resonance. Although one cannot state any simple general formulae for these quantities, it is possible to derive analytic estimates of the widths and libration time scales for particular resonances that are of interest (see e.g. Holman & Murray, 1996 for a concise description of the method).

Our focus up to this point on mean motion resonances should not blind the reader into forgetting that a planetary system has a large number of characteristic frequencies, and as a consequence the number of possible resonant arguments is very large. Other types of resonance that are of interest include:

- Secular resonances, where the commensurable frequencies involve the slow precession of the longitude of pericenter or the longitude of the ascending node due to mutual interactions between the bodies. A simple example of a secular resonant argument is $\theta = \varpi_{in} - \varpi_{out}$.
- Resonances that involve inclination.
- Three-body resonances. A Solar System example is the three-body resonance involving Jupiter, Saturn, and Uranus: $n_{Jup} - 7n_U = 5n_{Sat} - 2n_{Jup}$.
- So-called secondary resonances, which include cases where the libration frequency of two planets in resonance is itself commensurable with some other frequency of the system.

Although in the Solar System there are no simple mean-motion resonances among the planets, a plethora of these weaker resonances are nonetheless believed to have important dynamical effects. In particular, spatial overlap of different resonances is the accepted explanation for the apparent presence of chaos in the motion of the planets. The review by Lecar *et al.* (2001) provides a good starting point for the reader who wishes to explore this intricate subject in more detail.

7.2.1 Resonant capture

Since resonances have finite widths, there is some probability that two planets in a randomly assembled planetary system will happen to find themselves in a mean motion resonance. Even a cursory inspection of data on the Solar System, however, convinces one of the need for a causal rather than a probabilistic explanation for entry into resonance. Chiang *et al.* (2007), for example, find that in excess of 20% of Kuiper Belt Objects with well-determined orbits are in mean motion resonances with Neptune, with the 3:2 resonance occupied by Pluto being the most heavily populated. Among satellites too there are numerous known resonances, of which the most striking is the 4:2:1 resonance that involves three of the Galilean satellites of Jupiter – Io, Europa, and Ganymede. Mean motion resonances between planets themselves also appear to be common among extrasolar planetary systems. Resonances have been securely identified in approximately 20% of known multiple planet systems, and there are additional systems that are plausibly but not yet provably in resonance.[6] One well-known example is the GJ 876 system (Marcy *et al.*, 2001), in which two massive planets orbit in a 2:1 resonance with orbit periods of approximately one and two months. For this system both of the lowest-order mean motion resonant arguments

$$\theta_1 = \lambda_{in} - 2\lambda_{out} + \varpi_{in}, \tag{7.51}$$

$$\theta_2 = \lambda_{in} - 2\lambda_{out} + \varpi_{out}, \tag{7.52}$$

and the secular resonant argument

$$\theta_3 = \varpi_{in} - \varpi_{out}, \tag{7.53}$$

librate about zero degrees (Lee & Peale, 2001).

The likely origin of most of these mean-motion resonances lies in the phenomenon of resonant capture, which was first studied for spin-orbit resonances (Goldreich & Peale, 1966) and subsequently developed to explain resonances among planetary satellites (Sinclair, 1972; Yoder, 1979; Borderies & Goldreich, 1984). Resonant capture is possible when two bodies are driven towards resonance (i.e. their orbits converge) by the application of a (generally weak) external force – which could arise due to tidal effects on planetary satellites or torques from the gas disk in the case of extrasolar planets. As the bodies approach and then enter the resonance there is a secular exchange of angular momentum between them that acts so as to maintain the resonance despite the ongoing external forcing that would otherwise result in exit from the resonance. In the case of two planets whose orbits converge because the outer planet is subject to a torque causing inward migration, for example, the resonant coupling acts to remove angular momentum from the inner planet so that the two bodies move inward in lockstep. A heuristic (but still quite intricate) description of the dynamics that results in this angular momentum transfer is given by Peale (1976). An important general result of the theory is that capture can occur only for converging orbits. Two bodies whose orbits *diverge* such that they approach and enter a resonance still experience an increased strength of interaction – which may lead to significant excitation of eccentricity – but cannot be captured into resonance according to the classical theory.

Even when it is possible (in the case of converging orbits) capture is generally a probabilistic phenomenon whose likelihood depends (in part) upon the relationship between the libration time scale t_{lib} and the time scale on which the external forcing

---

[6] Identifying resonances in extrasolar planetary systems is generally difficult because in many cases the orbital solutions are of limited precision. Making an accurate statistical statement as to the true fraction of resonant planetary systems is even harder, because the selection function for detecting two planets of given masses will generally differ depending upon whether they are in a resonance or not.

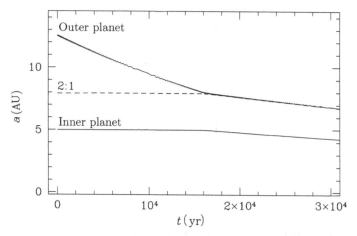

Fig. 7.5. Numerical example of resonant capture in a coplanar system of two initially well-separated Jupiter mass planets. For illustrative purposes the outer planet is subjected to a fictitious drag force which results in steady decay of its semi-major axis. The inner planet is almost unperturbed until the planets encounter their mutual 2:1 mean-motion resonance, at which point capture occurs and the two planets move inward in lockstep.

would result in resonance crossing if capture did not occur

$$t_{cross} = \frac{\Delta a_{res}}{|\dot{a}_{in} - \dot{a}_{out}|}. \tag{7.54}$$

Here Δa_{res} represents the width of the resonance under consideration. If $t_{cross} \gg t_{lib}$ then capture is guaranteed provided that the eccentricity of the body being captured is below some limit, while for faster crossing the probability of capture will depend upon the eccentricity, the rate of convergence of the orbit, and the particular resonance being encountered. Explicit expressions for the maximum drift rates that permit capture can be found in Quillen (2006, and references therein).

Resonant capture is a particularly appealing explanation for resonant extrasolar planets in short-period orbits, whose small semi-major axes provide circumstantial evidence for the existence of migration torques that could have both shrunk the orbits and driven the planets into resonance. Figure 7.5 shows a numerical calculation of resonant capture in a coplanar system of two Jupiter mass planets, where the outer planet is driven toward the inner one under the action of an external torque. For this choice of parameters capture occurs into the 2:1 resonance, which is then maintained as the orbits continue to shrink. Once in resonance, further decay of the semi-major axes of the planets is accompanied by growth of the eccentricity, though this may be damped (at least partially) by the interaction of the planets with the gas disk. Although existing theory is not fully predictive – in particular planets migrating through a turbulent gas disk can diffuse out of the resonance

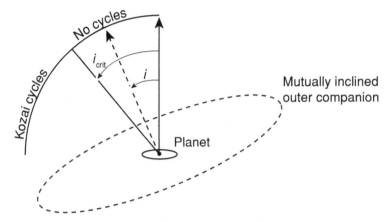

Fig. 7.6. Kozai cycles are possible in hierarchical triple systems in which the orbit of the inner body (here a planet) does not lie in the same plane as that of the outer body (here a stellar companion). Cyclic exchange of angular momentum occurs if the mutual inclination $i > i_{crit}$.

due to random variability in the torques they experience (Murray-Clay & Chiang, 2006; Adams *et al.*, 2008) – resonant capture is at least qualitatively consistent with the observation of a significant population of extrasolar planets in mean-motion resonances.

7.2.2 Kozai resonance

A quite different type of resonant behavior that is implicated in the evolution (and possibly formation) of some extrasolar planetary systems is Kozai resonance. A Kozai resonance is a remarkable type of secular resonance that was originally studied in the context of the motion of high inclination asteroids perturbed by the influence of Jupiter (Kozai, 1962).[7] For a model system that consists of a low mass body orbiting well interior to a massive perturber on a highly inclined orbit the resonant argument is simply the argument of pericenter of the inner body itself.

The interest in this particular resonance derives from the very unusual evolution that occurs when the resonance is active. The simplest situation to analyze (and also the one most relevant for extrasolar planetary systems) is that shown in Fig. 7.6. A planet of mass M_p and semi-major axis a_{in}, orbits the primary star (mass M_*) of a binary system. The secondary star of the binary system has mass M_s, semi-major axis a_s, and eccentricity e_s, and orbits in a plane inclined to that of the planet by an angle i. Triple systems of this kind are typically only stable if they are *hierarchical*

---

[7] Yoshihide Kozai's now celebrated paper languished in relative obscurity for some 30 years, no doubt in part because few inner Solar System bodies possess the large inclinations needed to trigger Kozai cycles.

(i.e. the outer binary has a much larger orbit than that of the inner binary, which is here the planet), so we consider the limit in which

$$a_s \gg a_{in}, \tag{7.55}$$

$$M_s \gg M_p. \tag{7.56}$$

The second of these conditions implies that essentially all of the angular momentum in the system resides in the orbit of the binary companion. Nothing that happens to the planet can affect the orbit of the binary, which thus remains fixed in space.

Given these approximations, Kozai (1962) showed that the behavior of the orbit of the inner planet depends upon the magnitude of the mutual inclination i between the orbital planes. In particular, if i exceeds a critical value

$$i_{crit} = \cos^{-1}\left[(3/5)^{1/2}\right] \simeq 39.2°, \tag{7.57}$$

then the eccentricity e and inclination i of the planet describe cyclic oscillations known as Kozai cycles. These oscillations can have a large amplitude. In the case where the planet has an initially circular orbit, the maximum value of the eccentricity during the cycle is

$$e_{max} = \left[1 - \frac{5}{3}\cos^2 i\right]^{1/2}, \tag{7.58}$$

where i is here the initial relative inclination. If i is large enough we find that $e_{max} = 1$ and the planet will be driven into collision with the star! Even more surprisingly **none of these results** depends upon the masses of the stars in the binary, the semi-major axis of the binary, or its eccentricity. At least in the model problem (where there are only three bodies, all of which can be described as Newtonian point masses) an extremely distant companion, whose perturbations one might assume would be negligible, can still excite dramatic evolution of the planetary orbit. In fact the only influence that the binary properties have on the cycle is to set its period. If the orbital period of the planet is P_{in}, the characteristic time scale of Kozai cycles is of the order of

$$\tau_{Kozai} \sim P_{in}\left(\frac{M_*}{M_s}\right)\left(\frac{a_s}{a_{in}}\right)^3 (1 - e_s^2)^{3/2}. \tag{7.59}$$

Binary companions that cause weaker perturbations, either on account of their lower mass or their greater distance, increase the time scale for cycles but do not otherwise alter their properties.

The simplest analytic treatment of the Kozai effect has known limitations, which are discussed for example by Ford *et al.* (2000). However, as one can readily confirm for oneself with the aid of a numerical integration such as the one whose results

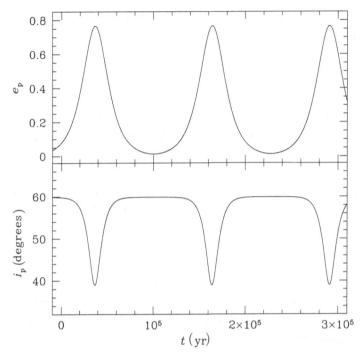

Fig. 7.7. Numerical illustration of a Kozai-type cycle in a marginally hierarchical system. The initial conditions assume that a Jupiter mass planet orbits a Solar mass star in a circular orbit at 5 AU. The star has a low mass binary companion (0.1 $M_\odot$) in a circular but inclined orbit at 50 AU.

are shown in Fig. 7.7, the basic picture is correct – a planet orbiting in a binary system does undergo Kozai cycles provided only that the mutual inclination is large enough.

From a theoretical perspective one would have to be dismayed if such a beautiful dynamical phenomenon remained an academic curiosity devoid of widespread application in nature. Such aesthetic considerations notwithstanding, the fraction of extrasolar planetary systems for which the Kozai effect is (or was) important remains poorly determined. Two considerations are of particular importance:

- The fraction of exoplanet host stars that are members of binary systems in which the plane of the planetary system is misaligned with the orbital plane of the binary. It is well established that Solar-type stars are frequently members of binary systems – Duquennoy & Mayor (1991) found that approximately half of the Solar-type (spectral type F7 to G9) primaries selected from an unbiased sample had one or more companions – and many of these systems are wide enough that a stable planetary system could form about the primary. The fraction of systems where the planetary orbital plane and the binary would be sufficiently misaligned to allow for Kozai cycles is, however, poorly determined.

- As a secular resonance, the Kozai effect is vulnerable to being washed out if there are additional sources of secular precession in the system, for example due to the presence of other planetary companions or due to General Relativistic effects. Approximately, the condition that must be fulfilled to allow Kozai cycles is that no other sources of precession operate on the time scale of the cycle.

In a number of cases a single highly eccentric planet is observed orbiting a star that is known to belong to a binary with properties that could plausibly drive Kozai cycles. The planet around 16 Cyg B is an example, and a strong argument can be made for the probable effectiveness of the mechanism in such specific cases (Holman *et al.*, 1997). Conversely the boldest hypothesis – that *all* eccentric extrasolar planets derive their eccentricity solely via the Kozai mechanism – can be ruled out on statistical grounds since it predicts substantially more near-circular planets than are observed (Takeda & Rasio, 2005). Most probably then, the Kozai mechanism plays a significant but not dominant role in the dynamics of extrasolar planets. One interesting idea is that some fraction of the hot Jupiters might have reached their present orbits as a consequence of the *combined* action of the Kozai effect and tidal dissipation (Eggleton & Kiseleva-Eggleton, 2001; Wu & Murray, 2003; Fabrycky & Tremaine, 2007). Although the Kozai mechanism on its own does not lead to orbital migration (the semi-major axis of the planet does not evolve), a large-amplitude cycle could lower the minimum pericenter distance $a_{in}(1 - e_{max})$ to the point where weak tidal interactions with the star occur to shrink the orbit. If this joint mechanism is at work to form hot Jupiters we would expect occasionally large misalignments between the stellar spin axis and the angular momentum vector of the planetary orbit, since the Kozai oscillations drive the planet out of its original orbital plane. This angle can be measured from radial velocity measurements taken during the course of planetary transits (using the Rossiter-McLaughlin effect; Rossiter, 1924), and the distribution of misalignments promises to provide at least circumstantial constraints on the theory.

7.3 Migration in planetesimal disks

A generic prediction of the core accretion model is that the time scale needed to assemble the core becomes longer in the outermost reaches of the protoplanetary disk. If giant planets form via core accretion we would therefore expect that outside a "giant planet zone" there ought to lie a region of debris that has been unable to form large bodies. This expectation is consistent with the fact that although planetesimals formed in the Solar Nebula out to a distance of ≈ 50 AU (as evidenced by the presence of the Kuiper Belt) there are no large bodies beyond Neptune in the Solar System. Our main goal in this section is to assess the possible dynamical influence that a disk of leftover debris (which we will dub a planetesimal disk, even though the sizes of the bodies involved can be much larger than the sizes usually considered for planetesimals) can exert on the orbits of neighboring giant planets.

Our first task is to estimate the amount of mass that might plausibly be present within a planetesimal disk in the outer region of a newly formed planetary system. This will depend upon the outer extent of the zone of giant planet formation (inside this radius the bulk of the planetesimals presumably end up within planets) and upon the assumed profile of the solid component of the protoplanetary disk. If we assume that the planetesimal disk tracks the surface density of the solid component of the minimum mass Solar Nebula[8] (Eq. 1.6), the integrated mass between radii of r_{in} and r_{out} is

$$M_{disk} = 4\pi \, \Sigma_0 \left(r_{out}^{1/2} - r_{in}^{1/2} \right), \qquad (7.60)$$

where Σ_0 defines the normalization (the fiducial value for the outer minimum mass Solar Nebula corresponds to 30 g cm^{-2} at 1 AU). As we will note later, there is considerable evidence to suggest that the Solar System's ice giants formed considerably closer to the Sun than their current locations. We therefore assume that the zone of giant planet formation in the early Solar System extended out to 20 AU, so that the disk of leftover debris ranged between 20 AU and 50 AU. With these parameters we estimate that

$$M_{disk} \simeq 40 \, M_\oplus. \qquad (7.61)$$

This estimate exceeds the *observed* mass of the present-day Kuiper Belt by more than two orders of magnitude. For it to be even remotely correct, dynamical processes must have removed almost all of the primordial material at some point during the lifetime of the Solar System.

The existence of a substantial debris disk in the outer reaches of a planetary system provides a long-lived (compared to the gas disk) reservoir of mass and angular momentum that can drive migration of any planets that are able to interact with it. Suppose, for example, that a planet of mass M_p orbiting interior to the disk scatters a mass of planetesimals δm into shorter period orbits at smaller radii. Elementary considerations suggest that the resulting change in the planetary semi-major axis ought to be of the order of

$$\frac{\delta a}{a} \sim \frac{\delta m}{M_p}. \qquad (7.62)$$

---

[8] Although we have previously warned of the dangers of placing too much trust in the minimum mass Solar Nebula profile its use here is justifiable, since it is derived empirically from estimates of the mass of heavy elements within the planets. Given only weak assumptions (primarily that the planetesimal disk had a continuous surface density distribution in the outer Solar System) it is reasonable to use it to estimate the mass of planetesimals that *did not* form planets beyond the orbit of the last giant planet.

Note that this cannot be the same process of small-angle scattering that we discussed in the context of gas disk migration (Section 7.1) since the sense of migration is reversed – we are imagining the planet to scatter planetesimals into lower angular momentum orbits and hence to migrate outward toward the exterior disk. For significant migration to occur we therefore require, at a minimum, that

$$M_p \lesssim M_{disk}. \tag{7.63}$$

True gas giants with masses (typically) of hundreds of Earth masses are therefore relatively impervious to planetesimal-driven migration, whereas this estimate suggests that lower mass ice giants such as Uranus and Neptune could suffer substantial migration given the existence of planetesimal disks with plausible masses. There is, however, an obvious complication. Even if the total disk mass is large enough to permit migration, the mass of material that the planet is able to interact with at any one time is much smaller. A planet migrating through a planetesimal disk may stall (or at least slow down dramatically) if the disk surface density is too small, since the planet will then scatter all of the bodies it is able to perturb without moving far enough to start interacting with a fresh population. This argument – which is a close cousin of the reasoning behind the existence of the isolation mass (Section 5.2.3) – suggests that not just the integrated disk mass but also the local surface density are important parameters that determine the behavior of planetesimal-driven migration.

An approximate analytic model for migration within planetesimal disks can be developed by borrowing ideas from the more complete treatments given by Ida *et al.* (2000) and by Kirsh *et al.* (2009). We consider an asymmetric configuration in which a planet of mass M_p and orbital radius a lies just inside an exterior disk of planetesimals of surface density Σ_p. At $t = 0$ there are no planetesimals at $r < a$. We assume that the planet migrates outward into the disk as a consequence of scattering planetesimals inward on to lower angular momentum orbits, and that its migration is fast enough that any individual planetesimal is scattered inward and does not then interact further with the planet.

An analysis of this model is straightforward. We first note that the planet will be able to perturb planetesimals strongly only within a radial zone whose width Δr is of the order of the radius of the planet's Hill sphere (cf. Section 5.1.2)

$$\Delta r \approx \left(\frac{M_p}{3M_*} \right)^{1/3} a. \tag{7.64}$$

The mass of planetesimals within this zone is

$$\Delta m = 2\pi a \Sigma_p \Delta r. \tag{7.65}$$

We now need to estimate the average change in the specific angular momentum of a planetesimal that results from scattering. This can be calculated accurately, but for now we simply assume that it is of the order of the difference in specific angular momentum of circular orbits across the scattering zone. If this is the case, then once all of the planetesimals within the scattering zone have encountered the planet they will have collectively lost an amount of angular momentum

$$\Delta J \approx \Delta m \left.\frac{dl}{dr}\right|_a \Delta r, \tag{7.66}$$

where $l = \sqrt{GM_*r}$ is the specific angular momentum for a circular orbit at distance r. The angular momentum lost by the planetesimals is gained by the planet, which (if it stays on a circular orbit) migrates outward a distance

$$\Delta a \approx \frac{2\pi a \Sigma_p \Delta r^2}{M_p}. \tag{7.67}$$

For rapid migration to occur, Δa must be large enough to move the planet into a region of the disk stocked with as yet unperturbed planetesimals. Requiring that $\Delta a \gtrsim \Delta r$ we obtain a condition on the planet mass

$$M_p \lesssim 2\pi a \Sigma_p \Delta r. \tag{7.68}$$

Fast planetesimal-driven migration requires that the planet mass be *smaller* than the mass of planetesimals within a few Hill radii of the planet. One may observe again that this behavior is the opposite of that predicted for Type 1 migration in a gas disk, which becomes more rather than less efficient as the planet mass increases.

The rate of migration can be determined by estimating how long it takes for all of the planetesimals within the scattering zone to encounter the planet and be scattered inward. The relevant time scale is that set by the shear across the scattering zone (Eq. 7.6)

$$\Delta t \sim \frac{2}{3}\frac{a}{\Delta r}P, \tag{7.69}$$

where P is the planetary orbital period. Combining this with the expression for the distance that the planet moves (Eq. 7.67), we conclude that the migration rate in the fast regime will be

$$\frac{da}{dt} \sim \frac{a}{P}\frac{\pi a^2 \Sigma_p}{M_*}. \tag{7.70}$$

Up to numerical factors of the order of unity this expression matches that derived by Ida *et al.* (2000), and is in good agreement with numerical results. An interesting feature of the result is that the migration of a low mass planet through a massive planetesimal disk occurs at a rate that is independent of the planet mass. Slower but still substantial migration can occur for planets that are up to an order of magnitude more massive than the rough limit defined by Eq. (7.68). Kirsh *et al.*

(2009) provide an empirical fit to numerical calculations of migration in the slow high-mass planet regime.

7.3.1 Application to the outer Solar System

The idea that scattering of planetesimals could have altered the orbits of planets in the outer Solar System is far from new. As with so many other fundamental results in the theory of planet formation, a brief discussion can be found in Safronov (1969). Safronov, however, along with other early investigators, was primarily interested in the ability of planets to *eject* bodies from the Solar System and thereby populate the Oort cloud (Oort, 1950). Ejection requires that the planet transfer energy to the planetesimal and hence results in inward planetary migration. The process is only efficient for Jupiter, whose large mass allows it to eject a substantial mass of material from the Solar System without moving very far inward.

Larger scale orbital migration is possible from scattering (without ejection) initiated by the ice giants. The first important development was the realization by Fernandez & Ip (1984) that the architecture of the outer Solar System lends itself to precisely the type of one-way scattering developed in our simple analytic model. If the giant planets formed within an exterior planetesimal disk, Neptune is massive enough to perturb bodies within the disk into orbits that lie between Uranus and Neptune, but is not massive enough to eject them from the Solar System. The same is true of Uranus and Saturn, and the overall result is that objects scattered inward from the disk by Neptune can be moved inward by successive encounters with Uranus and Saturn until eventually they reach Jupiter and are flung out of the Solar System. To conserve energy and angular momentum, the process results in the expansion of the orbits of Neptune, Uranus, and (to a lesser extent) Saturn, while the orbit of Jupiter shrinks slightly. An attractive feature of such a model is that it allows the giant planets to form in a more compact configuration closer to the Sun, ameliorating substantially the difficulty of trying to form Neptune via core accretion at its current location.

These theoretical ideas receive strong support from observations of the distribution of minor bodies in the outer Solar System (Fig. 1.3), which show that Pluto and a host of smaller Kuiper Belt Objects orbit in stable 3:2 mean motion resonance with Neptune. Malhotra (1993, 1995) showed that the properties of Pluto's orbit are consistent with those expected if it was resonantly captured by Neptune during the latter's slow outward migration, and was able to predict the distribution of the now well-observed population of resonant KBOs. Many – but not yet all – of the detailed properties of this population have subsequently been shown to be consistent with planetesimal–disk migration models (Chiang *et al.*, 2003; Hahn & Malhotra, 2005).

7.3.2 *The Nice Model*

The success of migration models in explaining the structure of the Kuiper Belt provides compelling evidence for the outward migration of Neptune, and has prompted investigation of more ambitious models which attempt to link the migration or rearrangement of the giant planets to other observed but poorly understood phenomena in the Solar System. One promising idea – known as the *Nice Model*[9] – is based upon the hypothesis that the giant planets formed within a compact configuration, with Saturn initially lying interior to the 2:1 mean motion resonance with Jupiter (Tsiganis *et al.*, 2005). The significance of these initial conditions is that subsequent planetesimal-driven expansion of the system results in divergent crossing of the 2:1 resonance between Jupiter and Saturn. This does not lead to capture, but does stimulate a sharp increase in the eccentricity of the planets and a rapid increase in the flux of planetesimals that are scattered into orbits that cross the terrestrial planet region. Morbidelli *et al.* (2005) have shown that capture of Jupiter's Trojan asteroids is dynamically possible during the immediate aftermath of the resonant crossing.

A dramatic increase in the influx of small bodies into the inner Solar System might be expected to leave geological evidence in the form of a clustering of crater ages on the Moon and other planetary bodies not subject to weathering processes, at least provided that it occurred well after the final assembly of the terrestrial planets had concluded. In fact, Tera *et al.* (1974) advanced evidence from radioactive dating of lunar rock samples for just such a "Late Heavy Bombardment" having occurred about 700 Myr after the formation of the Solar System, which Gomes *et al.* (2005) have more recently associated with the 2:1 resonance crossing within the Nice Model. This identification remains uncertain, not so much due to theoretical difficulties (the statistical outcome of a set of given initial conditions for the outer Solar System can be predicted reasonably well using N-body simulations), but rather as a consequence of ongoing debate as to whether the Late Heavy Bombardment actually happened. It is undisputed that the cratering rate on the Moon about 3.9 Gyr ago was extremely high, and that it then decayed rapidly over a short interval of 0.1 Gyr or less. What is debatable is whether we know enough of the early history of the Moon to say for certain whether the high cratering rate was a spike (what is normally meant by a Late Heavy Bombardment), or rather a sustained plateau of impacts stretching back to the Solar System's formation (see e.g. Chapman *et al.*, 2007 for a review of the evidence). Irrespective of the outcome of the debate, however, it is unquestionably exciting to think that further study of the Moon, together with additional investigations into the predictions that

---

[9] After the French city.

the Nice Model makes for other Solar System phenomena, may be able to shed light on the evolution of the outer Solar System during the first Gyr after planet formation.

7.3.3 Application to extrasolar planetary systems

Multiple giant planets and massive planetesimal disks are likely ubiquitous features of young planetary systems, and hence it is reasonable to expect that behavior analogous to that for which we have good Solar System evidence must also occur in extrasolar planetary systems. Observationally the bias of current radial velocity surveys toward finding massive planets at small orbital radii selects against systems in which planetesimal scattering could be important, and no existing observation is generally interpreted as providing evidence for planetesimal-driven migration. The theoretical considerations outlined above, however, suggest that the range of different outcomes possible from the interaction of several planets with a disk of debris is substantial, and will depend at a minimum on the masses of the planets involved, the mass and extent of the disk, and the architecture of the system (for example, whether the most massive planet is generically closest to the star, as in the Solar System).

As an illustration of just one possibility, Fig. 7.8 shows the predicted evolutionary track of a system of three giant (but sub-Jupiter mass) planets that form in a compact configuration within an exterior planetesimal disk. The outcome in this case is reminiscent of the Nice Model. The system expands slowly as the planetesimal disk is depleted via scattering before experiencing an accelerated phase of evolution that coincides with the eccentricity excitation accompanying a 2:1 resonance crossing. Once the resonance has been crossed, further interaction between the planets and the remaining disk material damps planetary eccentricities to small values.

7.4 Planetary system stability

In a colloquial sense the fact that the Earth has sustained life for a substantial fraction of the age of the Solar System provides compelling evidence for the basic stability of at least our planetary system. Similarly, it would be shocking to discover that any known extrasolar planetary system (most of which are also billions of years old) was about to suffer some catastrophic event such as a planetary collision or ejection. The interest in planetary system stability concerns not such gross questions, but rather two substantially subtler aspects. One is the mathematical question of whether (and if so when) it is possible to *prove* that an N-body system is stable for all time, especially in the practically important case that the motions of the bodies are chaotic

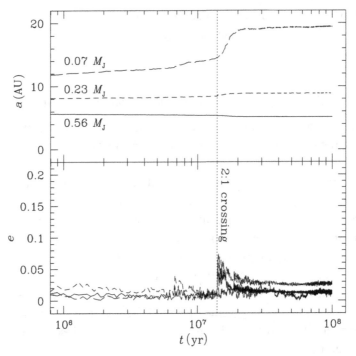

Fig. 7.8. Example of planetesimal-driven evolution of a system of three relatively low-mass giant planets, assumed here to form in close proximity to each other just interior to 10 AU. Beyond the planets lies a planetesimal disk containing 50 $M_\oplus$ of material distributed according to a $\Sigma_p \propto r^{-1}$ surface density profile between 10 AU and 20 AU. In this realization planetesimal scattering results in a slow expansion of the orbits of the outer two planets and an even more gradual contraction of the orbit of the innermost planet. Divergent crossing of the 2:1 resonance between the inner pair of planets briefly excites the eccentricities of the orbits, leading to enhanced scattering and more rapid outward migration of the outermost planet. The final values of the eccentricity are all small. Based on simulations by Raymond, Armitage, & Gorelick.

and thus intrinsically unpredictable over sufficiently long time scales. The second arises from the fact that nothing we have discussed about the *formation* of planetary systems requires that the outcome will be a stable system. On the contrary, since planet formation plays out in a highly dissipative environment – where either gas or planetesimal disks can frustrate the development of slow-growing gravitational instabilities – it is plausible to imagine that the typical planetary system forms in what turns out to be a long-term unstable state. If this is correct, features such as the chaotic orbits of Solar System planets and the eccentric orbits of some extrasolar planets may be endpoints of the evolution of initially unstable planetary systems, and it is important to understand how unstable systems evolve.

7.4.1 Hill stability

The trajectories of two point masses moving under Newtonian gravitational forces are stable and well known: elliptical orbits if the bodies are gravitationally bound, and parabolae or hyperbolae otherwise. For systems of three or more bodies, on the other hand, no general closed form analytic solutions to the motion of the bodies exist.[10] Nonetheless, it is sometimes possible to demonstrate the stability of a system of three bodies *despite our ignorance* of the actual trajectories, and although the stability criteria derived in this way are formally irrelevant for studies of more complex planetary systems they provide at least a framework for interpreting numerical results.

The only example of three-body motion for which it is easy to derive an analytic stability bound is the circular restricted three-body problem. We consider the motion of a test particle of negligible mass in the gravitational field of a circular star–planet system, and work in a frame that co-rotates with the orbital motion of the planet at angular velocity Ω. We have already derived the equations describing the motion of a test particle in this situation (the geometry is shown in Fig. 5.2 and discussed in Section 5.1.2). If the star and planet orbit in the (x, y) plane, then in a Cartesian coordinate system in which the star and planet lie along the $y = 0$ line at $x = -x_*$ and $x = x_p$ we have

$$\ddot{x} - 2\Omega\dot{y} - \Omega^2 x = -G\left[\frac{M_*(x + x_*)}{r_*^3} + \frac{M_p(x - x_p)}{r_p^3}\right], \tag{7.71}$$

$$\ddot{y} + 2\Omega\dot{x} - \Omega^2 y = -G\left[\frac{M_*}{r_*^3} + \frac{M_p}{r_p^3}\right]y, \tag{7.72}$$

$$\ddot{z} = -G\left[\frac{M_*}{r_*^3} + \frac{M_p}{r_p^3}\right]z. \tag{7.73}$$

Here r_* and r_p are the instantaneous distances between the test particle and the star and planet. No assumptions have been made as to the masses of the star M_* and planet M_p.

The acceleration due to the centrifugal force can be subsumed into a pseudo-potential. Defining

$$U \equiv \frac{\Omega^2}{2}\left(x^2 + y^2\right) + \frac{GM_*}{r_*} + \frac{GM_p}{r_p}, \tag{7.74}$$

---

[10] The exceptions to this statement include fascinating mathematical curiosities, such as the "figure of eight" orbit for three equal mass bodies that was discovered by Chenciner & Montgomery (2000).

we obtain

$$\ddot{x} - 2\Omega\dot{y} = \frac{\partial U}{\partial x},$$ (7.75)

$$\ddot{y} + 2\Omega\dot{x} = \frac{\partial U}{\partial y},$$ (7.76)

$$\ddot{z} = \frac{\partial U}{\partial z}.$$ (7.77)

The Coriolis terms on the left-hand-side of the first two equations can be eliminated by multiplying through by $\dot{x}$, $\dot{y}$, and $\dot{z}$ and adding. We then obtain,

$$\dot{x}\ddot{x} + \dot{y}\ddot{y} + \dot{z}\ddot{z} = \dot{x}\frac{\partial U}{\partial x} + \dot{y}\frac{\partial U}{\partial y} + \dot{z}\frac{\partial U}{\partial z},$$ (7.78)

$$\frac{d}{dt}\left(\frac{1}{2}\dot{x}^2 + \frac{1}{2}\dot{y}^2 + \frac{1}{2}\dot{z}^2\right) = \frac{dU}{dt}.$$ (7.79)

This equation integrates immediately to give

$$\dot{x}^2 + \dot{y}^2 + \dot{z}^2 = 2U - C_J,$$ (7.80)

$$C_J = 2U - v^2,$$ (7.81)

where v is the velocity and C_J, called the *Jacobi constant*, is the arbitrary constant of integration. Note that C_J is an energy-like quantity that is a conserved quantity in the circular restricted three-body problem.

The existence of this integral of motion is important because it places limits on the range of motion possible for the test particle. For a particle with a given initial position and velocity we can use Eq. (7.81) to compute C_J, and hence to specify *zero-velocity surfaces*, defined via

$$2U = C_J,$$ (7.82)

which the particle can never cross. If the volume enclosed by one of the zero-velocity surfaces is finite, then a particle initially within that region is guaranteed to remain there for all time. By this argument one can prove stability of the system without needing to derive an explicit form for the motion of the test particle.

The topology of the zero-velocity surfaces in the restricted three-body problem varies according to the value of C_J. An example is shown in Fig. 7.9. In this instance the zero-velocity surfaces define three disjoint regions in the (x, y) plane, one corresponding to orbits around the star, one corresponding to orbits around the planet, and one corresponding to orbits around the star–planet binary. A particle in any one of these states is stuck there – it cannot cross the forbidden zone between the different regions to move into a different state.

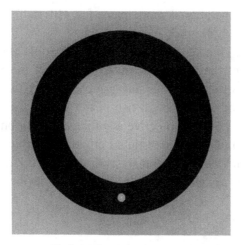

Fig. 7.9. Forbidden zones (dark regions) in the (x, y) plane in an example of the restricted three-body problem. For this particular choice of the Jacobi constant C_J, particles can orbit the star at small radii, the planet in a tight orbit, or the star–planet binary as a whole. The existence of zero-velocity surfaces, however, means that particles cannot be exchanged between these regions.

Given these results a test particle's orbit is guaranteed to be stable – in the sense that it can never have a close encounter with the planet – if its Jacobi constant is such that a zero-velocity surface lies between it and the planet. This condition can be written in terms of a minimum orbital separation needed for stability. As an example, consider a test particle in a circular orbit of radius a_{out} in a system where a planet has a circular orbit of radius a_{in}. We define a dimensionless measure of the orbital separation Δ via

$$a_{out} = a_{in}(1 + \Delta), \tag{7.83}$$

and write the mass ratio between the planet and the star as

$$q_{in} = \frac{M_p}{M_*}. \tag{7.84}$$

Stability is then assured (irrespective of the initial difference in the longitude between test particle and planet) provided that

$$\Delta > 2.4 q_{in}^{1/3}. \tag{7.85}$$

Since Δ is a measure of the orbital separation in units of (up to a small numerical factor) Hill radii, this is a version of a result used in Section 5.2.2, where we argued that a planet will rapidly perturb small bodies on to crossing trajectories if the separation is less than a few Hill radii. Here, we have demonstrated a more powerful inverse result: if the separation is more than a few Hill radii, close encounters are absolutely forbidden for all time.

Although Eq. (7.85) hints that the stability of a multiple planet system depends upon the separation of the planets measured in units of the Hill radius, nothing about the above derivation gives us any right to expect that a *formal proof* of stability can be derived for a general two-planet system. Surprisingly, however, for q_{in}, $q_{out} \ll 1$ it is possible to derive a sufficient condition for Hill stability[11] that can be applied to planets whose initial orbits have arbitrary eccentricity and inclination (Gladman, 1993; whose analysis draws on work by Marchal & Bozis, 1982). For two planets on initially circular coplanar orbits the system is Hill stable for separations

$$\Delta > 2 \cdot 3^{1/6} (q_{in} + q_{out})^{1/3}$$
$$+ \left[2 \cdot 3^{1/3} (q_{in} + q_{out})^{2/3} - \frac{11 q_{in} + 7 q_{out}}{3^{1/6} (q_{in} + q_{out})^{1/3}} \right] + \cdots, \qquad (7.86)$$

which simplifies to lowest order in the mass ratios to

$$\Delta > 2.4 (q_{in} + q_{out})^{1/3}. \qquad (7.87)$$

This criterion reduces to that deduced earlier for the equivalent restricted three-body problem (Eq. 7.85). Among the more general results derived and quoted by Gladman (1993) we note only the extension to the case where the planets have equal masses ($q_{in} = q_{out} = q$) and initially small eccentricities

$$\Delta > \sqrt{\frac{8}{3} \left(e_{in}^2 + e_{out}^2 \right) + 9 q^{2/3}}, \qquad (7.88)$$

which, as with Eq. (7.87), is also valid only to lowest order in the masses.

These stability criteria are formally only sufficient conditions for stability – they guarantee that a planetary system with larger Δ is Hill stable but say nothing about whether a more tightly packed system is actually unstable. For initially circular orbits (and only in that case), however, it is found empirically that the Hill limit is also a good general predictor of the onset of instability (with some exceptions such as systems with small Δ that are stable in resonant configurations). Another interesting empirical finding is that two-planet systems that are Hill stable can nevertheless exhibit chaos, which is attributed to the overlap of multiple resonances close to (but beyond) the minimum separation needed for stability.

---

[11] As in the circular restricted three-body problem, a system that is "stable" by this criterion is guaranteed to be free of close encounters between the planets for all time. This is not quite the common-sense definition of stability, since it leaves open the loophole that one planet might become unbound from the system via the cumulative action of many distant, weak encounters. A stronger definition of stability, known as *Lagrange stability*, requires that both planets remain bound. Lagrange stability cannot be proven in the same way as Hill stability, though for practical purposes the distinction is immaterial.

For systems of three or more planets there are no analytic guarantees of stability. The initial separation between planets in units of their mutual Hill radii

$$r_{\text{Hill,m}} = \left(\frac{M_{\text{p},i} + M_{\text{p},i+1}}{3M_*} \right)^{1/3} \frac{(a_i + a_{i+1})}{2}, \tag{7.89}$$

is still useful, however, as a guide to the time scale on which instability will result in crossing orbits (Chambers *et al.*, 1996). If we write the separation of the planets in the form

$$a_{i+1} = a_i + K r_{\text{Hill,m}}, \tag{7.90}$$

then it is possible to derive empirical expressions for the median time scale on which instability will develop. Chatterjee *et al.* (2008), for example, find that the instability time scale is well-represented by a three-parameter analytic fit of the form

$$\log_{10} t_{\text{instability}} = a + b \exp(cK), \tag{7.91}$$

where a, b, and c are constants. The instability time scale implied by this formula increases rapidly as the separation of the planets increases. As with two-planet systems, this general rule does not apply in the vicinity of strong resonances, but provided that this caveat is borne in mind it provides a good rule of thumb for assessing the stability of a complicated planetary system.

7.4.2 Planet–planet scattering

There is no evidence to suggest that the planets in the Solar System experienced close encounters with each other in the past, or that the early Solar System harbored additional planets that have subsequently been lost through collisions or ejections (though this latter possibility cannot be ruled out). The primary motivation for studying the evolution of unstable planetary systems comes from extrasolar planetary systems whose typically eccentric orbits immediately suggest that the observed planets may be survivors of violent planet–planet scattering events that occurred early on (Rasio & Ford, 1996; Weidenschilling & Marzari, 1996; Lin & Ida, 1997). In general, an initially unstable planetary system can evolve ("relax") via four distinct channels:

- One or more planets are ejected, either as a result of a close encounter between planets or via numerous weaker perturbations. An example of a system that displays this evolution is shown in Fig. 7.10.
- One or more planets have their semi-major axis and eccentricity changed in such a way that the system becomes stable.
- Two planets physically collide and merge.
- One or more planets impact the star.

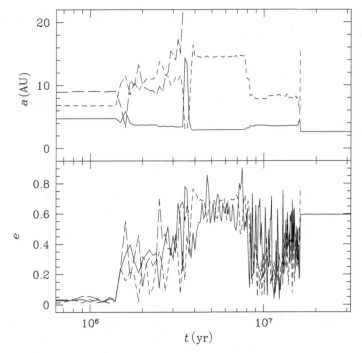

Fig. 7.10. Gravitational scattering in an initially marginally unstable system of three massive planets on circular coplanar orbits. The system relaxes via a complex set of interactions whose end result (in this particular realization) is the ejection of two of the planets. The lone survivor is left in a shorter-period orbit with an eccentricity $e \simeq 0.6$. Based on numerical simulations by Raymond, Armitage, & Gorelick.

The relative importance of the different channels is a function of the orbital radius at which scattering occurs (physical collisions obviously become more likely at smaller orbital radii) and of the mass distribution of the planets participating in the scattering.

Ideally perhaps, the distribution of planetary masses and orbital radii at some early epoch (when the gas disk has just been dissipated) would be a specified outcome of the theory of giant planet formation. The subsequent N-body evolution of an ensemble of planetary systems would then yield a single prediction for the distribution of final states of planetary systems after scattering. Although such calculations are possible, they require a possibly unwarranted degree of faith in the fidelity of the giant planet formation model, which as we have noted before is quite uncertain. Most authors have therefore followed a less ambitious path and studied the N-body evolution of unstable multiple planet systems starting from well-defined but essentially arbitrary initial conditions (Ford *et al.*, 2001; Chatterjee *et al.*, 2008;

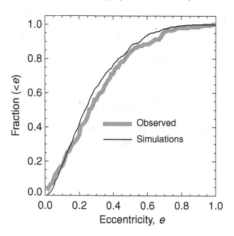

Fig. 7.11. The cumulative eccentricity distribution of known extrasolar planets is compared to the predicted distribution that results from scattering in three-planet systems. The simulation results were derived by numerically evolving an ensemble of unstable planetary systems made up of three planets whose masses were drawn from the observed mass function for extrasolar planets in the range $M_{Sat} < M_p < 3\,M_J$. Based on simulations by Raymond *et al.* (2008).

Jurić & Tremaine, 2008; Raymond *et al.*, 2008). The most important result from these studies is shown in Fig. 7.11, which shows the comparison between the cumulative eccentricity distribution of known extrasolar planets and that obtained theoretically as the end point of relaxation in an ensemble of unstable planetary systems. Good agreement with the observations – better in fact that one has any right to expect given the numerous other physical processes that can excite or damp planetary eccentricity – can be obtained starting from a variety of plausible initial conditions that vary in the assumed number and initial separation of giant planets within the system.

The agreement between the measured and predicted eccentricity distributions does not prove that strong scattering between planets is the only (or even the dominant) mechanism responsible for their typically significant eccentricities, but it is sufficiently persuasive as to lend support to the basic hypothesis that the typical outcome of giant planet formation is an unstable multiple planet system. One should bear in mind, however, that the observational sample is still biased toward massive planets at small orbital radii, and includes many systems that are so close in that significant prior gas disk migration is indicated. At larger radii planetesimal disk scattering is likely to become competitive with strong planet–planet scattering as a mechanism for dynamical evolution, especially for relatively low mass planets more akin to Uranus and Neptune than to Jupiter. It is thus probable that the eccentricity distribution of lower mass extrasolar planets at larger semi-major axis will differ from that currently measured, and it remains possible that the architecture and relatively low eccentricities of the Solar System's giant planets are typical, given the masses and orbital radii of the planets involved.

7.5 Further reading

"Planet formation and migration," by John Papaloizou and Caroline Terquem, 2006, *Reports on Progress in Physics*, **69**, 119.

Appendix 1

Physical and astronomical constants

Physical constants

Speed of light	$c = 2.998 \times 10^{10} \, \mathrm{cm \, s^{-1}}$
Newtonian gravitational constant	$G = 6.67 \times 10^{-8} \, \mathrm{cm^3 \, g^{-1} \, s^{-2}}$
Planck constant	$h = 6.626 \times 10^{-27} \, \mathrm{erg \, s}$
Proton mass	$m_\mathrm{p} = 1.673 \times 10^{-24} \, \mathrm{g}$
Boltzmann constant	$k_\mathrm{B} = 1.381 \times 10^{-16} \, \mathrm{erg \, K^{-1}}$
Stefan–Boltzmann constant	$\sigma = 5.670 \times 10^{-5} \, \mathrm{erg \, cm^{-2} \, K^{-4} \, s^{-1}}$
Molar gas constant	$\mathcal{R} = 8.314 \times 10^{7} \, \mathrm{erg \, mol^{-1} \, K^{-1}}$

Astronomical constants

Solar mass	$M_\odot = 1.989 \times 10^{33} \, \mathrm{g}$
Jupiter mass	$M_J = 1.899 \times 10^{30} \, \mathrm{g}$
Earth mass	$M_\oplus = 5.974 \times 10^{27} \, \mathrm{g}$
Solar radius	$R_\odot = 6.96 \times 10^{10} \, \mathrm{cm}$
Jupiter radius	$R_J = 7.15 \times 10^{9} \, \mathrm{cm}$
Earth radius	$R_\oplus = 6.38 \times 10^{8} \, \mathrm{cm}$
Solar luminosity	$L_\odot = 3.83 \times 10^{33} \, \mathrm{erg \, s^{-1}}$
Astronomical unit	$1 \, \mathrm{AU} = 1.496 \times 10^{13} \, \mathrm{cm}$
Parsec	$1 \, \mathrm{pc} = 3.086 \times 10^{18} \, \mathrm{cm}$

Appendix 2

N-body methods

Solving for the evolution of a system composed of N bodies interacting via their mutual gravitational force is a common problem in astrophysics. Although analytic or approximate numerical treatments (for example those based on the Fokker–Planck equation) can be useful, in many cases solution of N-body problems requires the explicit numerical integration of the trajectories traced by the bodies. A large body of technical literature (and at least one textbook, Sverre Aarseth's *Gravitational N-Body Simulations*[1]) is devoted to N-body methods, and codes that implement many of the most useful methods are readily available. This brief summary is intended to give an overview of the topic to help guide aspiring N-body practitioners to the methods and literature that may be most suitable for their problem.

Specification of the N-body problem

The state of a system of N point masses interacting only via Newtonian gravitational forces is fully specified once the masses m_i ($i = 1, N$), positions $\mathbf{r}_i$, and velocities $\mathbf{v}_i$ are given at some reference time. The evolution of the system is then described by Newton's Laws

$$\dot{\mathbf{r}}_i = \mathbf{v}_i, \tag{A2.1}$$

$$\dot{\mathbf{v}}_i = \mathbf{F}_i = -G \sum_{j=1; j \neq i}^{N} \frac{m_j(\mathbf{r}_i - \mathbf{r}_j)}{|\mathbf{r}_i - \mathbf{r}_j|^3}, \tag{A2.2}$$

where the dots denote time derivatives and we have defined $\mathbf{F}_i$ as the force per unit mass felt by body i due to the gravitational interaction with all the remaining bodies. The problem of numerically integrating an N-body system can then be split into two parts. First, how do we calculate the $\mathbf{F}_i$? Although it is trivial to perform the explicit sum indicated in Eq. (A2.2) this type of *direct* summation is computationally prohibitive for very large N. Approximate schemes which are much faster must then be used, but care is needed to ensure that the errors involved do not invalidate the results. Second, given the positions and velocities at some time t, together with an algorithm to compute the forces,

---

[1] Unfortunately Aarseth (2003), one of the pioneers of the field, does not discuss the methods of greatest interest for planetary integrations.

how do we update the state of the system to some later time $(t + \delta t)$? There exist an arbitrary number of finite difference representations that reduce to the ordinary differential equations (ODEs) as $\delta t \to 0$, and it is far from a trivial matter to decide which is best. Long term planetary integrations pose particular challenges since long term error control is paramount and standard schemes that are accurate for the integration of other ODEs can perform very poorly.

Exact and approximate force evaluation

Evaluation of the force by direct summation over all pairs of particles is straightforward and yields a force calculation that is exact up to the inevitable round-off errors that result from representing real numbers with a finite number of bits.[2] On a general purpose processor most of the computational work goes into computing the distances between particles (which requires evaluating a square root), and although this operation can be accelerated using special-purpose hardware (the video cards designed for avid PC gamers are surprisingly suitable for this purpose!), it is impossible to avoid the basic fact that the number of force evaluations per time step scales with the particle number as $\mathcal{O}(N^2)$. This scaling limits the use of direct summation schemes to relatively modest particle numbers.

If one is prepared to tolerate some inaccuracy in the evaluation of $\mathbf{F}_i$, a number of obvious ideas for accelerating N-body calculations suggest themselves. In a system with a large number of particles, for example, it is clear that the force felt by particle i due to its immediate neighbors will vary on a much shorter time scale than the force due to distant particles. We could therefore split the force into two pieces

$$\mathbf{F}_i = \mathbf{F}_{i,\text{close}} + \mathbf{F}_{i,\text{far}}, \tag{A2.3}$$

and save effort by recalculating the force from the distant particles less frequently than for the nearby bodies. This venerable approach, known as the Ahmad–Cohen (1973) scheme, results in a modest improvement in the scaling to $\mathcal{O}(N^{7/4})$ at the expense of a small decrease in accuracy.

More aggressive trade-offs of accuracy for speed are not only possible but also necessary if the system of interest has very large N (with current hardware a problem with $N = 10^5$ is challenging to solve using direct summation). The most powerful approach is to approximate the *spatial* distribution of the distant particles by grouping them together into ever larger clumps and evaluating the force from the center of mass of each clump rather than from the particles directly. There are many ways in which this could be done, of which the most elegant is the *tree code*, introduced to astrophysics by Josh Barnes and Piet Hut (1986). The basic principle is illustrated (in two dimensions) in Fig. A2.1. Starting with a cube which is large enough to enclose all of the particles within the simulation, space is first divided into 8 sub-cubes by splitting the original cube in half along each of the three coordinate axes. This division of space is continued recursively until every cell contains either one or zero particles. The lowest level cells that contain just one particle are known as the *leaves* on the tree. We then work backwards, calculating the total mass and center of mass of the cells at all higher levels of the tree, until eventually we return to

---

[2] Different implementations of floating point arithmetic can have different round-off properties even for a fixed number of bits (i.e. all "double precision" arithmetic is not created equal). Modern processors invariably implement standards for optimal or unbiased floating point arithmetic, but these desirable numerical properties can easily be lost if the user – in a desire for maximum performance – specifies the most aggressive compiler optimizations.

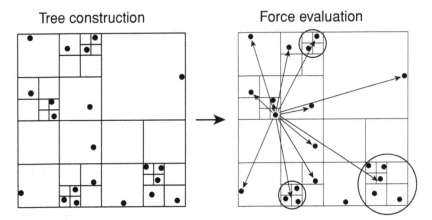

Fig. A2.1. Illustration of how a tree structure can be imposed upon a particle distribution by recursively partitioning space into cells (squares in this two-dimensional quad-tree example). Once the tree has been built, the force on any given particle is determined by summing contributions from nearby particles directly, while that from distant particles is evaluated approximately by grouping particles into larger clumps and calculating the force from the center of mass of each clump.

the original or *root* cube whose mass and center of mass are just the mass and center of mass of the whole N-body system. When this procedure has been completed the resulting data structure, known as an oct-tree, encodes a unique prescription for how particles can be clumped into successively larger groups by moving up from the leaves toward the root level of the tree. Unless the particle distribution is pathological, the recursive nature of the division scheme ensures that the depth of the tree (the number of subdivisions needed until each cell contains at most one particle) will scale with particle number as $\mathcal{O}(\log N)$.

With the tree in hand the force $\mathbf{F}_i$ on any individual particle can be calculated by traversing the tree from the top down. At each level in the hierarchy of nested cells we evaluate the opening angles θ subtended by the cells as seen from the location of particle i. If for a particular cell

$$\theta > \theta_{\text{open}}, \qquad\qquad (A2.4)$$

then the cell in question is too close for its contents to be treated as a single object and we proceed to the next level down in the tree. If, conversely, $\theta < \theta_{\text{open}}$ then it is acceptable to evaluate the force as if the contents of the cell were a single particle located at the center of mass of the true particle distribution within the cell.[3] Using this scheme it is possible to compute an approximate set of forces for the particles in $\mathcal{O}(N \log N)$ time, which for large N is a vast improvement over the N^2 scaling of direct summation methods. Moreover by varying the cell opening criterion θ_{open}, the accuracy of the approximation can be adjusted. The use of standard cell opening criteria (whatever the default value recommended by the code authors) will often suffice, but if the problem involves an unusual geometry or for some reason requires higher than usual accuracy one should be

---

[3] It is also possible to pre-compute and store higher order multipoles of the mass distribution within cells, and to use this information to improve the accuracy of the force calculation.

aware that the worst case force errors for tree codes can be quite different from the average errors (Salmon & Warren, 1994).

Tree codes have many advantages for large N simulations. The algorithm is very nearly coordinate independent, works well for arbitrarily clustered mass distributions that may have an unusual geometry, and can be parallelized with good efficiency (Salmon, 1991; Dave *et al.*, 1997; Springel, 2005). There is, however, a significant computational overhead inherent in the construction of the tree. As a result there are situations – particularly those where the particle distribution is not too strongly clustered – when it can be substantially more efficient to compute the forces from a *potential* which has been pre-computed on a regular lattice of points that encloses the mass distribution. This method (Hockney & Eastwood, 1981) involves three steps:

- Starting from the locations $\mathbf{r}_i$ and masses m_i of the particles, a density field $\rho(x_i, y_j, z_k)$ is defined on a uniformly spaced lattice by assigning the masses of the particles to the vertices of nearby lattice cells.
- The gravitational potential in Fourier space is calculated by taking the discrete Fourier transform of the density on the lattice, multiplied by the Green's function for Poisson's equation (basically this is just $-4\pi G/k^2$ for wavenumber k on a lattice with periodic boundary conditions, though small modifications are needed to remove artefacts that would otherwise arise from the mass assignment and force differencing schemes used). The real space potential on the lattice is then obtained by an inverse transform.
- The force at the location of each particle is found by finite differencing the potential on the lattice.

A discrete Fourier transform can be computed efficiently using readily available and highly optimized libraries. As a consequence grid methods of this kind are both fast and comparatively easy to implement on distributed memory systems. Currently, the most highly developed methods for *very* large N simulations (as of 2009, cosmological simulations with $N \sim 10^{10}$ have been completed) use nested grids to compute the long-range forces, with the grids being supplemented by local tree calculations to derive the short-range contributions.

Softening

Two classes of problem are commonly attacked using N-body methods. In the first class, the number of bodies in the physical system – which may be planets in a planetary system or stars in a small star cluster – is small enough that the individual particles in the N-body simulation can be considered to represent real bodies. Unless the bodies experience such close encounters that physical collisions or tidal effects become important, the only force that needs to be considered is Newtonian gravity with its $1/r$ potential, and we can aspire to integrate the orbits given this potential as accurately as possible.[4] A common problem is that binaries may form during the simulation even if none is present initially. This will often require the use of special methods (known as *regularization*) that can handle the large disparity in time scales between the internal motion of a single tight binary and the evolution time scale for the overall system (see e.g. Aarseth, 2003).

---

[4] Even in this case the presence of chaos may restrict the time frame over which the numerical integration can be considered to represent the "true" trajectory of the system. On very long time scales the integration of a chaotic N-body system should be considered as one sample of an ensemble of possible trajectories.

A second class of problems is composed of those where the number of bodies in the physical system is far too large to represent on a one to one basis in an N-body simulation. We cannot presently hope to simulate *every* planetesimal in the terrestrial planet forming region of the disk, and we will never be able to simulate galaxy formation using N-body particles that have the same mass as dark matter particles. The particles in a simulation of one of these problems should not then be thought of as real bodies within the physical system, but rather as tracers whose trajectories are intended to be a fair statistical sample of the dynamics of the real system. In this regime we may need to suppress effects – such as large angle gravitational scattering or binary formation – that would occur unphysically in the N-body system due to the fact that the particle mass is much larger than the mass of the physical bodies. This can be done by replacing the Newtonian potential $\Phi \propto 1/r$ with a *softened* potential, of which a simple example is

$$\Phi \propto \frac{1}{(r^2 + \epsilon^2)^{1/2}}. \tag{A2.5}$$

The parameter ϵ is a user-variable softening length, which needs to be chosen so as to suppress two-body interactions while leaving the large scale dynamics under the action of the smooth potential unaffected.

Time stepping

Once the algorithm for calculating the forces has been specified, it remains for us to decide how to advance the system in time. This requires settling on a discrete representation of the ordinary differential equations (Eq. A2.2). Advancing the discrete form of the equation will introduce errors, even if (as we assume henceforth) the forces have been computed exactly via direct summation. There are an infinite number of possible schemes, starting with the most naive

$$\mathbf{r}_i^{t+\delta t} = \mathbf{r}_i^t + \mathbf{v}_i^t \delta t, \tag{A2.6}$$

$$\mathbf{v}_i^{t+\delta t} = \mathbf{v}_i^t + \mathbf{F}_i^t \delta t. \tag{A2.7}$$

This scheme, known as Euler's method, updates the position and the velocity simultaneously using only information available at the initial time t. Euler's method is inaccurate and should never be used. By making just a small modification, however, and computing the new velocity at a time that is offset from the calculation of the new position, we arrive at the very useful leapfrog method

$$\mathbf{r}_i^{t+\delta t/2} = \mathbf{r}_i^t + \frac{1}{2}\mathbf{v}_i^t \delta t, \tag{A2.8}$$

$$\mathbf{v}_i^{t+\delta t} = \mathbf{v}_i^t + \mathbf{F}_i(\mathbf{r}_i^{t+\delta t/2})\delta t, \tag{A2.9}$$

$$\mathbf{r}_i^{t+\delta t} = \mathbf{r}_i^{t+\delta t/2} + \frac{1}{2}\mathbf{v}_i^{t+\delta t}\delta t. \tag{A2.10}$$

How can we compare different methods quantitatively? One way is to study how the error (the difference between the numerical solution to the discrete equations and the true solution) scales as the time step is reduced. This scaling is known as the *order* of the method. Higher order methods converge more rapidly toward the true solution with decreasing time step than low order schemes and, as a consequence, are often

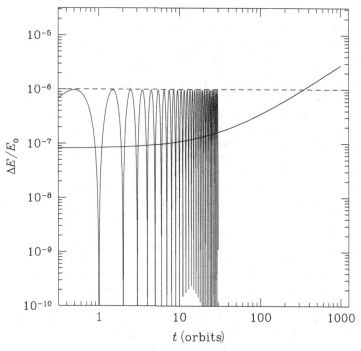

Fig. A2.2. Variation with time of the relative energy error $\Delta E = (E-E_0)$ from integrations of a Jupiter mass planet in a circular orbit about a Solar mass star. The oscillatory curve shows results obtained with a leapfrog integration scheme with 100 time steps per orbit (for clarity only the upper envelope of the errors is shown after 30 orbits) while the smoothly rising curve shows results obtained with a Hermite code.

recommended in numerical methods textbooks for problems whose equations admit smooth solutions.

Based on this reasoning, it might seem as if the choice of integration scheme is straightforward – surely all we need to do is to evaluate the single step error of the candidate methods and pick the one whose error at fixed computational work load is the smallest? This metric favors the use of very high order schemes, which were indeed used in most of the pioneering work on the dynamics of the outer Solar System (Cohen & Hubbard, 1965;[5] Applegate *et al.*, 1986). Matters are not so simple, however, because the manner in which errors accumulate over the long term is often unrelated to the magnitude of the single step error. This is illustrated in Fig. A2.2, which shows the drift with time of the total energy from integrations of a single massive planet on a circular orbit that were done using a leapfrog scheme and a Hermite scheme.[6] In the short term it is clear that the

---

[5] Cohen & Hubbard integrated the motion of Jupiter, Saturn, Uranus, Neptune, & Pluto for 120 000 yr using IBM's Naval Ordnance Research Calculator (NORC). The NORC had a memory of 2000 words and could execute 15 000 operations per second.

[6] The Hermite scheme is an example of a class of methods that make use of additional information in the form of time derivatives of the forces, which can be computed accurately via direct summation. Hermite schemes are often recommended and used with success for N-body integrations of star clusters.

Hermite scheme with this particular choice of integration options is superior – it has a smaller single step error than leapfrog and that advantage persists for several hundred orbits. In the long term, however, the leapfrog scheme's oscillatory error remains bounded, while the error of the Hermite scheme accumulates inexorably. It is clear that if we wish to integrate a planetary system for an extremely long period – perhaps for *billions* of orbits – the single step error or the order of the scheme may not be the best metric to use in comparing integration methods.

The realization that the order and short term accuracy of an integration scheme are poor guides to its performance over very long time scales led in the 1990s to the development and widespread use of a class of *geometric* integration schemes for planetary integrations (Kinoshita *et al.*, 1991; Wisdom & Holman, 1991). The central realization behind these methods is that for very long term integrations the best methods are those for which the discrete representation of the equations preserves as much as possible of the structure of the true equations. We know, for example, that the true dynamical system is both time reversible and energy conserving, and we might seek out integration schemes that share such properties. Schemes that exactly conserve energy can be designed, but global energy conservation is a relatively weak constraint on the dynamics and this avenue is not very profitable. Time reversibility, on the other hand, turns out to be an extremely desirable property of an integration scheme, since the dissipation that leads to drift in the total energy is a numerical analog for a physically irreversible phenomenon. The simple leapfrog scheme is an example of a time reversible algorithm.

In addition to time reversibility, altogether more subtle geometric properties of the true dynamics can serve as a further guide to constructing good integration schemes. In particular, the N-body problem is an example of a dynamical system that can be described by a Hamiltonian $H(\mathbf{p}, \mathbf{q}, t)$ that is a function of the generalized coordinates $\mathbf{q}$ and generalized momenta $\mathbf{p}$. The evolution of the state of a Hamiltonian system in the phase space described by these coordinates is given by

$$\dot{\mathbf{p}} = -\frac{\partial H}{\partial \mathbf{q}}, \tag{A2.11}$$

$$\dot{\mathbf{q}} = \frac{\partial H}{\partial \mathbf{p}}. \tag{A2.12}$$

Hamilton's equations describe how the state of the system at one time can be mapped into the state at some later time, and we can look for properties of this map that can be carried over to an integration scheme. Let us define the state of the system at time t to be $z = (\mathbf{q}, \mathbf{p})$. At some later time $(t + \delta t)$ the state is $z' = (\mathbf{q}', \mathbf{p}')$. One can then show that the map between these states $z \rightarrow z'$ has the property that it is a *symplectic transformation*, which means that it obeys the relation

$$\mathbf{M} \mathbf{J} \mathbf{M}^T = \mathbf{J}. \tag{A2.13}$$

Here the matrices $\mathbf{M}$ and $\mathbf{J}$ are defined as

$$\mathbf{M}_{ij} = \frac{\partial z_i'}{\partial z_j}, \tag{A2.14}$$

$$\mathbf{J} = \begin{bmatrix} \mathbf{0} & -\mathbf{I} \\ \mathbf{I} & \mathbf{0} \end{bmatrix}, \tag{A2.15}$$

with $\mathbf{0}$ and $\mathbf{I}$ being respectively the $N \times N$ zero and unit matrix. The important point is that we can apply this definition of a symplectic transformation not just to the real equations describing the evolution of an N-body system (which, since the system is Hamiltonian, are guaranteed to yield symplectic maps), but also to any discrete representation of those equations. Demanding that the numerical scheme generate a symplectic transformation of the state of the system from one time step to the next enforces a very powerful constraint on the trajectories that the system can describe in phase space, and ensures that at least qualitatively they must behave in the same way as do trajectories of the true system. For this reason symplectic integration schemes are often preferable for very long term integrations of planetary dynamics, where preserving the qualitative dynamics of the system is more important than obtaining the smallest error in the trajectory over the course of a few orbits. Somewhat amazingly the leapfrog scheme is not merely time reversible but also symplectic, and it is the simplest scheme that shares these two desirable traits.[7]

It is perfectly possible to use symplectic methods for arbitrary N-body problems, though care must be taken since many N-body problems require the use of variable time steps. These are needed, for example, to accurately resolve the large accelerations that occur during close encounters between bodies. It is in fact possible to construct time reversible schemes with variable time steps (Makino *et al.*, 2006), though it is by no means trivial, since in a variable time step code lack of symmetry in the *algorithm for selecting the time step* can very easily destroy the symplectic or time reversible nature of the underlying integration scheme. It is also unclear whether there is any real advantage to using a symplectic scheme if the force evaluation is done approximately, and so, although you will certainly find leapfrog recommended for use in cosmological simulations, the problem for which symplectic methods are best suited remains the small N direct integration of the Solar System or other planetary systems over very long time scales. In this situation, it is possible to make further optimizations by noting that the Sun is, by far, the dominant body in the system, and that as a result the system is *almost* integrable. Formally we could write

$$H = H_{\text{Kepler}} + \epsilon H_{\text{planets}}, \qquad (A2.16)$$

where H_{Kepler} is the Hamiltonian describing two-body motion of a planet around the Sun, and the perturbing term due to the mutual interactions between the planets is smaller by a factor that is of the order of the mass ratio between the planets and the star. To construct an integration scheme, we now approximate this Hamiltonian in the form (Wisdom & Holman, 1991; Saha & Tremaine, 1992)

$$H = H_{\text{Kepler}} + \epsilon H_{\text{planets}} \tau \sum_{n=-\infty}^{\infty} \delta(t - t_0 - n\tau), \qquad (A2.17)$$

where t_0 is some reference time and τ is some small fraction of an orbital period. Written this way, the motion of the body is *exactly* Keplerian except at a set of closely spaced discrete intervals when the planetary perturbations are applied. The resulting integration scheme has three steps:

- The bodies are evolved under the action of the Keplerian Hamiltonian for half a time step $\tau/2$. This amounts to translating the bodies along ellipses (if the orbits are bound)

---

[7] We have noted that it is possible to construct symplectic integration schemes – leapfrog being an example – and that it is also possible to construct schemes that exactly preserve the total energy (though we have given no explicit example of the latter). It may occur to the reader that it would be best of all to have a scheme that was *both* symplectic and exactly energy conserving. Alas such schemes do not exist, since it has been proved (Zhong & Marsden, 1988) that an energy conserving symplectic scheme would amount to a solution of the *exact* dynamical system, which is not possible if the system is nonintegrable.

and can be done analytically, either in terms of the normal Keplerian elements or in some other advantageous (non-Cartesian) coordinate system.

- The perturbing forces are evaluated in Cartesian coordinates, and each body receives an impulse that corresponds to the velocity change that would accumulate across a full time step.
- The step is completed with another half step under the Keplerian Hamiltonian.

Although this scheme requires continual conversion between coordinate systems, the extra work that this involves is more than compensated by the fact that it is possible to take a much longer time step than would be acceptable when using an unsplit scheme. Integrators of this class are known as *mixed variable symplectic* (MVS) integrators (Wisdom & Holman, 1991; Levison & Duncan, 1994; Saha & Tremaine, 1994; Chambers, 1999), and efficient implementations are available in several publicly released packages. One should note that the simplest MVS schemes are not well suited to handling close approaches between planets (since during flybys the "perturbation" due to the planets' mutual gravity can dominate over the force from the star), and a more sophisticated version must be employed if the problem might involve close encounters or collisions.

References

Aarseth, S. J. 2003, *Gravitational N-Body Simulations*, Cambridge: Cambridge University Press.

Adachi, I., Hayashi, C., & Nakazawa, K. 1976, *Progress of Theoretical Physics*, **56**, 1756.

Adams, F. C., Lada, C. J., & Shu, F. H. 1987, *Astrophysical Journal*, **312**, 788.

Adams, F. C., Laughlin, G., & Bloch, A. M. 2008, *Astrophysical Journal*, **683**, 1117.

Agol, E., Steffen, J., Sari, R., & Clarkson, W. 2005, *Monthly Notices of the Royal Astronomical Society*, **359**, 567.

Ahmad, A. & Cohen, L. 1973, *Journal of Computational Physics*, **12**, 389.

Alexander, R. D. & Armitage, P. J. 2007, *Monthly Notices of the Royal Astronomical Society*, **375**, 500.

Alexander, R. D., Clarke, C. J., & Pringle, J. E. 2006, *Monthly Notices of the Royal Astronomical Society*, **369**, 229.

Alibert, Y., Mordasini, C., Benz, W., & Winisdoerffer, C. 2005, *Astronomy & Astrophysics*, **434**, 343.

André, P. & Montmerle, T. 1994, *Astrophysical Journal*, **420**, 837.

André, P., Ward-Thompson, D., & Barsony, M. 1993, *Astrophysical Journal*, **406**, 122.

Andrews, S. M. & Williams, J. P. 2005, *Astrophysical Journal*, **631**, 1134.

Applegate, J. H., Douglas, M. R., Gursel, Y., Sussman, G. J., & Wisdom, J. 1986, *Astronomical Journal*, **92**, 176.

Arakawa, M., Leliwa-Kopystynski, J., & Maeno, N. 2002, *Icarus*, **158**, 516.

Armitage, P. J., Livio, M., Lubow, S. H., & Pringle, J. E. 2002, *Monthly Notices of the Royal Astronomical Society*, **334**, 248.

Artymowicz, P. 1993a, *Astrophysical Journal*, **419**, 155.

Artymowicz, P. 1993b, *Astrophysical Journal*, **419**, 166.

Artymowicz, P., Clarke, C. J., Lubow, S. H., & Pringle, J. E. 1991, *Astrophysical Journal*, **370**, L35.

Balbus, S. A. & Hawley, J. F. 1991, *Astrophysical Journal*, **376**, 214.

Balbus, S. A. & Hawley, J. F. 1998, *Reviews of Modern Physics*, **70**, 1.

Balbus, S. A. & Terquem, C. 2001, *Astrophysical Journal*, **552**, 235.

Baraffe, I., Chabrier, G., Allard, F., & Hauschildt, P. H. 2002, *Astronomy & Astrophysics*, **382**, 563.

Barnes, J. & Hut, P. 1986, *Nature*, **324**, 446.

Barranco, J. A. & Marcus, P. S. 2005, *Astrophysical Journal*, **623**, 1157.

Bate, M. R., Lubow, S. H., Ogilvie, G. I., & Miller, K. A. 2003, *Monthly Notices of the Royal Astronomical Society*, **341**, 213.

Beckwith, S. V. W. & Sargent, A. I. 1991, *Astrophysical Journal*, **381**, 250.
Begelman, M. C., McKee, C. F., & Shields, G. A. 1983, *Astrophysical Journal*, **271**, 70.
Bell, K. R. & Lin, D. N. C. 1994, *Astrophysical Journal*, **427**, 987.
Bell, K. R., Cassen, P. M., Klahr, H. H., & Henning, Th. 1997, *Astrophysical Journal*, **486**, 372.
Benz, W. & Asphaug, E. 1999, *Icarus*, **142**, 5.
Binney, J. & Tremaine, S. 1987, *Galactic Dynamics*, Princeton, NJ: Princeton University Press.
Blaes, O. M. & Balbus, S. A. 1994, *Astrophysical Journal*, **421**, 163.
Bodenheimer, P., Hubickyj, O., & Lissauer, J. J. 2000, *Icarus*, **143**, 2.
Borderies, N. & Goldreich, P. 1984, *Celestial Mechanics*, **32**, 127.
Boss, A. P. 1997, *Science*, **276**, 1836.
Bouchy, F., Bazot, M., Santos, N. C., Vauclair, S., & Sosnowska, D. 2005, *Astronomy & Astrophysics*, **440**, 609.
Britsch, M., Clarke, C. J., & Lodato, G. 2008, *Monthly Notices of the Royal Astronomical Society*, **385**, 1067.
Brownlee, D., Tsou, P., Aléon, J., et al. 2006, *Science*, **314**, 1711.
Butler, R. P., Marcy, G. W., Williams, E., et al. 1996, *Publications of the Astronomical Society of the Pacific*, **108**, 500.
Butler, R. P., Wright, J. T., Marcy, G. W., et al. 2006, *Astrophysical Journal*, **646**, 505.
Cameron, A. G. W. 1978, *Moon and the Planets*, **18**, 5.
Cameron, A. G. W. & Ward, W. R. 1976, *Abstracts of the Lunar and Planetary Science Conference*, **7**, 120.
Campbell, B., Walker, G. A. H., & Yang, S. 1988, *Astrophysical Journal*, **331**, 902.
Chambers, J. 2006, *Icarus*, **180**, 496.
Chambers, J. E. 1999, *Monthly Notices of the Royal Astronomical Society*, **304**, 793.
Chambers, J. E., Wetherill, G. W., & Boss, A. P. 1996, *Icarus*, **119**, 261.
Chandrasekhar, S. 1961, *Hydrodynamic and Hydromagnetic Stability*, International Series of Monographs on Physics, Oxford: Clarendon.
Chapman, C. R., Cohen, B. A., & Grinspoon, D. H. 2007, *Icarus*, **189**, 233.
Chapman, S. & Cowling, T. G. 1970, *The Mathematical Theory of Non-uniform Gases. An Account of the Kinetic Theory of Viscosity, Thermal Conduction and Diffusion in Gases*, 3rd edition, Cambridge: Cambridge University Press.
Chatterjee, S., Ford, E. B., Matsumura, S., & Rasio, F. A. 2008, *Astrophysical Journal*, **686**, 580.
Chenciner, A. & Montgomery, R. 2000, *Annals of Mathematics*, **152**, 881.
Chiang, E. 2008, *Astrophysical Journal*, **675**, 1549.
Chiang, E., Lithwick, Y., Murray-Clay, R., et al. 2007, in *Protostars and Planets V*, B. Reipurth, D. Jewitt, and K. Keil (eds.), Tucson: University of Arizona Press, p. 895.
Chiang, E. I. & Goldreich, P. 1997, *Astrophysical Journal*, **490**, 368.
Chiang, E. I., Jordan, A. B., Mills, R. L., et al. 2003, *Astronomical Journal*, **126**, 430.
Clarke, C. J. & Pringle, J. E. 1988, *Monthly Notices of the Royal Astronomical Society*, **235**, 635.
Clarke, C. J., Gendrin, A., & Sotomayor, M. 2001, *Monthly Notices of the Royal Astronomical Society*, **328**, 485.
Cohen, C. J. & Hubbard, E. C. 1965, *Astronomical Journal*, **70**, 10.
Cuzzi, J. N., Dobrovolskis, A. R., & Champney, J. M. 1993, *Icarus*, **106**, 102.
Damjanov, I., Jayawardhana, R., Scholz, A., et al. 2007, *Astrophysical Journal*, **670**, 1337.
D'Angelo, G., Lubow, S. H., & Bate, M. R. 2006, *Astrophysical Journal*, **652**, 1698.
Dave, R., Dubinski, J., & Hernquist, L. 1997, New Astronomy, **2**, 277.

Desch, S. J. 2004, *Astrophysical Journal*, **608**, 509.

Dominik, C., Blum, J., Cuzzi, J. N., & Wurm, G. 2007, in *Protostars and Planets V*, B. Reipurth, D. Jewitt, and K. Keil (eds.), Tucson: University of Arizona Press, p. 783.

Draine, B. T. 2006, *Astrophysical Journal*, **636**, 1114.

Draine, B. T., Roberge, W. G., & Dalgarno, A. 1983, *Astrophysical Journal*, **264**, 485.

Dubrulle, B., Morfiull, G., & Sterzik, M. 1995, *Icarus*, **114**, 237.

Dullemond, C. P. & Dominik, C. 2005, *Astronomy & Astrophysics*, **434**, 971.

Duquennoy, A. & Mayor, M. 1991, *Astronomy & Astrophysics*, **248**, 485.

Eggleton, P. P. & Kiseleva-Eggleton, L. 2001, *Astrophysical Journal*, **562**, 1012.

Eisner, J. A., Hillenbrand, L. A., Carpenter, J. M., & Wolf, S. 2005, *Astrophysical Journal*, **635**, 396.

Fabrycky, D. & Tremaine, S. 1997, *Astrophysical Journal*, **669**, 1298.

Feigelson, E., Townsley, L., Güdel, M., & Stassun, K. 2007, in *Protostars and Planets V*, B. Reipurth, D. Jewitt, and K. Keil (eds.), Tucson: University of Arizona Press, p. 313.

Feigelson, E. D. & Montmerle, T. 1999, *Annual Review of Astronomy & Astrophysics*, **37**, 363.

Fernandez, J. A. & Ip, W.-H. 1984, *Icarus*, **58**, 109.

Finkbeiner, D. P., Davis, M., & Schlegel, D. J. 1999, *Astrophysical Journal*, **524**, 867.

Fischer, D. A. & Valenti, J. 2005, *Astrophysical Journal*, **622**, 1102.

Fleming, T. & Stone, J. M. 2003, *Astrophysical Journal*, **585**, 908.

Font, A. S., McCarthy, I. G., Johnstone, D., & Ballantyne, D. R. 2004, *Astrophysical Journal*, **607**, 890.

Ford, E. B., Havlickova, M., & Rasio, F. A. 2001, *Icarus*, **150**, 303.

Ford, E. B., Kozinsky, B., & Rasio, F. A. 2000, *Astrophysical Journal*, **535**, 385.

Fromang, S. & Papaloizou, J. 2006, *Astronomy & Astrophysics*, **452**, 751.

Fromang, S., Terquem, C., & Balbus, S. A. 2002, *Monthly Notices of the Royal Astronomical Society*, **329**, 18.

Gammie, C. F. 1996, *Astrophysical Journal*, **457**, 355.

Gammie, C. F. 2001, *Astrophysical Journal*, **553**, 174.

Garaud, P. & Lin, D. N. C. 2004, *Astrophysical Journal*, **608**, 1050.

Garaud, P. & Lin, D. N. C. 2007, *Astrophysical Journal*, **654**, 606.

Ghosh, P. & Lamb, F. K. 1979, *Astrophysical Journal*, **232**, 259.

Gladman, B. 1993, *Icarus*, **106**, 247.

Godon, P. & Livio, M. 1999, *Astrophysical Journal*, **523**, 350.

Goldreich, P. & Peale, S. 1966, *Astronomical Journal*, **71**, 425.

Goldreich, P. & Sari, R. 2003, *Astrophysical Journal*, **585**, 1024.

Goldreich, P. & Tremaine, S. 1979, *Astrophysical Journal*, **233**, 857.

Goldreich, P. & Tremaine, S. 1980, *Astrophysical Journal*, **241**, 425.

Goldreich, P. & Ward, W. R. 1973, *Astrophysical Journal*, **183**, 1051.

Goldreich, P., Lithwick, Y., & Sari, R. 2004, *Astrophysical Journal*, **614**, 497.

Goldsmith, P. F., Bergin, E. A., & Lis, D. C. 1997, *Astrophysical Journal*, **491**, 615.

Gómes, R., Levison, H. F., Tsiganis, K., & Morbidelli, A. 2005, *Nature*, **435**, 466.

Gómez, G. C. & Ostriker, E. C. 2005, *Astrophysical Journal*, **630**, 1093.

Goodman, A. A., Benson, P. J., Fuller, G. A., & Myers, P. C. 1993, *Astrophysical Journal*, **406**, 528.

Goodman, J. & Rafikov R. R. 2001, *Astrophysical Journal*, **552**, 793.

Greenberg, R., Bottke, W. F., Carusi, A., & Valsecchi, G. B. 1991, *Icarus*, **94**, 98.

Greenzweig, Y. & Lissauer, J. J. 1992, *Icarus*, **100**, 440.

Guillot, T. 2005, *Annual Review of Earth and Planetary Sciences*, **33**, 493.

Guillot, T., Stevenson, D. T., Hubbard, W. B., & Saumon, D. 2004, in *Jupiter: The Planet, Satellites and Magnetosphere*, F. Bagenal, T. Dowling, & W. McKinnon (eds.), Cambridge: Cambridge University Press.

Gullbring, E., Hartmann, L., Briceno, C., & Calvet, N. 1998, *Astrophysical Journal*, **492**, 323.

Gutermuth, R. A., Myers, P. C., Megeath, S. T., *et al.* 2008, *Astrophysical Journal*, **674**, 336.

Hahn, J. M. & Malhotra, R. 2005, *Astronomical Journal*, **130**, 2392.

Haisch, K. E., Lada, E. A., & Lada, C. J. 2001, *Astrophysical Journal*, **553**, 153.

Hartmann, L., Calvet, N., Gullbring, E., & D'Alessio, P. 1998, *Astrophysical Journal*, **495**, 385.

Hartmann, L., Hewett, R., & Calvet, N. 1994, *Astrophysical Journal*, **426**, 669.

Hartmann, W. K. & Davis, D. R. 1975, Icarus, **24**, 504.

Hawley, J. F., Balbus, S. A., & Winters, W. F. 1999, *Astrophysical Journal*, **518**, 394.

Hayashi, C. 1981, *Progress of Theoretical Physics Supplement*, **70**, 35.

Hernández, J., Hartmann, L., Megeath, T., *et al.* 2007, *Astrophysical Journal*, **662**, 1067.

Higa, M., Arakawa, M., & Maeno, N. 1998, *Icarus*, **133**, 310.

Hirayama, K. 1918, *Astronomical Journal*, **31**, 185.

Hockney, R. W. & Eastwood, J. W. 1981, *Computer Simulation Using Particles*, New York: McGraw-Hill.

Hollenbach, D., Johnstone, D., Lizano, S., & Shu, F. 1994, *Astrophysical Journal*, **428**, 654.

Holman, M., Touma, J., & Tremaine, S. 1997, *Nature*, **386**, 254.

Holman, M. J. & Murray, N. W. 1996, *Astronomical Journal*, **112**, 1278.

Holman, M. J. & Murray, N. W. 2005, *Science*, **307**, 1288.

Hornung, P., Pellat, R., & Barge, P. 1985, *Icarus*, **64**, 295.

Hubickyj, O., Bodenheimer, P., & Lissauer, J. J. 2005, *Icarus*, **179**, 415.

Ida, S. 1990, *Icarus*, **88**, 129.

Ida, S. & Lin, D. N. C. 2004, *Astrophysical Journal*, **616**, 567.

Ida, S. & Makino, J. 1993, *Icarus*, **106**, 210.

Ida, S. & Nakazawa, K. 1989, *Astronomy & Astrophysics*, **224**, 303.

Ida, S., Bryden, G., Lin, D. N. C., & Tanaka, H. 2000, *Astrophysical Journal*, **534**, 428.

Ida, S., Guillot, T., & Morbidelli, A. 2008, *Astrophysical Journal*, **686**, 1292.

Igea, J. & Glassgold, A. E. 1999, *Astrophysical Journal*, **518**, 848.

Ikoma, M., Nakazawa, K., & Emori, H. 2000, *Astrophysical Journal*, **537**, 1013.

Ilgner, M. & Nelson, R. P. 2006, *Astronomy & Astrophysics*, **445**, 205.

Ilgner, M. & Nelson, R. P. 2008, *Astronomy & Astrophysics*, **483**, 815.

Inaba, S., Tanaka, H., Nakazawa, K., Wetherill, G. W., & Kokubo, E. 2001, *Icarus*, **149**, 235.

Ivanov, P. B., Papaloizou, J. C. B., & Polnarev, A. G. 1999, *Monthly Notices of the Royal Astronomical Society*, **307**, 79.

Ji, H., Burin, M., Schartman, E., & Goodman, J. 2006, *Nature*, **444**, 343.

Johansen, A., Oishi, J. S., Low, M.-M. M., *et al.* 2007, *Nature*, **448**, 1022.

Johns-Krull, C. M. 2007, *Astrophysical Journal*, **664**, 975.

Johnson, B. M. & Gammie, C. F. 2005, *Astrophysical Journal*, **635**, 149.

Johnstone, D., Hollenbach, D., & Bally, J. 1998, *Astrophysical Journal*, **499**, 758.

Joy, A. H. 1945, *Astrophysical Journal*, **102**, 168.

Jurić, M. & Tremaine, S. 2008, *Astrophysical Journal*, **686**, 603.

Kalas, P., Graham, J. R., Chiang, E., *et al.* 2008, *Science*, **322**, 1345.

Kenyon, S. J. & Hartmann, L. 1987, *Astrophysical Journal*, **323**, 714.

Kenyon, S. J. & Luu, J. X. 1998, *Astronomical Journal*, **115**, 2136.
Kinoshita, H., Yoshida, H., & Nakai, H. 1991, *Celestial Mechanics and Dynamical Astronomy*, **50**, 59.
Kippenhahn, R. & Weigert, A. 1990, *Stellar Structure and Evolution*, Berlin: Springer-Verlag.
Kirkwood, D., 1867, *Meteoric Astronomy: A Treatise on Shooting-stars, Fireballs, and Aerolites*, Philadelphia: J. B. Lippincott & Co.
Kirsh, D. R., Duncan, M., Brasser, R., & Levison, H. F. 2009, *Icarus*, **199**, 197.
Kokubo, E., Kominami, J., & Ida, S. 2006, *Astrophysical Journal*, **642**, 1131.
Königl, A. 1991, *Astrophysical Journal*, **370**, L39.
Korycansky, D. G. & Asphaug, E. 2006, *Icarus*, **181**, 605.
Korycansky, D. G. & Pollack, J. B. 1993, *Icarus*, **102**, 150.
Kozai, Y. 1962, *Astronomical Journal*, **67**, 591.
Krolik, J. H. & Kallman, T. R. 1983, *Astrophysical Journal*, **267**, 610.
Kuiper, G. P. 1951, *Proceedings of the National Academy of Sciences*, **37**, 1.
Kwok, S. 1975, *Astrophysical Journal*, **198**, 583.
Lada, C. J. & Lada, E. A. 2003, *Annual Review of Astronomy & Astrophysics*, **41**, 57.
Lada, C. J. & Wilking, B. A. 1984, *Astrophysical Journal*, **287**, 610.
Lada, C. J., Muench, A. A., Luhman, K. L., *et al.* 2006, *Astronomical Journal*, **131**, 1574.
Larson, R. B. 1981, *Monthly Notices of the Royal Astronomical Society*, **194**, 809.
Laughlin, G., Steinacker, A., & Adams, F. C. 2004, *Astrophysical Journal*, **608**, 489.
Lecar, M., Franklin, F. A., Holman, M. J., & Murray, N. J. 2001, *Annual Review of Astronomy & Astrophysics*, **39**, 581.
Lee, M. H. 2000, *Icarus*, **143**, 74.
Lee, M. H. & Peale, S. J. 2001, *Astrophysical Journal*, **567**, 596.
Leinhardt, Z. M. & Richardson, D. C. 2002, *Icarus*, **159**, 306.
Leinhardt, Z. M. & Stewart, S. T. 2009, *Icarus*, **199**, 542.
Levin, Y. 2007, *Monthly Notices of the Royal Astronomical Society*, **374**, 515.
Levison, H. F. & Agnor, C. 2003, *Astronomical Journal*, **125**, 2692.
Levison, H. F. & Duncan, M. J. 1994, *Icarus*, **108**, 18.
Lin, D. N. C. & Ida, S. 1997, *Astrophysical Journal*, **477**, 781.
Lin, D. N. C. & Papaloizou, J. 1979, *Monthly Notices of the Royal Astronomical Society*, **186**, 799.
Lin, D. N. C. & Papaloizou, J. 1986, *Astrophysical Journal*, **309**, 846.
Lissauer, J. J. 1993, *Annual Review of Astronomy & Astrophysics*, **31**, 129.
Lissauer, J. J. & Stevenson, D. J. 2007, in *Protostars and Planets V*, B. Reipurth, D. Jewitt, and K. Keil (eds.), Tucson: University of Arizona Press, p. 591.
Lissauer, J. J. & Stewart, G. R. 1993, in *Protostars and Planets III*, Eugene H. Levy & Jonathan I. Lunine (eds.), Tucson: University of Arizona Press, p. 1061.
Lodders, K. 2003, *Astrophysical Journal*, **591**, 1220.
Lubow, S. H., Papaloizou, J. C. B., & Pringle, J. E. 1994, *Monthly Notices of the Royal Astronomical Society*, **267**, 235.
Lynden-Bell, D. & Boily, C. 1994, *Monthly Notices of the Royal Astronomical Society*, **267**, 146.
Lynden-Bell, D. & Pringle, J. E. 1974, *Monthly Notices of the Royal Astronomical Society*, **168**, 603.
Makino, J., Hut, P., Kaplan, M., & Saygın, H. 2006, *New Astronomy*, **12**, 124.
Malhotra, R. 1993, *Nature*, **365**, 819.
Malhotra, R. 1995, *Astronomical Journal*, **110**, 420.
Marchal, C. & Bozis, G. 1982, *Celestial Mechanics*, **26**, 311.

Marcy, G., Butler, R. P., Fischer, D., *et al.* 2005, *Progress of Theoretical Physics Supplement*, **158**, 24.
Marcy, G. W., Butler, R. P., Fischer, D., *et al.* 2001, *Astrophysical Journal*, **556**, 296.
Marois, C., Macintosh, B., Barman, T., *et al.* 2008, *Science*, **322**, 1348.
Masset, F. S. & Ogilvie, G. I. 2004, *Astrophysical Journal*, **615**, 1000.
Mathis, J. S., Rumpl, W., & Nordsieck, K. H. 1977, *Astrophysical Journal*, **217**, 425.
Mayor, M. & Queloz, D. 1995, *Nature*, **378**, 355.
McKee, C. F. & Ostriker, E. C. 2007, *Annual Review of Astronomy & Astrophysics*, **45**, 565.
Militzer, B., Hubbard, W. B., Vorberger, J., Tamblyn, I., & Bonev, S. A. 2008, *Astrophysical Journal*, **688**, L45.
Mizuno, H. 1980, *Progress of Theoretical Physics*, **64**, 544.
Morbidelli, A., Chambers, J., Lunine, J. I., *et al.* 2000, *Meteoritics & Planetary Science*, **35**, 1309.
Morbidelli, A., Levison, H. F., Tsiganis, K., & Gómes, R. 2005, *Nature*, **435**, 462.
Murray, C. D. & Dermott, S. F. 1999, *Solar System Dynamics*, Cambridge: Cambridge University Press.
Murray-Clay, R. A. & Chiang, E. I. 2006, *Astrophysical Journal*, **651**, 1194.
Nelson, R. P. & Papaloizou, J. C. B. 2003, *Monthly Notices of the Royal Astronomical Society*, **339**, 993.
Nesvorný, D., Bottke, W. F., Jr., Dones, L., & Levison, H. F. 2002, *Nature*, **417**, 720.
Nettelmann, N., Holst, B., Kietzmann, A., *et al.* 2008, *Astrophysical Journal*, **683**, 1217.
O'Brien, D. P., Morbidelli, A., & Levison, H. F. 2006, *Icarus*, **184**, 39.
O'Dell, C. R., Wen, Z., & Hu, X. 1993, *Astrophysical Journal*, **410**, 696.
Ogilvie, G. I. & Lubow, S. H. 2003, *Astrophysical Journal*, **587**, 398.
Ohtsuki, K., Stewart, G. R., & Ida, S. 2002, *Icarus*, **155**, 436.
Oort, J. H. 1950, *Bulletin of the Astronomical Institutes of the Netherlands*, **11**, 91.
Papaloizou, J. C. B. & Pringle, J. E. 1984, *Monthly Notices of the Royal Astronomical Society*, **208**, 721.
Papaloizou, J. C. B. & Terquem, C. 1999, *Astrophysical Journal*, **521**, 823.
Papaloizou, J. C. B. & Terquem, C. 2006, *Reports of Progress in Physics*, **69**, 119.
Papaloizou, J. C. B., Nelson, R. P., & Masset, F. 2001, *Astronomy & Astrophysics*, **366**, 263.
Peale, S. J. 1976, *Annual Review of Astronomy & Astrophysics*, **14**, 215.
Perri, F. & Cameron, A. G. W. 1974, *Icarus*, **22**, 416.
Podolak, M. 2003, *Icarus*, **165**, 428.
Pollack, J. B., Hubickyj, O., Bodenheimer, P., *et al.* 1996, *Icarus*, **124**, 62.
Pollack, J. B., McKay, C. P., & Christofferson, B. M. 1985, *Icarus*, **64**, 471.
Pringle, J. E. 1981, *Annual Review of Astronomy & Astrophysics*, **19**, 137.
Pringle, J. E. & King, A. R. 2007, *Astrophysical Flows*, Cambridge: Cambridge University Press.
Pringle, J. E. & Rees, M. J. 1972, *Astronomy & Astrophysics*, **21**, 1.
Quillen, A. C. 2006, *Monthly Notices of the Royal Astronomical Society*, **365**, 1367.
Quirrenbach, A. 2006, in *Extrasolar Planets*. Saas-Fee Advanced Course 31, D. Queloz, S. Udry, M. Mayor, and W. Benz (eds.), Berlin: Springer-Verlag.
Rafikov, R. R. 2003, *Astronomical Journal*, **126**, 2529.
Rafikov, R. R. 2005, *Astrophysical Journal*, **621**, L69.
Rafikov, R. R. 2006, *Astrophysical Journal*, **648**, 666.
Rasio, F. A. & Ford, E. B. 1996, *Science*, **274**, 954.
Raymond, S. N. 2006, *Astrophysical Journal*, **643**, L131.

Raymond, S. N., Barnes, R., Armitage, P. J., & Gorelick, N. 2008, *Astrophysical Journal*, **687**, L107.

Raymond, S. N., Quinn, T., & Lunine, J. I. 2006, *Icarus*, **183**, 265.

Rice, W. K. M. & Armitage, P. J. 2003, *Astrophysical Journal*, **598**, L55.

Rice, W. K. M., Armitage, P. J., Bate, M. R., & Bonnell, I. A. 2003a, *Monthly Notices of the Royal Astronomical Society*, **339**, 1025.

Rice, W. K. M., Armitage, P. J., Bonnell, I. A., *et al.* 2003b, *Monthly Notices of the Royal Astronomical Society*, **346**, L36.

Rice, W. K. M., Lodato, G., & Armitage, P. J. 2005, *Monthly Notices of the Royal Astronomical Society*, **364**, L56.

Rossiter, R. A. 1924, *Astrophysical Journal*, **60**, 15.

Rybicki, G. B. & Lightman, A. P. 1979, *Radiative Processes in Astrophysics*, New York: Wiley-Interscience.

Safronov, V. S. 1969, *Evolution of the Protoplanetary Cloud and Formation of the Earth and the Planets*, English translation, NASA TT F-677 (1972).

Saha, P. & Tremaine, S. 1992, *Astronomical Journal*, **104**, 1633.

Saha, P. & Tremaine, S. 1994, *Astronomical Journal*, **108**, 1962.

Salmeron, R. & Wardle, M. 2005, *Monthly Notices of the Royal Astronomical Society*, **361**, 45.

Salmon, J. K. 1991, Parallel hierarchical N-body methods, Ph.D. thesis, California Institute of Technology, Pasadena.

Salmon, J. K. & Warren, M. S. 1994, *Journal of Computational Physics*, **111**, 136.

Sano, T. & Stone, J. M. 2002a, *Astrophysical Journal*, **577**, 534.

Sano, T. & Stone, J. M. 2002b, *Astrophysical Journal*, **570**, 314.

Sano, T., Miyama, S. M., Umebayashi, T., & Nakano, T. 2000, *Astrophysical Journal*, **543**, 486.

Saumon, D. & Guillot, T. 2004, *Astrophysical Journal*, **609**, 1170.

Scott, E. R. D. 2007, *Annual Review of Earth and Planetary Sciences*, **35**, 577.

Semenov, D., Henning, Th., Helling, Ch., & Sedlmayr, E. 2003, *Astronomy & Astrophysics*, **410**, 611.

Shakura, N. I. & Sunyaev, R. A. 1973, *Astronomy & Astrophysics*, **24**, 337.

Shen, Y., Stone, J. M., & Gardiner, T. A. 2006, *Astrophysical Journal*, **653**, 513.

Shlosman, I. & Begelman, M. C. 1989, *Astrophysical Journal*, **341**, 685.

Shu, F. H., Galli, D., Lizano, S., Glassgold, A. E., & Diamond, P. H. 2007, *Astrophysical Journal*, **665**, 535.

Sicilia-Aguilar, A., Hartmann, L., Calvet, N., *et al.* 2006, *Astrophysical Journal*, **638**, 897.

Simon, M. & Prato, L. 1995, *Astrophysical Journal*, **450**, 824.

Sinclair, A. T. 1972, *Monthly Notices of the Royal Astronomical Society*, **160**, 169.

Skrutskie, M. F., Dutkevitch, D., Strom, S. E., *et al.* 1990, *Astronomical Journal*, **99**, 1187.

Smak, J. 1984, *Acta Astronomica*, **34**, 93.

Smoluchowski, M. V. 1916, *Physikalische Zeitschrift*, **17**, 557.

Spitzer, L. & Tomasko, M. G. 1968, *Astrophysical Journal*, **152**, 971.

Springel, V. 2005, *Monthly Notices of the Royal Astronomical Society*, **364**, 1105.

Spruit, H. C. & Uzdensky, D. A. 2005, *Astrophysical Journal*, **629**, 960.

Stepinski, T. F. 1992, *Icarus*, **97**, 130.

Stevenson, D. J. 1982, *Planetary and Space Science*, **30**, 755.

Stewart, G. R. & Ida, S. 2000, *Icarus*, **143**, 28.

Stewart, G. R. & Wetherill, G. W. 1988, *Icarus*, **74**, 542.

Supulver, K. D., Bridges, F. G., & Lin, D. N. C. 1995, *Icarus*, **113**, 188.

Syer, D. & Clarke, C. J. 1995, *Monthly Notices of the Royal Astronomical Society*, **277**, 758.

Takeda, G. & Rasio, F. A. 2005, *Astrophysical Journal*, **627**, 1001.

Takeuchi, T. & Lin, D. N. C. 2002, *Astrophysical Journal*, **581**, 1344.

Tanaka, H., Takeuchi, T., & Ward, W. R. 2002, *Astrophysical Journal*, **565**, 1257.

Tera, F., Papanastassiou, D. A., & Wasserburg, G. J. 1974, *Earth and Planetary Science Letters*, **22**, 1.

Toomre, A. 1964, *Astrophysical Journal*, **139**, 1217.

Touboul, M., Kleine, T., Bourdon, B., Palme, H., & Wieler, R. 2007, *Nature*, **450**, 1206.

Trujillo, C. A. & Brown, M. E. 2001, *Astrophysical Journal*, **554**, L95.

Tsiganis, K., Gomes, R., Morbidelli, A., & Levison, H. F. 2005, *Nature*, **435**, 459.

Turner, N. J. & Sano, T. 2008, *Astrophysical Journal*, **679**, L131.

Turner, N. J., Sano, T., & Dziourkevitch, N. 2007, *Astrophysical Journal*, **659**, 729.

Umebayashi, T. & Nakano, T. 1981, *Publications of the Astronomical Society of Japan*, **33**, 617.

Velikhov, E. P. 1959, *Soviet Physics. JTEP*, **36**, 995.

Veras, D. & Armitage, P. J. 2004, *Monthly Notices of the Royal Astronomical Society*, **347**, 613.

Ward, W. R. 1986, *Icarus*, **67**, 164.

Ward, W. R. 1988, *Icarus*, **73**, 330.

Ward, W. R. 1997, *Icarus*, **126**, 261.

Wardle, M. & Ng, C. 1999, *Monthly Notices of the Royal Astronomical Society*, **303**, 239.

Weidenschilling, S. J. 1977a, *Astrophysics and Space Science*, **51**, 153.

Weidenschilling, S. J. 1977b, *Monthly Notices of the Royal Astronomical Society*, **180**, 57.

Weidenschilling, S. J. & Marzari, F. 1996, *Nature*, **384**, 619.

Wetherill, G. W. 1990, *Icarus*, **88**, 336.

Wetherill, G. W. 1996, *Icarus*, **119**, 219.

Wisdom, J. & Holman, M. 1991, *Astronomical Journal*, **102**, 1528.

Wolk, S. J. & Walter, F. M. 1996, *Astronomical Journal*, **111**, 2066.

Wolszczan, A. & Frail, D. A. 1992, *Nature*, **355**, 145.

Wu, Y. & Murray, N. 2003, *Astrophysical Journal*, **589**, 605.

Wurm, G., Paraskov, G., & Krauss, O. 2005, *Icarus*, **178**, 253.

Yoder, C. F. 1979, *Celestial Mechanics*, **19**, 3.

Youdin, A. N. 2010, *European Astronomical Society Publications Series*, **41**, 187.

Youdin, A. N. & Lithwick, Y. 2007, *Icarus*, **192**, 588.

Youdin, A. N. & Shu, F. H. 2002, *Astrophysical Journal*, **580**, 494.

Zharkov, V. N. & Trubitsyn, V. P. 1974, *Icarus*, **21**, 152.

Zhong, G. & Marsden, J. E. 1988, *Physics Letters A*, **133**, 134.

Zhu, Z., Hartmann, L., & Gammie C. 2009, *Astrophysical Journal*, **694**, 1045.

Index

[*]n = see footnote, f = see figure